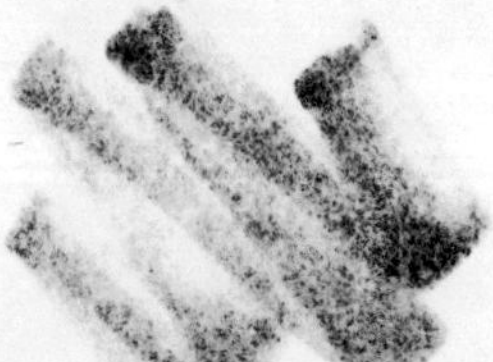

BIOCHEMISTRY AND PHYSIOLOGY OF PROTOZOA

BIOCHEMISTRY AND PHYSIOLOGY OF PROTOZOA

Edited by

ANDRÉ LWOFF

Head of the Department of Microbial Physiology
Pasteur Institute, Paris, France

VOLUME I

ACADEMIC PRESS INC., PUBLISHERS
NEW YORK · 1951

First Printing, 1951
Second Printing, 1958
Third Printing, 1966

PRINTED IN THE UNITED STATES OF AMERICA

CONTRIBUTORS

VIRGINIA C. DEWEY, *Amherst College, Amherst, Massachusetts.*
S. H. HUTNER, *Haskins Laboratories, New York, New York.*
GEORGE W. KIDDER, *Amherst College, Amherst, Massachusetts.*
ANDRÉ LWOFF, *Department of Microbial Physiology, Pasteur Institute, Paris, France.*
MARGUERITE LWOFF, *Pasteur Institute, Paris, France.*
RALPH W. MCKEE, *Department of Biological Chemistry, Harvard Medical School, Boston, Massachusetts.*
LUIGI PROVASOLI, *Haskins Laboratories, New York, New York.*
THEODOR VON BRAND, *Laboratory of Tropical Diseases, National Institutes of Health, Bethesda, Maryland.*

Foreword

Protozoa have been attracting the attention of biochemists to an increasing extent for some years. Of course, it is to be expected that the chemical and physiological features of Protozoa will be in harmony with our general biochemical concepts, but these organisms have evolved in many divergent directions and this evolution has led to a great variety of morphological and biological types. Some groups related to algae have been studied biochemically to a certain extent and the nutritional requirements of a few dozen species are known. But data concerning highly evolved groups such as the ciliates or the parasitic protozoa are scarce indeed. However, scarce though our data may be, it is already evident that evolution may lead to extreme biochemical specialization.

Zoologists, cytologists, geneticists, parasitologists, and pathologists have worked with Protozoa to a great extent but, compared with the biochemistry of bacteria and the lower fungi, the biochemistry of Protozoa has been neglected. Despite the technical difficulties encountered in the study of protozoan biochemistry, despite the fact that they grow slowly —a depressing feature in our swiftly-moving era—excellent work has been done and valuable results obtained.

This first volume of *Biochemistry and Physiology of Protozoa* is a series of contributions dealing with the nutrition, metabolism, and growth of the Phytoflagellates, the parasitic flagellates and amebae, the Trypanosomidae and the Bodonidae, the biochemistry of ciliates, and also the biochemistry of Plasmodium and the influence of antimalarials.

The second volume will deal with some groups that have been left aside in Volume I, for example, hypermastigines, ciliates of the rumen and other ciliates, phagotrophic flagellates, etc. Some chapters dealing with comparative biochemistry are also planned although this is a difficult task since our knowledge is very fragmentary, sometimes being limited to one aspect of one problem in one species.

The paucity of our knowledge is no excuse for inaction, however. All the data concerning the biochemistry of Protozoa, now scattered in various periodicals, should be made available in a single source. We hope that the material presented here will be useful both for students and for research workers and will stimulate interest in this important field.

André Lwoff

CONTENTS

Introduction to Biochemistry of Protozoa

ANDRÉ LWOFF

Head of the Department of Microbial Physiology
Pasteur Institute, Paris

CONTENTS

I. Protozoa

The term "Protist" was coined by Haeckel (1868) to designate what we may with Dobell (1911) call the noncellular organisms. These are generally referred to as uni- or monocellular, a most unfortunate designation, as was ably shown by Dobell. A protist has one or many nuclei. It lives generally as an isolated individual, but sometimes also forms filaments or aggregates or colonies in which no equivalent of the differentiated cells of multicellular organism are to be found. As recognized by Ehrenberg (1838), protists are organisms, and, as foreseen by Dobell (1911), all of them possess a nucleus. The noncellular state is found among plants, such as the lower fungi, schizomycetes (bacteria), and lower algae, as well as among animals of the group **Protozoa** (*sensu stricto*). However, difficulties arise when one tries to define in precise terms the differences between a plant and an animal, or when one attempts to differentiate satisfactorily yeasts, schizomycetes, and protozoa. R. J. Stanier and C. B. Van Niel (1941) in an excellent review have shown the impossibility "to frame a definition of the schizomycetes adequate to include all organisms which belong here, but sufficiently specific to exclude other groups of microorganisms." Morphology is sometimes useful for this purpose. For examples, budding is restricted to yeasts and certain protozoa. However, some species of yeast divide like bacteria. Furthermore, though the sexual process in yeast is characteristic, some species of yeast are apparently devoid of sexuality. Endospores are found only in schizomycetes, but some schizomycetes e.g., myxobacteriales exhibit external cystic membranes just as yeasts or protozoa, and many bacteria do not sporulate. Although bacteria are generally small, some are larger than the smallest protozoa. The bacterial nucleus is so small that it is not yet possible to show whether its structure is the same as that of the nuclei of yeasts or protozoa.

Certain groups of protozoa may be defined by the structure of their nuclear apparatus, by the pattern of their cilia, or by their life cycle; however, many protozoa have an ordinary nucleus, no cilia or flagella, and reproduce only by binary fission.

Thus, although error is better than confusion, it is impossible to give a satisfactory definition of Protozoa which would exclude lower fungi, schizomycetes, and lower algae. Protozoa could, of course, be defined as "animal protists." But what is an animal? Possession of chlorophyll and ability to photosynthesize could be considered as characteristic of the plant kingdom only. It is known, however, that a single mutation can induce the loss of chlorophyll. A chlorophyll-bearing flagellate can give rise to a colorless form either by losing the power to synthesize chlorophyll or by losing its plastids. If one were to decide arbitrarily that a chlorophyll-synthesizing flagellate belonged to the plant kingdom and a flagellate which could not synthesize chlorophyll to the animal kingdom, the first would be classified as a Protophyte, and the second as a true Protozoan (or Protozoan *stricto sensu*). If such an arbitrary definition is accepted, it must be remembered that, at the level of Protists, the distinction between plants and animals has no other significance than the presence or absence of chlorophyll and that a given plant and a given animal may be closely related. Every attempt to make such distinctions within natural groups involves such arbitrary definitions. As a matter of fact, a single order, e.g., *Euglenoidae,* may include green as well as colorless flagellates, which are studied as botanical and zoological material respectively.

It may be considered that the intake of particulate food is characteristic of animals. However, many photosynthetic organisms, such as the Chrysomonads, do engulf large particles, while many parasitic protozoa, such as the Trypanosomes, are able to absorb only dissolved food. Thus, differentiation between plants and animals on the basis of ability to ingest particles is meaningless. From a morphological point of view, the possession of pseudopods or of a cytostome is an important characteristic, but from a biochemical point of view the absorption and use of organic substances are more distinguishing features. The latter point of view leads, unfortunately, to difficulties with various organisms, of which an example is the photoautotrophic *Hematococcus.* This organism is able to live in a purely mineral medium in the light, but needs an organic energy source in the dark. In the light, it is a plant with potential animal behavior, while in the dark it is an animal with potential plant behavior. In addition, *Hematococcus* is able to combine both types of nutrition. Obviously, the terms "animal" and "plant" must be used merely for the designation of functional or physiological characteristics, and not for purposes of systematics. It should be noted that many chlorophyll-bearing organisms known today may become "animals" through loss of photosynthetic ability while the reverse process has never been observed.

The Protists as a whole should be considered as a monophyletic group. If this is done, it is not astonishing that so many features are common to lower fungi, schizomycetes, or bacteria, lower algae, and protozoa. If, nevertheless, we want to give a definition of "Protozoa" we may say:

1. They are relatively large protists, large, that is, compared to yeasts and bacteria.

2. They are generally motile, throughout the life cycle, or at least certain periods of the life cycle, a feature which they share with many bacteria and with some of the lower fungi such as the myxomycetes, but which clearly differentiates them from the yeasts;

3. Their cystic membranes, when they are formed, are formed outside the body, as in yeasts. Endospores are found only in schizomycetes, but some schizomycetes, e.g., myxobacteriales exhibit external cystic membranes just as yeasts on protozoa, and many bacteria do not sporulate.

4. "True Protozoa" are devoid of chlorophyll.

The necessary information concerning the morphology, cytology, life cycles, and the evolution of protozoa may be found in various treatises on protozoology, although, because of the specifically zoological or botanical interests of the author, it may be difficult to find in a single book all the desired information.

II. The Biochemistry of Protozoa

The present state of our knowledge of the biochemistry of Protozoa is very primitive as compared with that of our knowledge of bacterial biochemistry. Relatively little information is available concerning the chemical constitution of the cellular structures, the chemical nature of the food, and various aspects of the metabolism of Protozoa. There are many possible explanations of this situation. With a few exceptions, both free-living and parasitic forms of Protozoa generally occur in nature as isolated individuals. Therefore, to study their constitution and metabolism cultivation in bacteria-free cultures is necessary. Only a relatively small number of Protozoa have been obtained in pure culture. This is due to the fact that (1) highly evolved free-living protozoa generally live in the company of bacteria which grow more rapidly than the protozoa themselves, and (2) bacteria, or at least particulate complex food, is necessary for many Protozoa.

It is therefore not surprising that the first Protozoa to be grown in pure culture were certain green flagellates able to thrive in mineral media, and trypanosomes which, living in pure culture in the blood of their hosts and feeding on dissolved nutrients, were more or less easily cultivated in complex media such as broth enriched with blood. The problems of cultivation, however, do not in themselves explain entirely the relatively slow development of protozoan biochemistry.

Microbiological techniques were for many years used only by microbiologists and some botanists. The trypanosomes were studied by parasitologists, and the free-living Protozoa by zoologists. Since many Protozoa grow very slowly and do not develop abundantly, biochemists have quite naturally preferred studying the bacteria. In recent years, however, our knowledge of cellular and bacterial nutrition, especially with regard to growth factors, as well as our understanding of intermediary metabolism, have developed rapidly. By painstaking work, some free-living Protozoa have been freed from bacteria and obtained in pure culture. The biochemical knowledge of biologists has increased, and they have mastered the use of microbiological techniques. Moreover, the close cooperation of zoologists and parasitologists with biochemists has led to the constitution of balanced groups of workers able to attack many difficult problems, such as the biochemistry of malarial parasites. The fact that some cultures of Protozoa are now kept in many laboratories has facilitated their study considerably.

In 1932 Sandon wrote an excellent monograph entitled "The Food of Protozoa." According to the author, its scope was merely to review current knowledge of protozoan nutrition. The morphology of the feeding mechanism and the "choice of food" were also discussed. Most of the forms which had been studied at that time were those which need particulate food, the commonest types of this food being bacteria or other protista such as yeasts, algae, and protozoa. Most of the biochemical data available at that time were discussed in the book. This review is still very useful for protozoologists, since it provides abundant biological and ecological data, as well as an excellent bibliography.

The first review of our knowledge of protozoan metabolism is von Brand's "Stoffwechsel der Protozoen" which appeared in 1935. It seems to have been the only attempt to organize the data concerning the metabolism of water, lipids, carbohydrates, and nitrogenous compounds as well as the gas exchanges in Protozoa. The preparation of this review was a difficult task, since the data were scattered throughout zoological, parasitological, and microbiological journals. However, the author overcame this difficulty brilliantly, and his book retains great value up to the present time.

Since 1935, Protozoa have attracted increasing attention from biochemists, and numerous data in many periodicals eagerly await a new synthesis. Unfortunately, despite the progress of our knowledge, the number of species which have been studied is still very limited. Furthermore, the fact that it is much more difficult to obtain a few grams of Protozoa than a similar quantity of yeasts or bacteria continues to discourage most biochemists from using Protozoa as material for investigation of biochemical problems. There are, for example, no extensive papers dealing with the Keilin-Warburg system in the Protozoa. It would be of great importance to establish in at least one protozoan something like Meyerhof's scheme of sugar metabolism. The biosynthesis of essential metabolites, especially of amino acids has not been studied at all in these organisms. Some data are available on the nutrition, especially with regard to growth factors and sources of energy of Protozoa. In a few species of flagellates, amebae, and ciliates, certain enzymes have been identified. Because of the fundamental similarity of all living substances, because, on the whole, essential metabolites and mechanisms for their synthesis are the same in all organisms, and because some fundamental enzymes catalyzing similar reactions do exist in all protoplasm, though the intimate structure of such enzymes may be somewhat different in different biological types, one may reasonably expect that the biochemistry of protozoa will be in complete harmony with our concepts of general biochemistry.

The statement that Protozoa have many chemical features in common with other living beings may seem superfluous. However, despite the existence of life phenomena common to animals and plants and despite the fact that so far as we know the group Protozoa is monophyletic, it appears that certain types of Protozoa may exhibit peculiar features, some of which are shared with certain bacteria or yeasts, whereas others are shared with higher plants. Biochemical evolution has led in some cases to specialization, e.g., specialization of energy-yielding reactions. Specialization is particularly striking among parasites which have become perfectly fitted to their host, or even to some particular cells of their host; the biological significance of this adaptation is their inability to live anywhere except in a strictly limited environment.

It would have been desirable to include in this volume a section on the comparative biochemistry of Protozoa, a series of chapters, that is, describing their main biochemical characteristics, their chemical constitution, respiration and energy sources, intermediary metabolism of fats, carbohydrates, and amino acids, their coenzymes, pigments, encystment, sexuality, biochemical genetics, etc. Unfortunately, this would be a most difficult task, since our knowledge is very fragmentary, sometimes being limited to one aspect of one problem in one species. A comparative biochemistry of Protozoa must await the publication of more data. Systematic work on the diverse aspects of metabolism must be done on various species of all families. I wish to emphasize once more that close cooperation between protozoologists, botanists, parasitologists, and biochemists is a necessary condition of such progress.

During World War II, some biochemists were forced to cooperate with parasitologists in the study of certain parasitic protozoa. Some of them, whom I know personally, although they had made excellent achievements in their new field of research, were anxious to return to studies on muscle or yeast. For the future development of protozoan biochemistry, it seems of utmost importance that an atmosphere develop in which more biochemists may feel, without external pressure, that many problems of the biochemistry of Protozoa are now ripe for further investigations and that the Protozoa are quite ready to respond to their love and interest. This may happen quite naturally when scientists get the impression that muscle, yeasts, and bacteria have already delivered most of their messages and that Protozoa also offer a fruitful field for research. The painstaking work of establishing pure cultures is also a necessary condition for further progress.

The scarcity of our knowledge is no excuse for inaction. It can serve only as an excuse for the manner in which material is presented in this book, which is a series of monographs on species, genera, and families, with an occasional comprehensive discussion of certain features and of some of the higher groups. It is hoped that this material, such as it is, will be useful to students as well as to research workers. It is hoped that it will stimulate interest in the biochemistry of protozoa, and that, as a result of such interest, some of the further volumes will be enriched by chapters dealing with comparative biochemistry.

III. The Nutrition of Protozoa

A. Food Absorption

Two methods of food absorption may be described, at present, for protozoa. Some protozoa called the *compulsory osmotrophs* are devoid of cytostome and unable to produce pseudopods; thus they are able to utilize only dissloved food, either mineral or organic. The second group, which is able to ingest particles because its members possess a cytostome or produce pseudopods, is known as phagotrophs. Those protozoa which feed either on particles or on soluble substances are known as facultative phagotrophs or facultative osmotrophs, but those which feed only on particles are known as *compulsory phagotrophs.*

Many of the flagellates possessing photosynthetic pigments are strict osmotrophs. However, some of the photosynthetic flagellates, such as chrysomonadines, are able to produce pseudopods at certain phases of their life cycle and to ingest particles. Others, such as the dinoflagellates, are able to engulf even such large prey as copepods. Thus, phagotrophy may be combined with photosynthesis. It has been admitted that many chlorophyll-less "true protozoa" have been derived from flagellates possessing one or many plastids simply by the retarded division of the plastid. Thus, cytoplasmic division taking place when only one plastid is present would yield one pigmented flagellate with a plastid and another without plastid or pigment (for discussion see A. Lwoff, 1943). Such events have been observed among the chrysomonadines. We also know that it is very easy to suppress chloroplasts permanently in many Euglenae by treatment with streptomycin. Such organisms, lacking chloroplasts and chlorophyll, are, of course, no longer able to utilize light energy and need an organic energy source. Before discussing this question in detail, it is necessary to include a survey of the main energy sources known to be used among protozoa.

B. Energy-Yielding Reactions in Protozoa and Other Organisms

For a long period of time, living organisms were classified as either *autotrophic* (able to thrive on mineral food alone) or *heterotrophic* (needing organic substances for food).

Among the autotrophic organisms were the chemosynthetic bacteria, which derive their energy from the oxidation of inorganic substances, and the chlorophyll-bearing organisms, which utilize the energy of light. When Dusi in the Pasteur Institute discovered that certain chlorophyll-bearing Euglenae need organic substances in the form of growth factors, the old definition of an "autotroph" became meaningless, and it was necessary to distinguish between the use of organic substances as energy source and their use as growth factors. This thought has been expressed in a tentative nutritional system proposed by the author (1932). Later, it became clear that some photosynthetic bacteria need organic hydrogen donors for photoreduction of CO_2. During the Cold Spring Harbor Symposium of 1946, a group of workers proposed a nomenclature in which the available knowledge was coordinated in a system in which all terms were defined with maximal precision.

Two keys were given. The first utilized actions yielding energy, and the second the power of synthesis. We shall consider these keys as they are related to Protozoa, using the numerical system of the table.

As far as energy-yielding reactions are concerned, Protozoa are found among both great categories: chemotrophs and autotrophs.

I. Nomenclature based upon energy sources.
   - A. Phototrophy. Energy chiefly provided by photochemical reaction.
     1. Photolithotrophy. Growth dependent upon exogenous inorganic H-donors.
     2. Photoorganotrophy. Growth dependent upon exogenous organic H-donors.

		Autotrophy		Heterotrophy	Hypotrophy
		Autotrophy S.S.	Mesotrophy		
Phototrophy	Photolithotrophy	*Chlorella vulgaris*		*Rhodopseudomonas palustris*	
	Photoorganotrophy			*Rhodospirillum rubrum*	
Chemotrophy	Chemolithotrophy	*Thiobacillus denitrificans*			
	Chemoorganotrophy	*Pseudomonas fluorescens*	*Escherichia coli*	*Saccharomyces cerevisiae*	
Paratrophy	Schizomycetotrophy				Bacteriophages ?
	Phytotrophy				Plant viruses ?
	Zootrophy				Animal viruses ?

B. Chemotrophy. Energy provided entirely by dark chemical reaction.
   1. Chemolithotrophy. Growth dependent upon oxidation of exogenous inorganic substances.
   2. Chemoorganotrophy. Growth dependent upon oxidation or fermentation of exogenous organic substances.

C. Paratrophy. Energy apparently provided by the host cell.
   1. Schizomycetotrophy. Growth only in bacterial cells.
   2. Phytotrophy. Growth only in plant cells.
   3. Zootrophy. Growth only in animal cells.

II. Nomenclature based upon ability to synthesize essential metabolites.

A. Autotrophy. All essential metabolites are synthesized.
   1. Autotrophy *sensu stricto*. Ability to reduce oxidized inorganic nutrients.
   2. Mesotrophy. Inability to reduce one or more oxidized inorganic nutrients = need for one or more reduced inorganic nutrients.

B. Heterotrophy. Not all essential metabolites are synthesized = need for exogenous supply of one or more essential metabolites (growth factors or vitamins).

C. Hypotrophy. The self-reproducing units (bacteriophages, viruses, genes, and so on) multiply by reorganization of complex structures of the host.

a. Phototrophy in Protozoa. Chlorophyll-bearing protozoa are phototrophic and photolithotrophic. No phototrophic protozoan is known to need organic substances as hydrogen donor for the photoreduction of CO_2. This remains for the present a peculiar feature of bacteria of the group Athiorhiodaceae. Moreover, among the photolithotrophs, all protozoa belong to the "higher plant" group, in which water is the hydrogen donor. No photosynthetic "sulfur protozoan" is known in which sulfur compounds would replace water in van Niel's general equation for photosynthesis.

Protozoa have not been the subject of extensive studies of photosynthesis. Many investigations have been made on the alga *Chlorella.* The reader interested in photosynthesis should consult the excellent monographs of Rabinovitch (1945) and of Van Niel (1949).

When it is said that all chlorophyll-bearing protozoa are phototrophic, this means only that they are able to derive their energy from photosynthesis. For some of them, light may be the only energy source, but others are able to survive also from the oxidation of organic compounds. This will be considered later.

b. Chemotrophy in Protozoa. Chlorophyll-less protozoa must derive their energy from dark reactions. Not a single one is known as chemolithotrophic. A *Chilomonas* was at one time considered able to utilize the oxidation of inorganic substances as an energy source, but the claim did not withstand further investigation. All known protozoa which lack chlorophyll do need an organic energy source. Chemolithotrophic protozoa are, for the time being, a theoretical possibility. Among Protozoa, as among bacteria, the nature of the utilizable energy source varies according to species, though some features are peculiar to whole important groups.

In *the synthesis of essential metabolites* (II), protozoa are both autotrophic and heterotrophic. Examples of hypotrophs may be found when the nutrition of very parasitic intracellular protozoa is known.

Some photosynthetic flagellates, as well as some chemoorganotrophic are autotrophic *sensu stricto.* It seems that most photosynthesizers are able to use nitrates as nitrogen source, though this property is rare in chemolithotrophs; most of them must therefore be called mesotrophic. Some photosynthetic species, as well as numerous chemoorganotrophs, need growth factors. In *Euglena,* whether with or without chlorophyll, the main growth factor appears to be vitamin B_{12}. In Phytomonadines, it is the pyrimidine and thiazol parts of thiamine. The need for thiamine itself has not yet been found in Euglenoids or Phytomonadines. It has been described only in some trypanosomes and ciliates, which need many amino acids and, as far as their power of synthesis is concerned, resemble mammals more than they resemble most "primitive" flagellates.

c. Combination of nutritional features. It is indeed strange that photosynthesis, an energy-yielding reaction has been so long considered as, of necessity, associated with the synthesis of all essential metabolites. The author feels, as he stated in 1932, that energy-yielding reactions and the power of synthesis may have evolved independently and that different energetic types may be combined with different powers of synthesis. Any nomenclature based on nutritional types must allow for varied combinations of different energy-yielding reactions with different needs for growth factors.

Perhaps it is worth while to state that this form of nomenclature, because it is comprehensive and equilibrated, is practical for teaching purposes. The need for organic substances, previously called "heterotrophy" is now clearly defined. Knowing that even photosynthetic organisms may need growth factors, the student will no longer be tempted to use the requirement for organic substances as a criterion for grouping organisms which use organic substances as hydrogen donors or as a source of energy with those organisms which need a specific organic growth factor.

This is, of course, only a nomenclature of nutritional types. Different species of one group, such as *Phytomonadines,* may be found in many subdivisions. The nomenclature does not, by any means, pretend to coincide with the subdivisions of systematics.

IV. Obligate Phototrophic Flagellates and Specialized Energy Sources

As stated above, some photosynthetic flagellates are able to combine photosynthesis with the absorption of organic food which may be absorbed in one of two states: (1) as particulate food as in chrysomonadines and some dinoflagellates; (2) as dissolved food as in *Euglena.* Organic food may represent a source of nitrogen, of essential metabolites, or an energy source. Are photosynthetic flagellates able to utilize simultaneously photosynthesis and the oxidation of an organic substance as energy source? This seems possible but has apparently never been demonstrated. It would be interesting to compare the metabolism and growth rates of photosynthetic flagellates in the dark with an organic energy source and in the light under controlled CO_2 pressure and light intensity with and without an organic energy source. Whatever the case may be, photosynthetic flagellates are able to utilize light as their only energy source. But it happens that some species require light as the only possible energy source while others are able to grow in the dark if provided with an adequate organic substance acting as energy source. Their situation is similar to that of some phototrophic bacteria or to that of some chemolithotrophic bacteria able to derive their energy either from the oxidation of one specific inorganic substance or from the oxidation of various organic substances.

Euglena gracilis, for example, is able to derive its energy from photosynthesis or from the oxidation of acetic acid. *Euglena pisciformis,* on the other hand, is able to grow only in the light and not in the dark even in complex organic media supplemented and enriched with vitamins and acetic acid. Euglenes of the *E. gracilis* type are therefore able to lose their chlorophyll and actually do so in some cases, and survive as colorless *Astasia* whereas euglene of the *E. pisciformis* type are sentenced to death if maintained in the dark or if the assimilatory pigment is lost, unless a mutation permits the utilization of an organic energy source.

It is obvious that a mechanism permitting the utilization of the energy from organic substances is a prerequisite for the formation of viable true "protozoa" from photosynthetic organisms. Such irreversible evolution has taken place very often. In many groups of flagellates certain species differ only, as far as can be seen, in their ability to produce chlorophyll, the green and the colorless forms being classified in two different genera.

In *Euglenoida* and *Phytomonadina* the green forms manufacture paramylon or starch. So far as we know, the colorless "homologues" synthesize the same polysaccharide as do the green forms. Thus, in one case polysaccharide is formed with the energy of light, in the other with the energy derived from the oxidation of acetate. The interesting situation is that neither the *Phytomonadina* nor *Euglena* are able to utilize any sugar as energy source. They synthesize and utilize a glucose-polysaccharide and yet are unable to metabolize either glucose, maltose, or saccharose.

There are three known mechanisms of starch synthesis: by the classical phosphorylase.

$$n\text{-glucose-1-phosphate} \rightleftharpoons \underset{\text{starch}}{(\text{glucose})_n} + n\text{-phosphate}$$

by amylomaltase

$$n\text{-}[4(\alpha\text{-glucosido}) - \text{glucose}] \rightarrow \underset{\text{starch}}{(\text{glucose})_n} + n\text{-glucose}$$

and by amylosucrase

$$n\text{-(glucosido-fructoside)} \rightarrow \underset{\text{starch}}{(\text{glucose})_n} + n\text{-fructose}$$

Nothing is known about the intermediary metabolism of paramylon (see p. 11) which, being soluble only in formalin and concentrated KOH, is difficult to study in biological systems.

A study of starch synthesis has been made only in the colorless Phytomonadine *Polytomella coeca.* The polysaccharide of the starch granules has been thoroughly investigated by Bourne, Stacey, and Wilkinson (1950). It has a starch-like constitution. It can be fractionated with thymol into two fractions: fraction (A) (*ca.* 30%) resembles amylose in that it has a blue value of 1.180 (potato amylose B.V. = 1.20) and is converted into maltose with β-amylase to the extent of 88.7%, while fraction (B) (*ca.* 70%) resembles amylopectin with a blue value of 0.125 (potato amylopectin: B.V. = 0.200) and a β-amylase conversion of 52.4%. These data seem to be the only ones on the structure of the starch of flagellates.

The flagellate contains, beside an amylase, a phosphorylase: glucose-1-phosphate has been obtained from the action of enzyme of the flagellates on starch in the presence of mineral phosphate. As the flagellate is unable to utilize either glucose, sucrose, or maltose, it seems highly probable that it synthesizes starch from glucose-1-phosphate with its phosphorylase. If a hexokinase and a phosphoglucomutase were present, the synthesis of glucose-1-phosphate from glucose would be possible. The absence of hexokinase explains the fact that glucose is not utilizable. Except during exponential growth, the flagellate accumulates a considerable number of starch granules which disappear upon starvation. It is therefore certain that starch represents a reserve of energy rich bonds and of organic carbon. There can be no doubt that the flagellate is able to synthesize and utilize starch.

We have no precise data on the first step of CO_2 reduction and CO_2 metabolism in phototrophic organisms.

Van Niel's general equation for photosynthesis is

$$CO_2 + 2H_2A \rightarrow (CH_2O) + H_2O + 2A$$

According to Van Niel whose views are generally accepted, the photochemical reaction is only responsible for the production of special hydrogen donors.

The formation of a complex carboxyl phosphate $R-\overset{\overset{\displaystyle O}{\|}}{C}-OH_2PO_3$ has been postulated as an early step in photosynthesis. Ruben (1943) has suggested that photosynthesis plays a role in the synthesis of high-energy phosphate compounds for the phosphorylation of a reduction product of CO_2, or perhaps the phosphorylation of an intermediary carboxylated substance. This is in good agreement with Lipmann's concept (1941) concerning the role of high-energy phosphate bonds and the specific hypothesis of Lipmann and Tuttle (1945) concerning the utilization of hydrogen produced by photolysis for chemosynthesis. The hypothesis that in photosynthesis the bulk of the light energy is first converted into energy-rich phosphate bonds has been critically discussed by Rabinowitch (1945) who reaches the conclusion that the high-energy phosphate bonds required for reactions other than initial CO_2 reduction are supplied by nonphotochemical oxidations. If this conclusion is true, the substrate of these nonphotochemical oxidations, obviously resulting from CO_2 reduction, should be able to substitute for photosynthesis. These compounds may, of course, be highly labile substances, or substances unable to penetrate through the cell membrane, and their existence may well be difficult to prove. For the time being, the existence of obligate phototrophs is a serious challenge to students of photosynthesis.

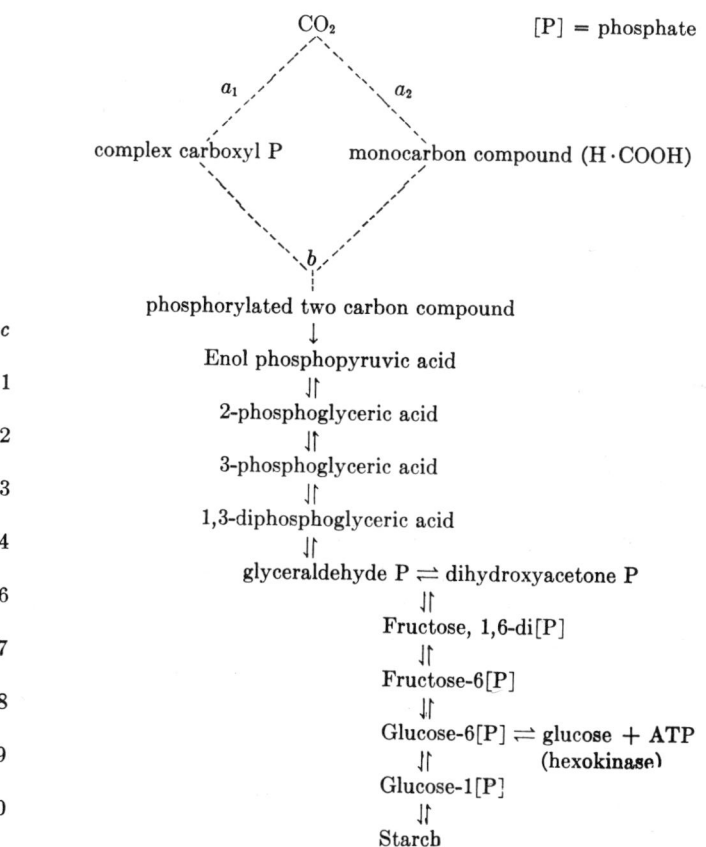

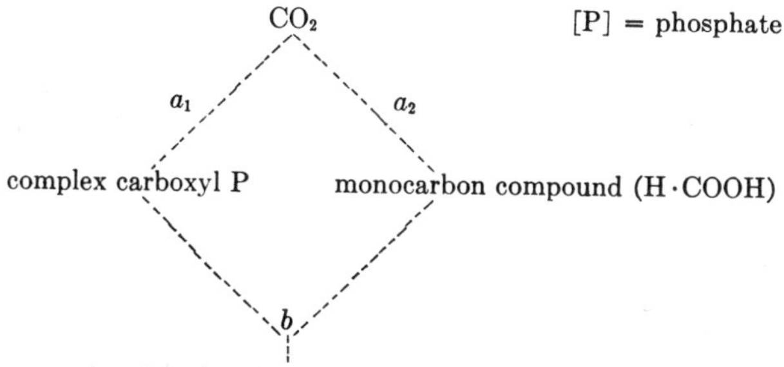

phosphorylated two carbon compound

c ↓

Enol phosphopyruvic acid

1 ⇅

2-phosphoglyceric acid

2 ⇅

3-phosphoglyceric acid

3 ⇅

1,3-diphosphoglyceric acid

4 ⇅

glyceraldehyde P ⇌ dihydroxyacetone P

6 ⇅

Fructose, 1,6-di[P]

7 ⇅

Fructose-6[P]

8 ⇅

Glucose-6[P] ⇌ glucose + ATP

9 ⇅ (hexokinase)

Glucose-1[P]

10 ⇅

Starch

Let us consider now the pathway of starch synthesis. The best-known reaction involves glucose-1-phosphate. It is highly probable that in starch-synthesizing organisms, glucose-1-phosphate is synthesized from triosephosphates and those from phosphoenol-pyruvic acid. Albaum *et al.* (1949) have identified glucose-1-phosphate, fructose-6-phosphate, fructose-1,6-phosphate, phosphoglyceric acid in *Euglena gracilis.* At least, some of the reactions 1 to 10 must be valid for photosynthetic organisms. Reactions a_1, a_2, b, and c are hypothetical unknown reactions leading to the synthesis of phosphoenol pyruvic acid from CO_2. It is obvious that any intermediate compound in such a chain should act as a substitute for photosynthesis provided this compound can pass through the cell membrane. Most phosphorylated compounds are devoid of this property. On the other hand, if any substrate, e.g., pyruvic acid or glucose, could be phosphorylated, pyruvic acid or glucose should be able to substitute for photosynthesis. In general, a substance able to act as energy source must either (1) be able to be phosphorylated or (2) yield, as product of its metabolism, a substance which can be phosphorylated or which can provide energy-rich bonds. When glucose substitutes for photosynthesis, e.g., in *Chlorella,* it is impossible to know a priori whether it is glucose itself or a substance derived from its oxidation which is phosphorylated. But we may expect some information when we study simpler substances.

Some phototrophic flagellates are able to live in the dark, and it is possible to compare their nutrition in the presence or absence of light and also with the nutrition of their colorless homologues. It happens that (with the possible exception of one species of *Euglena* which is perhaps able to utilize glucose) all flagellates which have been studied to date are unable to utilize any carbohydrate as exogenous energy source. Among the energy sources which have been found to be suitable for various organisms are lower fatty acids, some alcohols, lactic acid, and pyruvic acid. Some flagellates may utilize many carbon sources, others a small number, while still others can use only one. In those cases in which only one organic compound can be used, as in *Chlorogonium* and *Hyalogonium,* this unique energy source is acetic acid. And in all known cases in which a variety of organic compounds can be utilized, acetic acid is among them.

Although very few data are available, it is hoped that a discussion of the situation of obligate phototrophic flagellates and of closely related forms able to utilize acetic acid may lead to some working hypothesis.

As obligate phototrophs do not utilize acetic acid as energy source, it seems clear that acetic acid itself—provided it can be absorbed—is not on the pathway between CO_2 and starch. When acetic acid is the only utilizable energy source, in the absence of light or in the absence of chlorophyll, this means either that acetic acid itself or a substance derived from its metabolism can be phosphorylated. If it is assumed that the biosynthesis of starch in photosynthetic organisms goes through the phosphopyruvic step, the hypothetical phosphorylated substances originating from the metabolism of acetic acid would probably occur in the $CO_2 \rightarrow$ starch sequence of reactions between CO_2 and phosphoenol pyruvic acid.

Let us now consider the hypothesis that phosphopyruvic acid is synthesized from a phosphorylated two-carbon compound, which could be acetyl phosphate or phosphoenol acetic acid and a monocarbon compound such as formic acid, and let us discuss the possible reason why acetic acid cannot substitute for CO_2 in the photosynthesis of obligate phototrophs. In these organisms, acetyl phosphate would certainly not be formed by phosphorylation of acetic acid; otherwise acetic acid would be utilizable as energy source. It follows that the hypothetical phosphorylated acetic acid is synthesized in another way, probably by the combination of two one-carbon compounds, one of them being a phosphorylated substance.

Any of the hypothetical monocarbon compounds intermediate between CO_2 and phosphoacetic acid should be able to substitute for photosynthesis provided (1) they are able to be phosphorylated and (2) their oxidation provides enough energy. No monocarbon compound, nor any dicarbon compound more oxidized than acetic acid is known to act as energy source for chemoorganotrophic flagellates, any more than of course as for obligate phototrophs. The hypothesis that in obligate phototrophs phosphorylation has to take place at a very early step must therefore be considered.

This hypothesis is in agreement with Lipmann's and Ruben's views. It is also in agreement with Brown, Fager and Gaffron's demonstration (1948) of a substance "B (light)" found in illuminated *Scenedesmus,* which is stable in unilluminated cells. This substance, according to the authors, seems to combine in one molecule the properties of a first photosynthetic intermediate with those of a dark acceptor for CO_2. For starch synthesis CO_2 reduction has to take place (1) for the formation of two one-carbon compounds which will react to build a two-carbon compound which might be phosphoenol acetic acid and (2) for the formation of phosphoenol pyruvate from a two-carbon compound.

One of the first steps in photosynthesis would be the formation of the complex carboxyl phosphate postulated by Ruben $R{-}\overset{\overset{O}{\|}}{C}{-}OH_2PO_3$. This would correspond to substance "B(light)" and would act as acceptor of a monocarbon compound for the synthesis of phosphorylated acetic acid. Obligate phototrophy could be the result of the inability to oxidize any organic substance, at least at a sufficient rate to provide energy for growth. But as obligate phototrophs do perform oxidations, the hypothesis seems therefore more plausible that obligate phototrophy is due to the inability to phosphorylate any intermediate between the complex carboxyl phosphate and starch. In the pathway between CO_2 and starch, only one compound could be phosphorylated and this has to be an intermediary between CO_2 and the complex carboxyl phosphate. It could be a monocarbon compound or a complex carboxyl compound and represents the first photosynthetic intermediate. As a corollary we have to admit that the synthesis of this compound is intimately involved in the photolytic process.

Thus obligate phototrophy appears as the result of a lack of phosphorylases. Obligate phototrophs could perform only two phosphorylations, one at each end of the main energy highway: (1) the phosphorylation of one of the product of CO_2 reduction and (2) the phosphorylation of starch itself.

To reach this conclusion, we had to postulate:

1. That phosphopyruvic acid is an intermediate in the synthesis of glucose-1-phosphate.

2. That phosphopyruvic acid is synthesized from a two-carbon compound perhaps a phosphorylated acetic acid.

3. That the inability to utilize organic substances as energy sources is not a mere question of permeability.

The importance of acetic acid is well stressed by Hutner and Provasoli who further on put forward the hypothesis that the utilization of many organic substances as energy sources may be a question of penetration. They consider that the rapid passage of acetate across the cell membrane from neutral or alkaline media is probably an activated transport, perhaps a phosphorylation of acetate at the plasma membrane. According to this hypothesis, the inability to utilize acetate would be bound to the inability of acetate phosphorylation. The same conclusion is therefore reached by two different modes of reasoning. Van Niel (1949) has discussed the utilization of acetate in the light and in the dark and concluded that the decomposition of acetate must be accomplished essentially in the same manner in the darkness with oxygen as it is in light with carbon dioxide. That one or some enzymatic systems are common to dark and light metabolism of acetate seem highly probable. But one would except some differences when acetate is acting either as energy source either as hydrogen donor. And as long as the case of obligate phototrophs have not been cleared, it is difficult to know if Van Niel's very interesting conception applies to all photosynthetic organisms.

If the unsolved problem of obligate phototrophy has been discussed, it is to attract the attention on photosynthetic flagellates with the hope that they can be utilized to help solve the problems of photosynthesis.

V. Biochemical Evolution of Protozoa

The problem of nutritional evolution in Protozoa has been discussed in these terms by A. Lwoff (1943).

A. The Argument for Nutritional Subordination

"Zoologists and protozoologists have for a long time considered the Rhizopods, the prototypes of which are the amebae, as the simplest and the most primitive of the protozoa. These organisms, without a differentiated membrane, without definite form, with variable contours, do in fact seem to be primitive beings from which, by a sort of continuous perfectioning, the 'higher' protozoa with a definite form and complicated structure have arisen.

"This idea, which had a lasting success, is now completely abandoned following especially the work of Pascher. It is to be noted that in 1889 Bütschli derived the amebae and flagellates from the rhizomastigina, the amebae by predominance of the flagellate phase over the ameboid phase. Pascher described precisely Chrysomonads possessing cells which appeared alternately in an ameboid form and a flagellated form, or in which there coexisted the flagellated and the ameboid state, then predominance of the latter.

"The problem which interests us particularly here is that of the ancestors of the protozoa. This question has been approached both from a morphological point of view and a physiological point of view. In 1882 Bergh introduced into the phylogenetic discussion the argument of nutritional subordination, which may be summarized as follows: organisms with chlorophyll are the only ones able to support themselves upon inorganic substances and, particularly, alone possess the ability to assimilate carbon dioxide from the atmosphere owing to the radiant energy fixed by the pigment, chlorophyll; now, organisms lacking assimilatory pigment support themselves upon organic substances; their existence therefore depends upon that of the autotrophic organisms and they could not have appeared until after the latter. However, in 1888, Winogradsky described the chemotrophic bacteria, which do not fix radiant energy but are nevertheless able to maintain themselves upon inorganic materials, obtaining their energy by oxidation of inorganic compounds, the great diversity of which was later recognized. Then bacteria with assimilatory pigment were discovered, also capable of supporting themselves on inorganic materials owing to the energy furnished by photosynthesis. These chemotrophic and phototrophic bacteria synthesize living substance from inorganic compounds and are thus capable of insuring, at least qualitatively, the cycle of carbon and nitrogen compounds indispensable to the 'allotrophic'* organisms for the synthesis of their constituents. Orla-Jensen (1909) remarked that the bacteria served effectively as food for the protozoa.

"Thus the value of the argument of nutritional subordination was compromised. However, Reichenow (1929) in his 'Lehrbuch der Protozoenkunde' which was authoritative, wrote: 'If one considers the protozoa from the point of view of their metabolism, it seems clear that forms the most closely related to the organisms which appeared to have been first are the forms whose nutrition depends only upon inorganic materials, for these are the autotrophic organisms which have given the possibility of life to the heterotrophs' an exact conclusion, as we shall see, although based on an argument which no longer deserves to be considered correct because the existence of autotrophic flagellates is not a condition *sine qua non* for the existence of protozoa, many species of which support themselves effectively on bacteria. The fact that the protozoa require organic substances does not lead automatically to the conclusion that the protozoa are derived from the Chlorophytes.

"The argument of nutritional subordination, since it has no absolute value, must be categorically rejected from phylogenetic discussions. It is a morphological study of the eukaryote protista which will furnish data permitting the establishment of the origin of the leucophytes and the protozoa from the chlorophytes. It is their biological study which will give the experimental proof in the reality of this relationship."

The problem of origin of life has been considered by Oparin who

* (= organotrophic)

put forward the hypothesis that organic substances have preceded life. Haldane (1932) and Dauvillier and Desguin (1942) have developed similar views. According to these authors, organic substances have preceded life. This view is today accepted by all those who have discussed the question of origin of life (Bernal, 1949; Van Niel, 1949). It was accepted too by Horowitz (1945) to whom we owe an important discussion of the problem of evolutionary biosynthesis of essential metabolites. The synthesis of an essential metabolite supposes a series of enzymes and of genes, one for each step. The first living entity was a completely heterotrophic unit, reproducing itself at the expense of prefabricated organic molecules. As these organic molecules were exhausted, the organism developed mechanism for their synthesis. This happened by a stepwise process. For example, the species was heterotrophic for a sub stance A. Substances B and C were also present in this environment, and they were able to combine with the help of a catalyst to give substance A. If A became exhausted, organisms could survive only if they developed by mutation an enzyme able to perform the reaction $B + C = A$. Thus the primitive heterotrophic living substance developed gradually into an autotrophic organism. As pointed out by Horowitz, this process requires an environment in which all the intermediates for the biosynthesis of essential metabolites were present. This is of course necessary to explain the passage from heterotrophy to autotrophy. Horowitz's concept is very attractive, and the proposed mechanism of evolution is biosynthesis has probably really taken place, although it is difficult to conceive of such complex essential metabolites as cozymase or hematin and all their biological precursors being present in the environment. It is perhaps to avoid these future theoretical consequences of Horowitz's concept that I have proposed (1943, p. 278) another hypothesis.

''Essential metabolites are today products of living substances. Have they always been so? Dauvillier and Desguin have developed a concept according to which life appeared at the expense of organic substances previously formed by the action of short waves of solar radiation. It is in fact not possible to understand how living matter with all its enzymatic systems would have been formed from inorganic substances. But it is also very difficult to imagine how substances as complex and rigidly specific as certain essential metabolites, such as thiamine or lactoflavine (riboflavin), may have arisen outside of life. Perhaps, initial living matter was much simpler than the simplest of present-day organisms and capable of perpetuating itself in the absence of substances now indispensable. If this hypothesis is accepted, the intervention of certain metabolites, of certain enzymes, would be secondary. Thus, the necessity for growth factors would not necessarily be the consequence of losses of functions. The indispensable characters of exogenous supply of essential metabolites could be the manifestation of an evolution of more and more complex organic substances which was made necessary for the functioning of the living machine. The tendency toward specialization of certain highly differentiated cells of multicellular organisms is known; their activity depends on one or many hormones, exogenous with respect to the cell. An analogous tendency could have manifested itself at the protoplasmic or micellar level. In whatever fashion the evolutionary development of living substances occurred in earlier times, one fact remains without doubt: in the free living or parasitic organisms which we know, the requirement for growth factors is the consequence of losses of functions."

As noted by Van Niel (1949) "it will be necessary to accept the occurrence of a physiological evolution characterized by a loss of synthetic ability, and it is even probable that this is the chief if not the exclusive route by which the vast majority of now living organisms have originated. But acceptance of this view does not rule out the possibility that the earliest stages of physiological evolution have been marked by events along the lines suggested by Haldane and Oparin, and this seems to me the most plausible hypothesis."

B. Photosynthesis and Evolution

Since the problem for organic substances in general has been discussed, the problem of the energy source and of mechanisms allowing their utilization may be considered. Evolution of photosynthesis and of photosynthetic machines has been critically reviewed by Van Niel (1949). Photosynthetic flagellates not having provided any data for the solution of this problem, it does not seem necessary to add anything to Van Niel's excellent discussion. We have only to remember that phototrophy does not appear as a compulsory primitive condition. And chemo-organotrophy may well have been a property of some primitive types of organisms.

The fact that protozoa *sensu stricto* have arisen from photosynthetic flagellates by the loss of chlorophyll or of chlorophyll and chloroplasts does not mean therefore that obligate phototrophy is a primitive condition. Some phototrophic organisms could have kept the alleged power of their ancestor to utilize organic energy sources. As already stated, only those photosynthesizers able to utilize an organic energy source are able to survive the loss of chlorophyll. Therefore, in order to allow the formation of viable Protozoa, a mechanism has to be present, whether maintained or developed, which allows intake of either particulate or dissolved organic food. Unfortunately, information concerning nutrition and enzymatic equipment in Chrysomonadines and Dinoflagellates, the richest groups in evolutionary series leading to true protozoa, is scarce indeed, if not totally absent.

The only consistent data concern *Phytomonadina* and *Euglenoida.* The utilization of sugar as an exogenous energy source, is, if it occurs at all, exceedingly rare among those flagellates, whether green or colorless. Most of them, so far as we know, seem to be unable to utilize amino acids as energy source. The number of possible energy sources is, as already stated, sometimes restricted to one substance which is acetic acid. The role of acetic acid for flagellates has been emphasized by Treboux (1903) and rediscovered by Pringsheim (1921) for *Polytoma uvella* who designated this flagellate as "Azetat-Flagellat." It was shown later that organic acids were able to compensate for the absence of photosynthesis of many green flagellates, *Chlamydomonas, Hematococcus,* and *Euglena,* and played an important role in the nutrition of *Astasia,* the colorless equivalent of *Euglena* and for green Euglenas deprived of photosynthesis (A. Lwoff and H. Dusi, 1936). It was shown also that not only fatty acids but also lactic and pyruvic acid (Lwoff and Dusi, 1936) could be utilized. (Alcohols and glutamic acid were subsequently added to the list of energy sources.) Those organisms with a more or less specialized energetical nutrition, unable to utilize sugars and amino acids, and utilizing organic acids with a tendency to thrive only on lower fatty acids were called *oxytrophic.* They were opposed to the polytrophic types, such as most schizomycetes, molds, and yeasts, called haplotrophic. Oxytrophy is the condition of organisms corresponding to Pringsheim's acetate flagellates and haplotrophy to nonacetate organisms.

The comparative study of different types of protozoa showed a definite relation between starch or paramylon synthesis and oxytrophy. Whereas ciliates, trypanosomids, and, in general, protozoa which do not synthesize starch granules are haplotrophic, it turned out that starch or paramylon synthesizers are "oxytrophic." This is, for the time being, true only for protozoa. A similar condition to oxytrophy exists in some specialized schizomycetes where it has not been connected with other special biochemical features.

It is known that in the plant kingdom starch granules are synthesized only in the plastidic line, in chloroplasts or in leucoplasts. Only two exceptions are known, *Cyanophycees* and *Rhodophycees.* Starch granules are never formed in mitochondria. The fact that plastids may lose their chlorophyll and maintain their originality and specificity, so far as starch synthesis is concerned, must have a signification. (For the discussion of this problem see Lwoff, 1950.) It seems possible that evolution of cytoplasmic organelles has led to a situation in which the chloroplast is no more—if it ever was—a mere substrate for chlorophyll.

It is known that some species of Euglena, such as *Euglena gracilis* may lose the power to synthesize chlorophyll. This condition, as shown by Hutner and Provasoli, is easy to obtain by streptomycin treatment. The colorless form survives and remains colorless. As no plastid is seen in colorless Euglenas, Hovasse and Pringsheim (1948) have concluded that the whole chloroplast has disappeared. Starch is manufactured in chloroplasts in the green form. If we admit that the whole chloroplast has disappeared in colorless flagellates, this would mean that no essential enzymatic system is bound to chloroplasts. But when *Euglena gracilis* is subcultured in the dark and when chloroplasts are disappearing, some chloroplasts establish intimate connections with the mitochondrial system. One can question whether the totality of the chloroplast is really lost in colorless *Euglenas* and whether the colorless part, the leucoplast, does not persist, intricated with the mitochondrial system. Such situations are known at certain phases of the cycle of some plants. It is known also that leucoplasts are sometimes indistinguishable from mitochondria, except for their property to synthesize starch. It is therefore possible that a part of the chloroplast is maintained in the colorless *Astasia.* (For discussion see Lwoff, 1950.)

If *Euglena mesnili* is treated with streptomycin, it loses its chloroplasts. The chloroplastless flagellate is unable to live in the dark, whereas *E. mesnili,* as long as it keeps some chloroplasts is able to grow in the absence of light. Things happen as if photosynthesis were not essential for life, whereas chloroplasts are. One may question whether evolution has not led in some organisms to a situation where some essential enzymatic systems are bound to chloroplasts. The brilliant experiments of Ephrussi and his co-workers (1949) have proved the existence in yeast of enzymes bound to cytoplasmic particle endowed with genetical continuity.

Thus we have been led to conceive that colorless flagellates synthesizing starch of paramylon must possess a leucoplast or its biochemical equivalent. Things happen as if some enzymatic system of the leucoplasts did persist in leucophytes, probably as remainder of the wreckage of the chloroplasts. It is also possible to conceive that in some flagellates the enzymatic systems responsible for starch synthesis have been maintained in two different lines of cytoplasmic particles. For the time being, the available data do not permit a clear-cut conclusion.

Whatever the case may be, those leucophytes or colorless flagellates synthesizing starch or paramylon are oxytrophic. A peculiar mode of nutrition is in protozoa bound to the synthesis of starch or starch-like material. The problem of the cause of this interrelation still await its solution. It may have something to do with the evolution of enzymatic system in plastid possessing organisms or with the fact that energetics of these leucophytes are bound to starch or paramylon and that biochemical evolution has led in some groups to a situation in which only a small number of phosphorylases are present and especially phosphorylases for lower fatty acids. The systematic study of enzymatic system in evolutionary series should bring to light data allowing approach to the solution of this problem.

REFERENCES

Albaum, H. G., Ogur, M. and Hirshfield, A. (1949). *Federation Proc.* **8**, 179.

Bergh, R. S. (1882). *Morph. Jahrb.* **7**.

Bernal, J. D. (1949). *Proc. Phys. Soc. Am.* **62**, 537.

Bourne, E. J., Stacey, M., and Wilkinson, I. A. (1950). *J. Chem. Soc.* 517, 2694.

Brown, A. H., Fager, E. W., and Gaffron, H. (1948). *Arch. Biochem.* **19**, 407.

Bütschli, O. (1889). *Protozoa,* **I** and **II**.

Dauvillier, A., and Desguin, E. (1942). La genèse de la vie. Phase de l'évolution géochimique. Hermann, Paris.

Dobell, C. C. (1911). *Arch. Protistenk.* **23**, 269.

Doflein, F., and Reichenow, E. (1929). Lehrbuch der Protozoenkunde. G. Fischer, Jena.

Ehrenberg, C. G. (1838). Die Infusionsthierchen als vollkommene Organismen. Leipzig.

Ephrussi, B., Hottinguer, H., and Chimenès, A. M. (1949). *Ann. inst. Pasteur* **76**, 351.

Ephrussi, B., Hottinguer, H., and Tavlitzki, J. (1949). *Ann. inst. Pasteur* **76**, 419.

Ephrussi, B., L'Héritier, PH, and Hottinguer, H. (1949). *Ann. inst. Pasteur* **77**, 64.

Haeckel, E. (1868). *Jena. Z. Naturw.* **4**.

Haldane, J. B. S. (1932). Pelican Books.

Horowitz, N. H. (1945). *Proc. Natl. Acad. Sci. U. S.* **31**, 153.

Lipmann, F., and Tuttle, L. C. (1945). *J. Biol. Chem.* **158**, 505.

Lwoff, A. (1932). Recherches biochimiques sur la nutrition des Protozoaires. Le pouvoir de synthèse. Masson, Paris.

Lwoff, A. (1943). L'évolution physiologique. Étude des pertes de fonctions chez les microorganismes. Hermann, Paris.

Lwoff, A. (1950). *The New Phytologist* **49**, 72.

Lwoff, A., and Dusi, H. (1936). *Compt. rend.* **202**, 248.

Orla-Jensen, S. (1909). *Zentr. Bakt., Abt. II* **22**, 305.

Pascher, A. (1931). *Beihefte zum. Bot. Centr.* **48**, 317.

Pringsheim, E. G. (1921). *Beit. allg. Bot.* **11**, 88.

Pringsheim, E. G., and Hovasse, R. *In* Pringsheim, E. G. (1947). *The New Phytologist* **47**, 52.

Rabinowitch, E. I. (1945). Photosynthesis and Related Processes, Vol. I. Interscience Publishers, New York.

Ruben, S. (1943). *J. Am. Chem. Soc.* **65**, 279.

Sandon, H. (1932). The Food of Protozoa, Misr-Sokkar Press, Cairo. 187 pp.

Slonimski, P. P. (1949). *Ann. inst. Pasteur* **76**, 510; **77**, 774.

Slonimski, P. P., and Ephrussi, B. (1949). *Ann. inst. Pasteur* **77**, 47.

Stanier, R. Y., and Van Niel, C. B. (1941). *J. Bact.* **42**, 437.

Tavlitzki, J. (1949). *Ann. inst. Pasteur* **76**, 497.

Treboux, O. (1903). *Flora* **92**, 49.

von Brand, T. (1935). *Ergeb. Biol.* **12**, 161.

Van Niel, C. B. (1949). *In* Photosynthesis in Plants, pp. 437–495. J. Franck and W. E. Loomis, ed., The Iowa State College Press, Ames, Iowa.

Winogradsky, S. (1888). Beiträge zur Morphologie und Physiologie der Bakterien. I. Zur Morphologie und Physiologie der Schwefelbakterien. Arthur Felix, Leipzig.

The Phytoflagellates

S. H. HUTNER AND LUIGI PROVASOLI

Haskins Laboratories, New York

It is better to ask some of the questions than to know all the answers.

JAMES THURBER

CONTENTS

I. Introduction

The phytoflagellates are a heterogeneous group occupying a systematic position at intersections of plant and animal lines of descent. The presentation of their biochemistry requires a closer attention to evolutionary principles and to the supporting bichemical evidence than is necessary in the treatment of a more specialized, more clearly defined natural group. For the sake of orientation, the affinities of the plant flagellates are sketched in Fig. 1.

The elaboration of phylogenetic trees preoccupied nineteenth century biologists feeling the impact of the Darwinian revolution. But the mor-

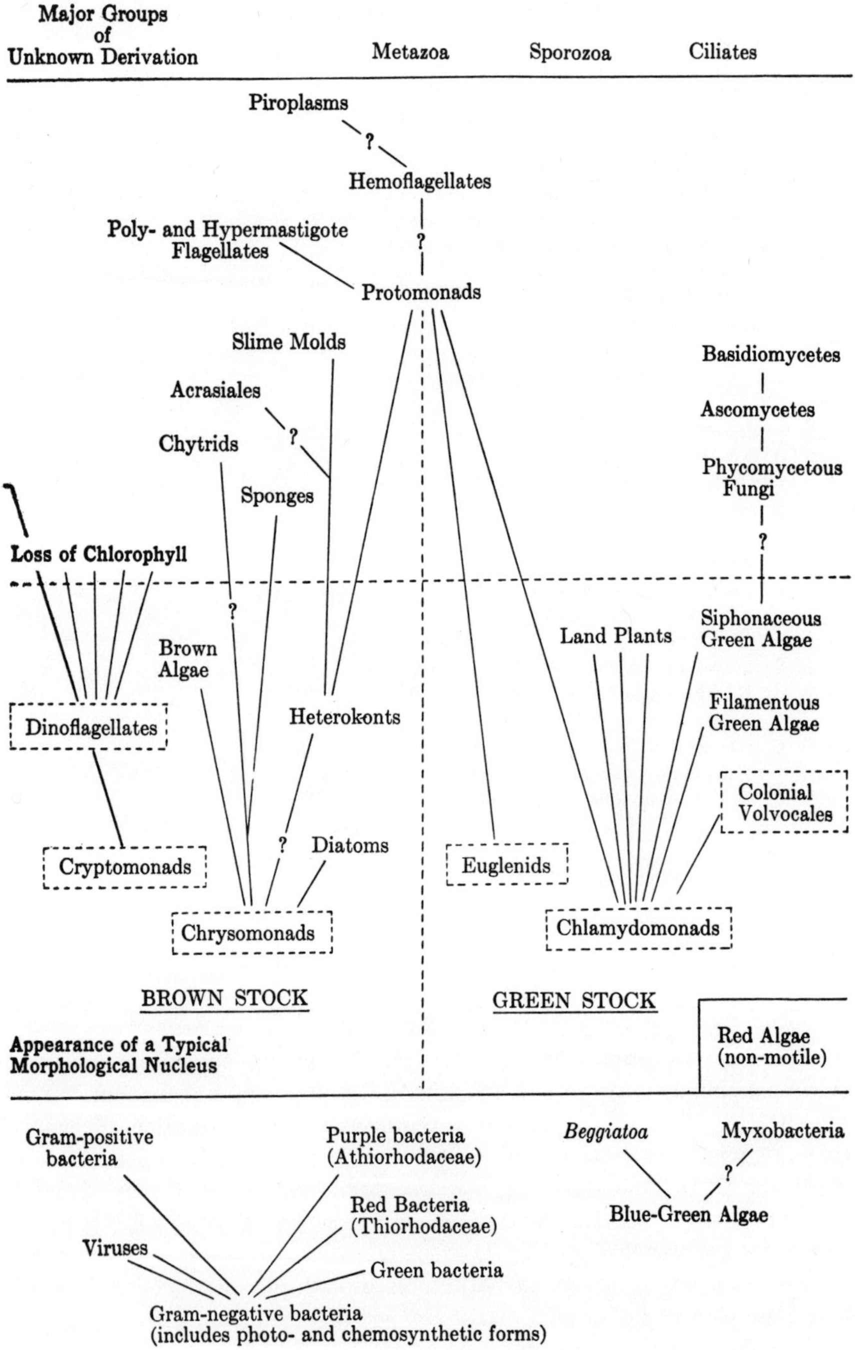

FIG. 1. Phylogenetic relationships of phytoflagellates. Highly speculative!

phological data lying at hand were inadequate for bridging the wide gaps in many evolutionary sequences; elaborate genealogies were constructed on overspeculative premises. A broader point of view is emerging: classical morphology is joined to molecular morphology, that is to say biochemistry, and powerful new tools such as the electron microscope connect the two morphologies. Comparative biochemistry, as exemplified in the school of Kluyver and Van Niel (Van Niel, 1949a), has shown that the myriad chemo- and photosynthetic processes and dissimilations of foodstuffs are reducible to mere differences in hydrogen donors and hydrogen acceptors in oxidation-reduction systems. Since the salient physiological attribute of the phytoflagellates is photosynthesis, and their most interesting direction of biochemical development is a lessened dependence on light as a source of energy for the assimilation of carbon, the teachings of comparative biochemistry are needed if one is to discern which components of the photosynthetic mechanism participate in heterotrophic reactions. The tenets of comparative nutrition as expounded by Lwoff (1932, 1943) and Knight (1936), combined with the comparisons of reaction chains in the intermediary metabolism of metazoa (Baldwin, 1948; Florkin, 1949), lead to the belief that many cellular structures and functions are conservative, even stereotyped in pattern—phylogenetically archaic—in contrast with the evolutionary plasticity of much of gross morphology. Comparative biochemistry offers the hope of putting to test the relationships postulated in classical taxonomy and of detecting hitherto overlooked kinships. Relevant data have now accumulated to the extent that a logical and expanded body of theory is required to preserve valuable findings from being lost in a welter of detail. An enduringly useful organization of biological data is attainable, as emphasized by Van Niel (1946) and Copeland (1947), only in the measure that taxonomy, i.e., the man-made classificatory system, approximates the natural, the phylogenetic, system. And, reciprocally, physiologists must be aware of taxonomic principles, either tacit or openly expressed, in selecting the best available experimental object for inquiry into a function.

A. General and Phylogenetic Considerations

Figure 1 is intended as a frame of reference; of course no claim of definitiveness is made. Nonetheless an outstanding feature of any comprehensive taxonomic system is the position of the phytoflagellate at the base of important plant and protozoan lines of descent.

The phytoflagellates are treated as a group because they combine three characteristics: (1) possession of a morphological nucleus in the conventional sense (nuclear membrane and detectable chromosomes; phytoflagellates also have a centriolar mitosis); (2) motility by means of flagella; and (3) photosynthetic pigments localized in plastids. For a detailed treatment, principally along morphological lines, of the connections of the phytoflagellates with other algae, the reader is referred to Fritsch (1935), Smith (1938), and the numerous publications written or edited by Pascher (for references see Fritsch and Smith). An incisive evaluation of current ideas on the affinities of the phytoflagellates to protozoa and metazoa is provided by Hyman (1940).

The taxonomic limits of the phytoflagellates, as recognized here, are shown enclosed in broken lines in Fig. 1.

Many of the bars to the understanding of the nature of the phytoflagellates are needless and arise from careless usage of the simple-seeming terms, *cell, plant,* and *animal.* These designations are unambiguous when applied to land plants and metazoa, but certain connotations are irrelevant and misleading if applied to Protista. A major confusion stems from misapprehension of the limits of the cell theory. As defined by Dobell (1911) in a trenchant essay, the protists, among which are numbered the phytoflagellates, are not built on the same plan as metazoa and metaphyta: protists, the name for all the organisms not compartmented into *cells,* are *non*-cellular. A single phytoflagellate is not homologous with the cell of a compartmentalized organism; it is equivalent to the whole metazoan or metaphyte. Certain attributes of protists—absence of cellularity, minute size—are regarded by students of cellular organisms as primitive or undifferentiated. But evolution within protistan lines has gone on throughout the same geological periods as has the development of cellular organisms. In filling their countless environmental niches, extreme specializations in morphology and physiology, without detectable compartmentation except for nucleus from cytoplasm, have developed among protists. While the protists are uncommonly as obviously complicated as larger organisms, their complexities and specializations soon become real to the investigator. Wonder at protistan complexity within minuteness of size is naive: one tiny fertilized ovum is large enough to determine the make-up of the whole man. It is to be deplored that "microbiology" with its emphasis on size is commonly employed where the older term "protistology" is more appropriate. Regrettably, the science of fundamental biochemical reactions is called "cell physiology," and the bodies of protists are referred to as "cells." These expressions are too firmly embedded in biological parlance to warrant their elimination from the aforementioned inappropriate contexts, but the pitfalls in their use should be recognized.

As indicated in Fig. 1, cellularity has arisen independently several times: in the Rhodophyceae and in the Phaeophyceae, and in the green algal lines that led to the land plants ("metaphyta"); again independently in the metazoan line, and perhaps *de novo* in the sponges. It would seem more than fortuitous that in all flagellated nucleated organisms, whether mammalian spermatozoan or zoospore of a filamentous alga, the basal granule of the flagellum—the blepharoplast or kinetosome—assumes the function of, or is connected with, the centriole in mitosis. The flagellate is one of nature's basic patents.

B. "Plant" *vs.* "Animal"

Defined simply, an animal *ingests particulate food; it is phagotrophic; it eats.* This entails deep-reaching adaptations: flexibility of the cell wall to permit the surrounding of particles, and secretion of digestive enzymes. Predation may be furthered by the elaboration of food-gathering and locomotory devices. The pseudopodial and rhizopodial habits represent two solutions to the problem of the ingestion of solids; other types of animal develop special sites for ingestion and digestion. And, as noted in later discussion, phagotrophy allows an increased molecular complexity of food. All this is not to be taken to mean that, when photosynthesis is abandoned, a stage where nutrition is entirely at the expense of dissolved nutrients ("saprophytism") must intervene between the holophytic and the holozoic ways of life. Several groups of chlorophyll-bearing flagellates display a conspicuous ability to ingest solids, a phenomenon disquieting to botanists and zoologists with fixed ideas, derived from experience confined to metazoa and metaphyta, as to orthodoxy in plant and animal behavior; also uncomfortable are microbiologists victimized by academic fragmentations of biology. A plant may be defined, negatively, as an organism unable to utilize particulate food. Certain forms, unquestionably animals, are not phagotrophic, e.g., astomatous ciliates and tapeworms, but these have undoubtedly undergone secondary simplification in response to parasitism. Overlapping of plant and animal nutrition is of common occurrence; *plant* and *animal* are not mutually exclusive designations in the Protista.

The criterion of non-phagotrophy permits application of the term *plant* to non-photosynthetic organisms, such as heterotrophic bacteria and the fungi, without awkward qualifications. The coexistence of plant and animal modes of feeding in the same individual is by no means a makeshift arrangement if the number of such individuals in nature be taken as the index of evolutionary success. Phagotrophic chlorophyllous chrysomonads are ubiquitous in fresh water and soils, e.g., *Chromulina* and *Ochromonas*. The nannoplankton of the ocean consists in large part of flagellates of the brown stock: dinoflagellates, chrysomonads, and cryptomonads; mostly chlorophyllous (Gross, 1937). Even should only a small proportion of these strangers to the experimentalist prove to be phagotrophic, the number of photosynthetic phagotrophs would still be vast. A sidelight is cast on the success of phagotrophy side-by-side with photosynthesis by the frequent symbiosis between algae and water-dwelling animals; Yonge (1945) estimates that the symbiotic algae in coral-building coelenterates, in sea anemones, and in radiolaria may in number of individuals surpass free-living algae. It remains for experiment and ecological observation to decide the relative importance to an organism of the two modes of nutrition, and whether phagotrophy, where it occurs, is obligate. Photosynthetic phagotrophs may become important enough objects of investigation to justify coining a less cumbersome appellation: "phytozoan" may serve.

Animals (in the sense of phagotrophs) appear to have sprung from several, perhaps many, stocks. Signs of this tendency are unmistakable in existing chrysomonads and euglenids. Among the least studied from the biochemical standpoint of the available plant → animal chains of species are those contained in the marine dinoflagellates; Kofoid and Swezy's (1921) magnificently illustrated monograph depicts forms almost certainly photosynthetic, lacking obvious adaptations for predation; these intergrade with an abundance of colorless species, some bizarrely equipped for predation.

1. LOSS OF CHLOROPHYLL

In many existing groups chlorophyll appears to be lost rather readily; for the older literature see Lwoff (1943). It may be, as proposed by Oparin (1938) and Horowitz (1945), and considered by Van Niel (1949b), that heterotrophic bacteria were on earth before autotrophs (autotrophs in the sense of obtaining the bulk of their carbon from CO_2), and that chemoautotrophic and photoautotrophic bacteria came later. But there can be little question that photosynthesizers were the first *nucleated* organisms. Artificially induced loss of chlorophyll is discussed farther on (p. 79).

As indicated in Fig. 1, loss of chlorophyll among nucleate organisms has its parallels in non-nucleate groups. Any comprehensive theory of the factors shaping this transition must take into account the capacities for heterotrophism of chemoautotrophic and photosynthetic organisms; the peculiarities of heterotrophism in phytoflagellates cannot, by themselves, adequately be appreciated.

2. MOTILITY

Motility is common to plants and animals. The land plants and the Phaeophyceae have lost it completely except for their ultra-conservative sex cells, and two major plant lines, the red algae and the blue-green algae, do not seem ever to have developed specialized locomotory organelles. With these qualifications, motility is not a distinction between plants and animals. Predation obviously is aided by locomotion, and as animals are distinguished predators indeed, the association of animals with locomotory specialization is justified. In phytoflagellates other functions besides predation are served by motility: distribution in space and photo-orientation.

The adoption of the ameboid habit has undoubtedly occurred repeatedly in different lines. Many small amebae often pass through a flagellate phase; the same organism may be at once ameboid, flagellated, and provided with plastids. Pascher has described remarkable organisms suggestive of ways in which amebae may have originated directly from algae. The heterokont alga *Perone,* containing chromatophores and oily food reserves, has ameboid swarmers. As the swarmers age, they develop the characteristic two unequal flagella. After a time the amebae round up into heterococcoid vegetative cells (Pascher, 1932). Another ameba has a stigma and is phototactic (Pascher, 1930). The various chlorophycean series, most conventionally plant-like, are not innocent of ameboid tendencies: not only are colorless counterparts of chlorophyllous genera of common occurrence (Pascher, 1931), but at least one of these colorless forms (*Collodictyon*) has pseudopodial feeding habits (Pascher, 1927a). The frequency of the rhizopodial habit (extension of long, delicate, food-catching filaments) in Chrysophyceae (Pascher, 1940), coupled with their close relationships to common colorless genera such as *Monas* and *Oikomonas,* suggest many linkages of chrysomonads to colorless phagotrophic groups, particularly to the radiolaria and heliozoa; the occurrence of silification in these groups (p. 44) is another token of affinity.

C. Evolutionary Gaps; What Are Protozoa?

The morphological record is woefully incomplete. The ancestry of the metazoa is a baffling mystery. Equally obscure and vast is the gulf dividing the phytoflagellates from their presumably non-nucleate ancestors. The notion expressed by some bacteriologists that the actinomycetes, with their filamentous habit and production of conidia from aerial hyphae, bridge the gap between bacteria and fungi, finds little support from mycologists: Bessey (1942) for instance proposes a flagellate ancestry for the fungi, and Stanier and Van Niel (1941) look upon the resemblance between actinomycetes and filamentous fungi as crude convergence. It is prudent, in seeking a phylogenetic relationship between nucleate and non-nucleate protists, to appreciate the position of the phytoflagellates.

Another gap lies between the brown and the green flagellate: there is no evidence of true transitional forms between the brown and the green stocks viewed in their entireties, although examples of convergence are plentiful, particularly in the Heterokontae (brown) and Protococcales and Ulotrichales (green).

At what point does a phytoflagellate become a protozoan? The term *protozoa* may embody over-arbitrary and superficial assumptions; unless these are stated explicitly, the term cannot be used without risking entanglement in verbal difficulties. Lwoff (1950) recognizes two kinds of "protozoa": (1) colorless forms, *leucophytes,* which still synthesize grains of starch or starch-like paramylum, and "protozoa *sensu strictu*" which do not. If phagotrophy were to be made an essential feature of protozoa, most leucophytes and important groups such as sporozoa would be left out, and many chlorophyll-bearing species admitted. A less troublesome course would appear to be to include all permanently colorless forms as well as *all* nucleated phagotrophic protists. The usefulness of the term *protozoa* can be maintained only if there is an awareness of the fact that the lumping together of unrelated forms under the heading Protozoa, in the name of didactic expediency, carries the disadvantage of implying relationships where none exists. Clear definitions are needed to avoid sterile disputation, and to avoid the relegation of important phytoflagellate groups to a quiet no-man's-land lying between the preserves of botanists, zoologists, and protozoologists.

II. Evolutionary Aspects of Photosynthesis

A. The Overall Equation

An understanding of the physiology of the phytoflagellates rests on an appreciation of the principles of photosynthesis; a detailed discussion of photosynthesis is beyond the scope of this chapter, but as photosynthesis forms the background against which other assimilatory and energy-yielding processes are examined, a précis at least of current concepts is necessary. The task of organizing in a critical and coherent manner the vast literature on photosynthesis has been accomplished by Rabinowitch (1945) in a treatise hailed as ". . . the outstanding recent event in the field" (French, 1946). The subsequent symposium volume edited by Franck and Loomis (1949) brings the subject to 1949.

The photosyntheses found in bacteria and in other organisms (lumped as "green plants"—a term unambiguous in practice if not in taxonomy) are reducible to the general or overall equation:

$$CO_2 + 2H_2A \xrightarrow{\text{Light}} (CH_2O) + H_2O + 2A \qquad (1)$$

"wherein H_2A represents any one of a variety of oxidizable compounds and A their corresponding oxidation (dehydrogenation) products" (Van Niel, 1949b).

Two molecules of the reductant H_2A are required: one pair of H-atoms reduces CO_2 to the *level* of the photosynthetic product (CH_2O); the oxygen atom thus removed from CO_2, and this first pair of H-atoms reappears in the molecule of water. The second pair of H-atoms reappears in the photosynthetic product having the structurally noncommittal but stoichiometrically accurate designation (CH_2O).

In *green plants*—blue-green, red, brown, and green algae, and land plants—the sole H-donor is water. The equation of photosynthesis then takes the form:

$$CO_2 + 2H_2O \xrightarrow{\text{Light}} (CH_2O) + H_2O + O_2 \qquad (2)$$

On subtracting one molecule of water from each side of the equation, equation (2) reduces to the familiar expression:

$$CO_2 + H_2O \xrightarrow{\text{Light}} (CH_2O) + O_2 \qquad (3)$$

Compared with the variety of photosyntheses encountered in bacteria, where the H-donor may be H_2S, organic compounds, or even molecular hydrogen, photosynthesis in green plants is stereotyped. *The bacterial photosyntheses are all anaerobic in type.* Thus in green sulfur bacteria (Thiorhodaceae):

$$CO_2 + 2H_2S \xrightarrow{\text{Light}} (CH_2O) + H_2O + 2S \quad (4)$$

Free oxygen here is not evolved; elementary sulfur appears instead. Atmospheric oxygen is intensely inhibitory to green and to red sulfur bacteria; however, many non-sulfur purple bacteria (Athiorhodaceae) are readily adapted to aerobic growth. *Only green plants evolve oxygen in photosynthesis.* As yet no exceptions are known.

The primary photochemical process in all photosyntheses may be the photodecomposition of water (Van Niel, 1949b): water serves as the *immediate* source of hydrogen for the reduction of CO_2. An oxidized product is then considered to accumulate and to bring photosynthesis to a stop unless it is disposed of. In bacteria, this hypothetical oxidized product is reduced with the aid of external H-donors, which cannot come from water. In green plants the oxidized product is reduced by the elimination of molecular oxygen. The sequence of reactions in bacteria takes the form:

$$4H_2O + 4Z \xrightarrow{\text{Light}} 4OH + 4HZ \quad (5)$$

$$CO_2 + 4HZ \longrightarrow (CH_2O) + H_2O + 4Z \quad (6)$$

$$4\{OH\} + 2HA \longrightarrow 4H_2O + 2A \quad (7)$$

In green plants, equation (7) is replaced by (8):

$$4\{OH\} \longrightarrow O_2 + 2H_2O \quad (8)$$

In accordance with Rabinowitch's conventions, Z is a hypothetical H-acceptor, and the reactants enclosed in braces do not react in the free condition. Equations (5), (6), and (7) reduce, on addition, to the general equation (1). The crucial point in distinguishing between the bacterial and the green plant photosyntheses is seen to be the fate of the primary oxidation product {OH}. In bacterial photosynthesis the dependence on external H-donors to reduce the {OH} appears inescapable. The green plants, to Van Niel, represent a line of evolution beginning with chemoautotrophs dependent on external H-donors for the reduction of CO_2, followed by the development of a photosynthetic form using water as the reducer, but with additional H-donors still necessary, and culminating, in the green plants, in independence of external H-donors. The presence in green plants of a photosynthetic apparatus capable of using radiant energy of shorter wavelengths than can be utilized by bacteria is the basis of the ability of green plants to carry out an extensive photolysis of water since greater energy per quantum is found at shorter wavelengths. The principal argument for photolysis of water being the primary photochemical process is that otherwise it would be necessary to postulate a separate photochemical activation for each of the many substrates that can be utilized in the bacterial photosyntheses; the purple bacteria can use virtually the same substrates for photosynthesis that ordinary heterotrophic pseudomonads use as oxidation substrates. Moreover, in purple bacteria, the rate of dehydrogenation of the substrates when used as H-donors in photosynthesis, and the rate of oxidation of the same substrates in the dark with molecular oxygen, are the same.

B. Photoreduction and H_2S Utilization

If any of the phytoflagellates were to exhibit the ability to reduce CO_2 with external H-donors, without evolution of O_2, this could be interpreted as a vestige of the more versatile bacterial photosyntheses. Work in this direction with phytoflagellates is hardly started, but because these experiments promise important contributions to future ecological and photosynthetic theory, a detailed account of these beginnings is worth while.

Most of the careful recent work on non-bacterial photosyntheses has been on *Chlorella* and *Scenedesmus,* both protococcoid (sessile, non-filamentous) green algae. How much can this rich fund of information be drawn on for phytoflagellates? On morphological grounds there would appear little objection to the assumption that photosynthesis, at least in the chlorophycean flagellates, would prove similar to photosynthesis in *Chlorella* and *Scenedesmus.* The Protococcales are a polyphyletic assemblage derived from sundry phytomonads; the phytomonads (= Volvocales) are, in the strict taxonomic sense, the flagellates belonging to the Chlorophyceae or green algae. The members of the catch-all genus *Chlamydomonas* approximate the morphology of the archetypal chlorophycean flagellate. Certain species of *Chlamydomonas* form zoospores by vegetative divisions of a single individual and show tendencies for these daughter cells to remain within the parent cell wall, and to lose motility. Permanent loss of motility in this situation would result in multiplication by means of *Chlorella*-like autospores, as suggested by Pascher (1927b). Probably many of the phytomonads are more closely related to protococcoid species than they are to one another. Hence it is to be anticipated that patterns of photosynthesis found in *Chlorella* and *Scendesmus* will be rediscovered among the chlamydomonads.

1. PHOTOREDUCTION

Frenkel (1949) has reported preliminary observations on the reduction of CO_2 by H_2 in *Chlamydomonas moewusii.* This reaction

$$CO_2 + 2H_2 \xrightarrow{\text{Light}} (CH_2O) + H_2O \tag{9}$$

discovered and studied by Gaffron, is reminiscent of the many purple bacteria which can use H_2 as the external H-donor in photosynthesis. Previously only *Scenedesmus* and two other protococcoid algae (but not *Chlorella*), and a blue-green alga, had been shown to exhibit this behavior. The following discussion of photoreduction is based largely on Rabinowitch's (1945a) summary. Photoreduction is induced by keeping the alga for several hours in an atmosphere of hydrogen, in light; darkness is less effective. In the *dark,* adapted algae *absorb* hydrogen from an atmosphere rich in hydrogen and *evolve hydrogen* in an atmosphere of nitrogen. If oxygen is present also, there may be simultaneous uptake of H_2 and O_2 by adapted algae; CO_2 may be fixed as a result, as in hydrogen bacteria.

In *light* there may be:

(*a*) Enhanced hydrogen absorption in an atmosphere of hydrogen and enhanced *hydrogen evolution* in an atmosphere of nitrogen.

(*b*) In an atmosphere of H_2 and CO_2, reduction of CO_2 by hydrogen or by hydrogen donors, e.g., glucose.

Gaffron and his associates have not reported any support of growth by photoreduction, or in the dark by the oxidation of H_2. Since their experiments were carried out over a period of no more than a few hours, at most 30 hours (A. H. Brown, 1948), observations of growth were limited. Moreover in most of these studies the media used for suspension of the cells for respirometric measurements were inadequate for growth, e.g., the buffer solutions used by Rieke (1949). This absence of growth is therefore not as yet to be taken very seriously. If, by either photoreduction in light or hydrogen oxidation in darkness, growth of a green plant were permitted, i.e., if CO_2 were not merely *fixed,* but further *assimilated for growth,* one gap between photosynthesis in bacteria and in green plants would be abridged. In *Scenedesmus,* de-adaptation to photoreduction, as measured by the evolution of oxygen, takes place when the light intensity exceeds 600 lux. But for *Chlamydomonas moewusii* an intensity of at least 1,200 lux is required to restore O_2 production and normal photosynthesis; from this standpoint *C. moewusii* has a more stable anaerobic metabolism than does Gaffron's *Scenedesmus* D_3. A priori one might expect to find photoreduction best developed, perhaps geared to growth, in algae found growing in complete anaerobiosis in nature. The existence of such phytoflagellates is touched on in the next section.

2. H_2S UTILIZATION

Where waters or soils are anaerobic, H_2S is always present, produced, sometimes in large amount, by the bacterial reduction of sulfate. In these circumstances all three types (green, red, and purple) of photosynthetic bacteria live together. This appears to be a suitable environment for certain algae also.

Frenkel *et al.* (1949) observed that *Synechococcus,* a blue-green alga, carried out an ordinary photosynthesis. Under anaerobic conditions in the presence of H_2, the organism could be adapted to carry on photoreduction. When the H_2 was replaced by N_2, and Na_2S added, the sulfide disappeared in light with the simultaneous reduction of CO_2. In agreement with the overall equation for photosynthesis, for each molecule of CO_2 reduced, 2 molecules of sulfide (presumably H_2S) were oxidized; as soon as the sulfide was consumed, the cells reverted to normal photosynthesis. A culture of *Scenedesmus* adapted to photoreduction also reduced CO_2 with simultaneous oxidation of sulfide in light. Nakamura (1937, 1938) had claimed earlier that the diatom *Pinnularia* and the blue-green alga *Oscillatoria* reduced CO_2 with H_2S, sulfur globules being deposited in the cells. The phytoflagellates living at the interface between pond water and a black mud bottom should, by all appearances, be equipped for survival in this milieu that is devoid of, or very low in, oxygen, and high in CO_2. Black muds commonly owe much of their color to ferrous sulfide, and free H_2S (an excellent oxygen-remover) and H_2 are likely to be present in abundance, as in the Dutch canal muds which the Dutch school of microbiology, from Beijerinck onwards, exploited so successfully as a source of obligately anaerobic bacteria. In one suggestive experiment (Lindemann, 1942) lake bottom mud was sealed in a jar: all oxygen quickly disappeared. At the end of 120 days *in darkness* at 10° C., cryptomonads were more abundant and varied than before; the green *Euglena deses* initially present had disappeared, and the colorless *Heteronema acus* survived. Von Dach (1940) noted that the colorless euglenid *Astasia klebsii,* after an initial lag, grew almost as well anaerobically as aerobically. This implies, inescapably, that colorless euglenids can make use of an unidentified anaerobic energy-yielding reaction, since the medium used was sugar-free. Another group of flagellates whose ecology suggests that they should be examined for ability to carry out bacteria-style photosyntheses are those flourishing in waters laden with sewage or unusually rich in organic matter. Euglenids are common in these surroundings (Lackey and Smith, 1940) along with certain chrysomonads and members of the Volvocales (Lackey, 1944).

Because of the greater ease in regulating the gas phase, all experiments on anaerobic photosyntheses and photoreductions have been per formed in respirometric vessels rather than culture vessels. With the usual culture conditions it would be technically difficult to establish the anaerobiosis and regulated atmosphere necessary for photoreduction. The metabolism of facultatively anaerobic phytoflagellates appears to be an attractive field awaiting exploration. One may ask what happens to these facultative anaerobes at night: do they forsake the anaerobic bottom waters and migrate to the oxygen-rich surface? Certain species of *Cryptomonas* are commonly encountered in proximity to black muds and would therefore seem worthy of trial as photoreducers, but, as will be shown later, the mass culture of *Cryptomonas* species may be hindered by an unusual set of difficulties.

III. Descriptive Chemistry and Phylogenetic Relationships

A. Introduction

The aforementioned works on photosynthesis contain detailed descriptions of the carotenoids and chlorophylls associated with photosynthesis, and, as the chemistry of these compounds is an extensive subject in itself, these pigments are dealt with here only as they pertain to phytoflagellates. However, as the distribution of these pigments is of extraordinary taxonomic value, these substances and certain other constituents of phytoflagellates will be discussed from the standpoint of systematic biology rather than of organic chemistry.

The phytoflagellates fall into two great groups: those with predominantly brownish green plastids, and those with grass green plastids containing the same pigments in about the same proportions as that found in the plastids of land plants. This is not an arbitrary taxonomic distinction dependent solely on a difference in distribution of xanthophylls: differences in gross morphology, type of flagellation, silicification, occurrence of unusual chlorophylls, and differences in reserve food and cell wall materials, accompany the difference in xanthophylls (Table I).

TABLE I. Some Chemical Characteristics of Certain Plant Groups

Plants	Chlorophylls	Principal xanthophylls	Storage products	Cell wall constituents
Seed plants, ferns, mosses, stoneworts, most Chlorophyceae	$a + b$	Lutein Neoxanthin Violaxanthin	Starch (fat)	Cellulose Pectins
Brown algae (Phaeophyceae)	$a + c + e$*	Fucoxanthin Violaxanthin	Free and polymerized mannitol	Polymannuronic acids Fucose sulfate esters
Diatoms	$a + c$	Fucoxanthin Diadinoxanthin Diatoxanthin	Fat	Silicified
Dinoflagellates	$a + c$	Peridinin Diadinoxanthin Dinoxanthin	Fat Starch	Cellulose (pectin)
Chrysophyceae	?	Lutein Fucoxanthin	Leucosin	Silicified cell walls sometimes present

* Granick (1949a).

The distribution of photosynthetic pigments and other constituents of the phytoflagellates and related groups is outlined in Table I, compiled largely from Heilbron (1942), Cook (1945), Strain (1949), and J. H. C. Smith (1949). In nearly all species β-carotene is the principal carotene, accompanied by a smaller amount of α-carotene; this also holds for blue-green and for red algae, but the carotenoid pigments of purple bacteria are quite different from those of the green plants—yet another sign of the cleft between the bacteria and the phytoflagellates.

Too little work has been done with representatives of the brown algal stock to determine whether they embody peculiarities in photosynthesis other than those directly resulting from their difference in pigmentation vis-a-vis the green forms. In respect to photosynthetic efficiencies in different regions of the spectrum, there are differences, to some extent predictable from their difference in total absorption spectrum. The storage of oil or fat by many of the brown-pigmented algae does not imply a difference in photosynthesis: storage products have nothing directly to do with photosynthesis.

B. Morphology

The brown flagellates are all asymmetric, with two (sometimes only one) unequal flagella inserted subapically, in a few forms apically. The green flagellates, on the other hand, typically have a symmetrical body, with two equal apical flagella. In general, one flagellum of the brown flagellates, where two flagella are present, bears delicate lateral processes along both (or all?) sides ("flimmer" or "tinsel" type); flagella of the Volvocales lack these processes. The euglenids (*Euglena* and *Astasia*) bear a single row of delicate filaments on one side only (H. P. Brown, 1945); the euglenids may be regarded as derived from the green flagellate stock (p. 49).

C. Silicification

Silicification is a biochemically unexplored yet fundamental feature encountered in the brown stock. Silica skeletal structures are prominent in many chrysomonads: indeed one flagellate group related to them is known as the silicoflagellates. The relationship of the chrysomonads (wherein are placed, tentatively, the somewhat heterogeneous silicoflagellates) to diatoms, heliozoa, radiolaria, Heterokontae, and, more speculatively, the sponges, is postulated in part on the occurrence in all these groups of more or less elaborate siliceous skeletal structures and the possession of the rhizopodial tendency (prominent in some groups; rare or absent in others; see the previous discussion of *amebaė*). This hypothesis is founded on the existence of pigmented chrysomonads bearing silica spicules which are disposed in a manner similar to that of the simplest heliozoa; certain silicoflagellates (Ebriaceae) bear a strong resemblance to radiolaria and to certain dinoflagellates (Hovasse, 1934). Deflandre (1937) has described fossil dinoflagellates with silicified thecal plates.

Spiculate chrysomonads have never been used as experimental objects. yet certain genera, e.g., *Mallomonas,* are fairly common in freshwater; they reach comparative abundance in the ocean. The physiology of silicification is an almost unopened chapter of natural history. Hendey (1946) reported that cultures of a diatom (*Navicula*) possessing the usual silica frustules, lost their frustules in a silicate-deficient medium, and could be subcultured as a naked form. Upon addition of silicate the frustules were regenerated. An entrancing vision thus dawns of putting to the test of experiment the belief that the chrysomonads are exceptionally prone to give rise to amebae. Recent advances in the technology of translucent autoclavable plastics, polytetrafluoroethylene in particular, opens the prospect of doing away with silicate-shedding glassware for the compounding of Si-free culture media, an essential for an inquiry into the metabolism of silicates. With silicate-free culture media at one's disposal, and with cultures of silica-walled flagellates on hand, the transformation of a rigid-walled flagellate into a viable naked ameboid form might be accomplished.

Much the same interest attached to siliceous phytoflagellates attends another group of chrysomonads, the all-marine Coccolithophoridae, which have a cell membrane studded with calcareous bodies. Braarud and Fagerland (1946) succeeded in maintaining several coccolithophorids in impure culture in the laboratory at temperatures up to 25° C.; pure cultures were not attempted and the upper temperature limit was not determined. Judging from the high calcium requirements found for the common inshore (neritic) diatom *Nitzschia closterium* (Hutner 1948), it may prove a delicate enterprise so to regulate calcium supply as to hamper calcification of cell walls without killing the cells by internal calcium deficiencies.

D. Fucoxanthin

The yellow carotenol fucoxanthin, $C_{40}H_{54\text{-}60}O_2$ (the exact structure is in question) is the main xanthophyll of diatoms and Phaeophyceae, and, according to Heilbron (1942), is present also in Chrysophyceae. Strain (1949) notes that in diatoms "this yellow color of the plastids seems to result from the physical state or condition of the pigments or from their geometric arrangement rather than from a preponderance of the yellow pigments." A thorough investigation of the xanthophyll pigments of dinoflagellates, cryptomonads, and heterokont algae would be a help both to taxonomy and to the comparative biochemistry of photosynthesis.

E. Chlorophyll *c* and Porphyrins

Like fucoxanthin, the distribution of chlorophyll *c* is a pledge of the ties linking the brown-pigmented algae: chlorophyll *c* was reported, in small amount, in Phaeophyceae, diatoms, and dinoflagellates, and in none of the green stock. Green algae have chlorophyll *b* which is absent from brown-pigmented algae.

Chlorophyll *c* may aid in elucidating the biosynthesis of the chlorophylls: heme compounds and chlorophylls have the same tetrapyrrolic nucleus (for a stimulating review of this branch of biochemistry see

Granick and Gilder, 1947) and there has been much speculation about a common pathway of synthesis. This likelihood has been raised to a virtual certainty: Granick (1948a) found an induced mutant of *Chlorella* that formed brown colonies on glucose media; the brown pigment proved to be hemin by chemical and spectroscopic examination and by growth tests with the hemin-requiring bacterium *Hemophilus influenzae*. The accumulation of protoporphyrin 9 (the exact designation of Fe-free heme) was attributed to a block in the synthesis of chlorophyll. Another mutant accumulated magnesium (not iron!) protoporphyrin (Granick, 1949a). Chlorophyll *c* may be the near-embodiment of another putative stage between porphyrins and chlorophylls. First, Granick (1949) found that the magnesium-free derivate of chlorophyll *c* from the brown alga *Laminaria* was so soluble in water as to suggest that chlorophyll *c* lacked the phytol side chain, absent in porphyrins. Second, the spectrum of chlorophyll *c* resembled that of a porphyrin—a sign of the resemblance of its resonating ring system to that of protochlorophyll, a pale green compound which may be a precursor of chlorophyll (Rabinowitch, 1945).

It would be most desirable for these investigations to be extended to phytoflagellates and to some of the photosynthetic bacteria which form colonies on aerobic agar plates, thus facilitating the selection of mutant colonies. By using heterothallic strains of sexual phytoflagellates in studying chlorophyll-deficient mutants, it should be possible to subject these mutations to Mendelian analysis, as has been done with the fungi *Neurospora* and *Ophiostoma* (Cerastomella). The phytoflagellates are uniquely suitable for this undertaking: they are probably haploid except for the zygote. Certain of the technical difficulties impeding this type of experimentation are described in the sections on growth of phytoflagellates in darkness and on sexuality.

F. Food Reserves and Cell Wall Materials

Cellular plants contain several kinds of macromolecules functioning as food reserves or as materials of the cell wall. It is no great problem to obtain land plants or seaweeds directly from nature in sufficiently uncontaminated state for one to proceed directly to the isolation and characterization of particular cell constituents. This is seldom possible with the phytoflagellate representatives of the brown and the green stocks: they rarely occur naturally in abundance unmixed with other organisms. Laboratory pure cultures on a heroic scale, approaching the magnitude of a pilot plant industrial fermentation, are generally required for chemical isolation work. Ideas of the composition of the high polymers in phytoflagellates are therefore, for the most part, extrapolations from work with macroscopic plants. Yet from the point of view of evolution these problems are brimful of interest: each major plant group contains substances relatively abundant in, or peculiar to, its own members, and the question then arises as to the presence of rudiments of these substances in the flagellate members (where they exist) in the group. To illustrate: the cell walls of land plants have three characteristic constituents, cellulose, lignin, and pectins. Cellulose has repeatedly been detected in green and brown algae with the aid of fairly specific microchemical tests and x-ray diffraction studies. But it is not known whether lignin is present in algae; and the pectic substances of phytoflagellates have not been investigated chemically. Many such substances in bacteria, because of their importance as surface antigens, have received considerable attention, from which the conclusion may be drawn that pectin is essentially a bacterial and not a phytoflagellate invention. In Phaeophyceae, pectin is replaced by the similar alginic acid and also by polysaccharides derived from L-rhamnose and L-fucose residues (Hirst and Jones, 1946).

1. STARCH, PARAMYLUM, AND LEUCOSIN

Starch in *Polytoma* (the colorless counterpart of *Chlamydomonas*) yielded glucose on hydrolysis (Bréchot, 1937); glucose was characterized as the phenylosazone. *Polytoma* starch was digested by several amylases effective on ordinary starch. The structure of *Polytomella* starch has been investigated by I. A. Wilkinson (unpublished data).

The common occurrence of anisotropic bodies staining blue with iodine would appear to be acceptable proof of the presence of typical starch. *Paramylum* (also spelled *paramylon*), the characteristic reserve carbohydrate of euglenids, is, as its name suggests, different from typical starch: it is not stained by iodine. That paramylum has the same empirical formula as starch was known a century ago (Gottlieb, 1850). In unpublished observations (Hutner and Hockett, 1936), paramylum obtained from mass cultures of *Euglena gracilis* was found after hydrolysis to yield mostly glucose, in accordance with the evidence from preceding studies (Habermann, 1874; Kutscher, 1898). The reason for re-examining paramylum was that *E. gracilis* cannot utilize exogenous, i.e., externally supplied, glucose as a substrate, and this seemed to cast some doubt on the belief that paramylum was built of glucose units. (See p. 74 for a discussion of the metabolism of glucose in flagellates.)

The *leucosin* of cryptomonads has been characterized adequately (Hutchens *et al.*, 1948). The availability of *Chilomonas paramecium*, the colorless counterpart of the photosynthetic *Cryptomonas,* made this achievement possible; *C. paramecium* grows very densely in certain chemically defined simple media. The leucosin granules gave the starch-iodine reaction only after boiling. It was estimated, from color reactions with iodine, that leucosin consists of approximately equal parts of amylopectin and amylose. The sole source of carbon used in this instance for growing the organisms was acetate.

2. CARBOHYDRATES AND CARBOHYDRATE-LIKE SUBSTANCES OF PHAEOPHYCEAE

The polysaccharide reserves of brown algae (briefly discussed by J. H. C. Smith, 1949), differ sharply from those in the green stock. *Laminarin,* a common polysaccharide of fucoids, is a polyglucose composed of β-type 1,3-glycosidic linkages (glucose in starch has the α-configuration). The cell wall material of *Laminaria* consists mostly of alginic acid, probably a poly-D-mannuronic acid (Stacey, 1946). Polymerized sulfate esters of fucose (a methylpentose) also occur in the gums of Phaeophyceae. Free mannitol is frequently present in abundance, perhaps serving here as the equivalent of glucose or sucrose in other plants.

These constituents of Phaeophyceae direct attention to neglected areas in biochemistry. The carbohydrates awaiting discovery in the food reserves and cell wall constituents of brown flagellates are likely to belong to classes of importance in mammalian physiology as well as in seaweeds. Studies of the nutritional capabilities of marine phytoflagellates veering toward saprophytism and phagotrophism should not neglect the peculiar sulfuric acid esters composing a large part of brown and red algae. These sulfuric derivatives, constituting an important factor in the marine environment, are potential substrates for marine heterotrophs.

Organisms utilizing a polymer should also utilize the *repeating unit* (the simplest recurring molecular grouping of the polymer) and the monomeric unit or building unit of the polymer. In the best-known syntheses among carbohydrates—starch and glycogen—the starting point or monomer is not the simplest *hydrolytic* product, glucose, but the simplest *phosphorolytic* product, glucose-1-phosphate. The true monomer is glucose-1-phosphate notwithstanding the sloughing off of the phosphate groups during polymerization (Bell, 1948). An omnivorous marine phagotroph might utilize polymeric carbohydrate sulfuric esters. Failure to grow on a postulated monomer would be evidence that the assumed monomeric unit was too simple, in the same sense that glucose is simpler than glucose-1-phosphate. Polymeric sulfate esters would be new objects of investigation from this standpoint. Sulfuric derivatives of carbohydrates include several kinds of compounds: sulfate esters of glucosamine, uronic acids, D- and L-galactose or methylpentoses. Examples of sulfate-containing substances are alginic acid in Phaeophyceae, agar in Rhodophyceae, and heparin and chondroitin sulfate in vertebrates (Stacey, 1946). Sulfuric derivatives of carbohydrate derivatives have not been reported in bacteria or blue-green algae.

Phagotrophic marine members of the brown stock seem especially suitable for such investigations since their own cellular constituents are undoubtedly similar to unique polymeric components in their substrates. It seems likely, therefore, that they have an ability to utilize these similar substrates to synthesize their own polymers. Hardin's (1942) success in isolating from soil a pure culture of the phagotrophic *Oikomonas termo,* a colorless counterpart of the plastidic chrysomonad *Chromulina,* strengthens the hope that the culture of similar marine phagotrophs will not prove excessively difficult. Unfortunately, Hardin's culture, which, conveniently, flourished in autoclaved media, is no longer available.

G. Phylogeny of the Euglenids

Euglenids are abundant in fantastic variety in nature and have been investigated so often that their uncertain phylogenetic setting is an unusually troublesome problem to protozoologists and phycologists alike, for the euglenids include many colorless phagotrophs as well as green species. Their morphology, ecology, and physiology were concisely reviewed by Jahn (1946). Fritsch (1935) hesitated to affiliate the euglenids with any other algal group. Recent findings suggest chlorophycean affinities. The open- or appressed-spiral body of many if not all euglenids may be derived by torsion from a bilaterally symmetrical ancestor, and the reservoir may be an invagination of the anterior end. The stout flagellum of the uniflagellate genera (which include most of the green species) with its unique bifurcate root is interpreted as derived by incomplete fusion from a biflagellate ancestor (Pringsheim, 1948a). The electron microscope studies of H. P. Brown (1945) indicate that the flagellum of the uniflagellate forms has an axoneme composed of two major and equal fibers. The absorption spectrum of *Euglena* suspensions resembles that of land plants (Baas-Becking, 1926), and hematochrome in *Euglena* and *Haematococcus* is similar, as indicated by Tischer's studies. Incorporating the euglenids in the green stock seems the simplest way of coordinating these morphological and biochemical findings.

H. The Stigma and Hematochrome

The stigma or eye-spot of phytoflagellates has long interested physiologists seeking the simplest basis of phototropism and vision. The stigma does not arise *de novo* in the phytoflagellates. Its phylogenetic *anlage* is almost certainly the system responsible for phototaxis in photosynthetic bacteria and blue-green algae, where definite phototactic structures are lacking. The stigma may represent a morphologically observable concentration of phototactic units, in the same sense that the plastid is a concentration of previously diffuse photosynthetic units, and the chromosome of genes. The evolutionary conservatism of the stigma is shown by the fact that the zoospores of cellular sessile algae are provided with a stigma as also with plastids and flagella.

The substance responsible for phototaxis in photosynthetic bacteria and in blue-green algae was concluded to resemble β-carotene, as observed from the phototactic action spectrum as distinct from the photosynthetic action spectrum. Regions of the spectrum active in photosynthesis were comparatively inactive in inducing photo-orientation, and vice versa (Manten, 1948). This is in harmony with the view expressed by Wald (1943) in a comprehensive review of photoreceptor function (photoreception = phototaxis + vision): "It seems possible now that photoreceptor systems in general are linked with a common dependence upon carotenoids." The yellow color of the β-carotene type of carotenoid is the basis of the phototactic effectiveness of blue light.

The stigma of many euglenids is especially conspicuous. Furthermore there are red species of *Euglena* whose pigment, distributed throughout most of the body, seems indistinguishable from what would result from hypertrophy of the stigma. The connection between the stigma and the body hematochrome calls for clarification, since much chemical work has been done with the hematochrome extracted from red euglenas, but not with extracts of euglenas where the red pigment is almost entirely restricted to the stigma. Many euglenids, e.g., *Euglena gracilis,* contain flecks of hematochrome throughout the cytoplasm which become more noticeable as cultures age.

The strikingly rapid morphological changes exhibited in response to variations in light intensity by certain hematochrome-reddened euglenas, coupled with the report of similar behavior of a red dinoflagellate, suggest a relationship between the hematochrome of red flagellates and the stigma. Johnson (1939) found that *Euglena rubra* changed at dusk from bright red to green. In the morning, as the sun rose, the euglenas reverted to red. The response was so sensitive that a cloud shading the area for 15 minutes induced the change. A change from red to green was accomplished by the migration of pigment granules from a central to a peripheral or a more uniformly distributed position (Johnson and Jahn, 1942). The green → red change was induced by heating cultures to 30° to 40° C. and also by irradiation with infrared or visible light. This rather high temperature range was in agreement with observations in the field, for the red organism became abundant each year when the water temperature reached between 35° to 45° C. During cooler weather this species might exist, it was supposed, in a prolonged green phase. Remarkably similar independent observations were made on a red dinoflagellate (*Glenodinium sanguineum*) from an Alpine lake (Baldi, 1941). It had a red carotenoid-like pigment also, and red and green phases varying with the height of the sun. Chilling transformed the red form into a green form, and the change had the same morphological basis of an alteration in the distribution of pigment masses from a peripheral to a central position.

The happy accident of finding a pond with a water-bloom consisting of red euglenas created the opportunity for isolating hematochrome by the conventional methods of carotenoid chemistry (early references in Tischer, 1944). The pigment was provisionally considered to be isomeric with the ketonic carotenoid astacene, the familiar red pigment of the boiled lobster. Later work was carried out with massive laboratory cultures of *Haematococcus pluvialis*. From 130 l. of media, Tischer obtained 118.5 g. (dry weight) of aplanospores; 100 g. of aplanospores yielded 1.03 g. of free astaxanthin, the parent substance of astacene. The similarity to the lobster pigment was close: astaxanthin in the aplanospores was present in the form of fatty acid esters. Astaxanthin is assigned the following structure (reviewed by Goodwin and Srisukh, 1949):

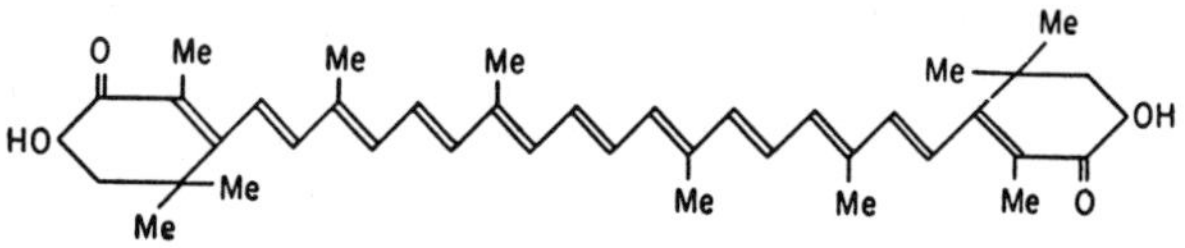

Strictly speaking, the compound is a double α-ketole; spontaneous oxidation of the ketole groups to double ketonic groups gives astacene. Following the current convention for numbering carotene carbons, astaxanthin is 3,3′-dihydroxy, 4,4′-β-carotene. It is noteworthy that Goodwin and Srisukh isolated astaxanthin from the eyes of insects. Tischer observed that astaxanthin was accompanied in *Haematococcus* by the xanthophylls lutein and zeaxanthin and by small amounts of other xanthophylls; α- and β-carotene were present as usual.

There is little evidence that the hematochrome of *Haematococcus* exhibits the same phototactic behavior shown by organisms with the labile red-green transformation. Hematochrome in *Haematococcus* and in the cytoplasm of ordinary green species of *Euglena* and many other algae seems to function differently: when cell division is checked by nutritional deficiencies of one kind or another, *Haematococcus* and many other algae, some of these sessile, pile up hematochrome. An investigation of the phytoflagellates of "red snow" should yield further interesting results.

Evidence for a participation by astaxanthin in the visual purple cycle of the mammalian retina or in the metabolism of vitamin A is lacking. However, astaxanthin occurs in the retinal globules of birds and turtles, but none has been found in chicken tissues other than the retina (Wald, 1943). This, coupled with the finding of astaxanthin in insect eyes mentioned previously, tempts one to ascribe the ostensible absence of astaxanthin in the mammalian retina to the lability of astaxanthin. These are problems of extraordinary difficulty; thousands of eyes must be collected for a single experiment—a mammalian counterpart of the technical difficulties experienced in collecting enough euglenas devoid of cytoplasmic hematochrome for examination of the pigments of the stigma.

I. Miscellaneous Cell Constituents

1. dinoflagellate toxin

Outbreaks of "red water" in the ocean are frequently mass growths of dinoflagellates. Kofoid and Swezy (1921), in a compilation of other reports and their own observations, noted that the repeated appearance of red water off the coast of California was due most often to *Gonyaulax polyhedra* and *Gymnodinium flavum*. Elsewhere, *Cochlodinium catenatum* and species of *Amphidinium* were responsible. Mass mortality of fish and other marine animals was frequently an accompaniment of red water. The earlier workers blamed the mortality on the excessive crowding of the water by the dinoflagellates and on their decomposition products. This deadliness to animals may in part be attributed to a toxin contained in or secreted by the dinoflagellates. Circumstantial evidence links outbreaks of poisoning in humans, with symptoms of paralysis, following eating of mollusks and fish, with the appearance of red water in the areas where the shellfish and fish was obtained. Riegel *et al.* (1949)

isolated a toxin from *Gonyaulax catenella;* these flagellates were so abundant (20–40 million organisms per milliliter) that the ocean (Monterey Bay) was a deep rust red by day and brilliantly luminescent by night. The toxin was extracted with acidified ethanol, and, after removal of the ethanol, fractionated on acid-washed charcoal. The poison had the properties of a strong base. The most active preparations were comparable in potency with the toxin of *Clostridium botulinum:* a dose of 0.6 μg. was lethal for the mouse. The symptoms of paralysis, coupled with the basicity, suggested a quaternary base with a curare-like pharmacological activity. *Botulinum* toxin, while paralytic, has a different site of action in the mammalian body.

The catastrophic mortality of fish and crustacea, associated with an outbreak of red water off the coast of Florida, was graphically described by Gunter *et al.* (1948). Fish, turtles, *Balanus,* oysters, shrimp, and crabs (*Callinectes* and *Penaeus*) were killed. This water-bloom was caused by the unarmored dinoflagellate *Gymnodinium brevis.* The sea water became oily or viscous, and yellow rather than red: it killed fish in the laboratory, whereas sea water taken from just outside the affected area did not. Death was not caused by lack of oxygen.

From the parochial viewpoint of the protistologist, these spectacular consequences of phytoflagellate multiplication are mitigated disasters: they emphasize the prevailing ignorance of the factors controlling the proliferation of marine phytoplankton. Analyses of sea water for the principal mineral nutrients, phosphate, nitrogen compounds, iron, and manganese, are being carried out in government and academic laboratories in an attempt to predict outbreaks. These efforts are hampered by the present complete lack of pure cultures of marine unarmored dinoflagellates necessary for working out exact nutritional requirements. Reported studies on the physiology of marine dinoflagellates with the aid of pure cultures (Barker, 1935) are so scarce as to be countable on the fingers of one thumb, to borrow a phrase.

2. PHYTOHORMONES (AUXIN)

The discovery of auxin in seed plants has had but few reverberations in protistology. In seed plants, appropriate concentrations of 3-indoleacetic acid (the classic "heteroauxin") bring about cell elongation in root and shoot. The effect of light upon the local auxin concentration is responsible for the phototropic bending of the stem. Higher concentrations of auxin inhibit cell elongation: concentrations which are auxetic in the shoot are commonly inhibitory in the root. Auxin has, in addition, profound morphogenetic effects. The development of lateral buds, flowering, and the development of the vascular system in flowering plants, are influenced by natural and synthetic auxins. The biochemical mechanisms of auxin action (which vary with concentration and with the tissue involved) are essentially unknown in spite of the industry with which the problem has been attacked and the success met in the use of auxins as weed-killers.

Considering this situation, it is not surprising that trial of the natural auxin, 3-indoleacetic acid, or its simple homologues has not led to clear results in protococcoid algae (Algéus, 1946). Stimulation of *Euglena gracilis* in light, but not in darkness, and absence of effect on *Astasia* (the colorless counterpart of *Euglena*) were actions exerted by several auxins including naphthaleneacetic acid (Elliott, 1939). The behavior of phytoflagellates, if they are susceptible to the hormonal action of 3-indoleacetic acid (which might exert effects resulting from its involvement in tryptophan metabolism, possibly unrelated to effects on auxin), cannot be predicted on the basis of present knowledge of the mode of action of auxin in flowering plants. The phytoflagellate is, in one philosophic sense, both root and shoot; the effects of auxin might cancel out. Auxin has effects on other cellular, differentiated plants, such as Phaeophyceae; these studies might guide further work on the effects of auxin on protists.

As a prelude to studies of auxin activity in protists, it would be interesting to know whether indoleacetate is a metabolite in phytoflagellates. This question is raised by the likelihood that pseudomonad bacteria may synthesize or degrade tryptophan by a pathway which does not go through indoleacetate (Stanier and Tsuchida, 1949). Auxetic activity in culture fluids of protists has been demonstrated in *Rhizopus* and a few gram-negative bacteria (Schopfer, 1943).

IV. The Metabolic Pool and "Acetate Flagellates"

A. Introduction

Acetate has replaced glucose as the center of interest in studies of the *metabolic pool*. Acetate, in the context of external substrate, means ionic acetate plus a varying proportion of the free acid; where intracellular processes are being considered, it is understood that the true reactant is closely related phosphorylated derivative, most likely an "acetylphosphate" of undetermined structure (Lipmann, 1948).

The intense activity in this field presages the rapid obsolescence of much of the present recital of phytoflagellate activities. For a detailed summary of the principles of intermediary metabolism, the reader is referred to the lucid monographs by Baldwin (1947) and by Work and Work (1949). The University of Wisconsin symposium volume (Lardy, 1949) has much detail on the pertinent enzymes. The yearly reviews are indispensable in coping with the paper blizzard of research publications. For the sake of a more coherent presentation of certain features of phytoflagellate metabolism in need of alignment with this central body of knowledge, the writers have drawn freely on hitherto unpublished data.

The concept of the *metabolic pool,* in which acetate is a key substance, was formulated to explain the rapid spread of isotopic carbon atoms, initially incorporated in the form of various small molecules, throughout many cell constituents previously thought to be off the main stream of metabolism. Conversely, tracer atoms in certain large molecules were detected in small molecules or fragments after a remarkably short incubation period. Two-carbon (C_2) compounds, together with adenosine triphosphate and several C_3 to C_6 fragments, are at the vortex of the metabolic whirlpool. These fragments may, depending on the need of the moment, serve either as fuel or for building blocks in construction of the fats, carbohydrates, proteins, nucleic acids, and the other constituents of protoplasm. C_1 compounds, particularly CO_2, also participate to an important, even indispensable, extent in heterotrophic as well as in autotrophic reactions. "Pool" therefore connotes a stock pile freely added to or drawn upon in furtherance of the activities of the organism.

Certain flagellates, some green, others colorless, are called "acetate flagellates" (Pringsheim, 1937). They grow well with acetate as sole organic carbon source and utilize poorly or not at all such traditional substrates as glucose. The term "acetate flagellate" was proposed for *Polytoma* by Pringsheim at a time when he thought that these flagellates could use only acetic acid. When it was later learned that some related organisms could utilize pyruvic and lactic acids, Lwoff (1938) proposed the designation "oxytroph." Because of the emphasis placed here on the $C_2 + C_2$ combination and because of the recent work of Foster *et al.* (1949), the term "acetate flagellate" is retained in this discussion; it should be understood therefore that this usage differs somewhat from that originally intended by Pringsheim. Earlier knowledge of these flagellates is adequately summarized with respect to respiration (Jahn, 1941) and nutrition (Hall, 1941; and Trager, 1941); other peculiarities they share as a group will be discussed following the consideration of the role of acetate as a substrate in other types of oxidative organisms.

B. Acetate Nutrition

Acetic acid has the ratio $C:H_2O$ found in carbohydrates. Consequently a respiratory quotient (CO_2/O_2) of 1.0 does not indicate whether acetate or carbohydrate is being oxidized. Carbon for carbon, acetate has the same energy content as glucose: it is potentially a good fuel. Oxidation of glucose appears to be the chief source of energy for many organs and organisms, but the assignment of glucose to a preeminent role in the nutrition of aerobic Protista is not justified: acetate has the stronger claim. It is difficult to find a protist which, while dissimilating glucose and showing during the process a substantial uptake of O_2, cannot also utilize acetate. Trypanosomes have an unusual type of oxidative glycolysis in which there is an accumulation of products of splitting such as succinate and acetate (see the chapter by von Brand on Trypanosomes); they seem to be the only obligately aerobic non-acetate utilizers that have been studied in some detail.

Many bacteria which utilize acetate avidly are nearly or totally indifferent to glucose; most strains of non-sulfur purple bacteria fall into this category. The failure of these bacteria and of many phytoflagellates to utilize external sugars has a fairly satisfactory explanation (p. 75). The phytosynthetic phytoflagellates and autotrophic bacteria, which when deprived of light or the appropriate inorganic hydrogen donor do not grow on acetate or any other organic substrate, pose much thornier problems.

1. Acetate Utilization by Organisms Other than Phytoflagellates

As the utilization of acetate appears general in nature, the more significant become acetate flagellates as tools in cell physiology. It may be instructive, therefore, to examine two well-known exemplars of glucose-powered energy systems and then to describe a dissection of the energy system of a superlatively aerobic sugar utilizer; this last example (*Azotobacter*) provides the clearest evidence that acetate is indispensable in the oxidative energy cycle.

a. Mammals. So much of the information and insight required for formulation of the theory of the aerobic energy cycle was derived from studies of mammalian and avian tissue, that the question of the acetate metabolism of the traditionally glucose-energized mammals is of special interest. Acetate is readily oxidized by man (Buchanan *et al.*, 1943). It is not ordinarily considered a good bulk substrate because the alkaline residue left on oxidation places a heavy demand on the pH-buffering systems of the body, just as the consumption of acetate in culture media leads to the accumulation of alkali. In one instance where direct comparisons of the rates of oxidation of acetate and glucose were possible—in perfusates of rabbit heart—acetate was oxidized at nearly twice the rate of glucose (Barcroft, 1943).

b. Yeast. This classical example of a glycolytic organism is not an exception to the rule that aerobic organisms can utilize acetate. Bakers' yeast, which has a more complete system of cytochromes than brewers' or distillers' yeast, had exhibited an O_2 uptake in the presence of acetate that was a little over half that obtained with sucrose (Meyerhof, 1925). Skoog and Lindegren (1947) described a bakers' yeast that was unable to ferment glucose and sucrose except after a period of adaptation; it grew on lactate; acetate was not tried. Reports from other laboratories indicate that external acetate is readily swept into the energy cycle of yeasts.

c. Azotobacter and the Significance of Acetate in the Tricarboxylic Acid Cycle. One of the profound discoveries in the field of cellular oxidations is that few if any compounds are burned to completion by the *direct* addition of oxygen. Complex compounds must first be broken down into C_2–C_6 fragments; for the purpose of exposition, one such fragment is here assumed to be acetate. Acetate in turn must be *condensed* with another compound (also derivable, under certain circumstances, from substrates or complex compounds). The condensed product then undergoes stepwise dehydrogenations and decarboxylations until the equivalent of the original acetate is fully converted to CO_2 and H_2O, and the condensing agent reappears, ready for union with another equivalent of acetate. It may be conjectured that there are circumstances where acetate is to be considered the constantly renewed condensing agent, and that it is the other partner which is cyclically consumed. In the conventional scheme based, as mentioned, largely on experiments with such tissues as pigeon breast muscle, C_6 acids such as citric figure prominently (Fig. 2), hence the term "tricarboxylic acid" cycle. In one current version of the tricarboxylic acid cycle, acetate condenses with oxaloacetate, i.e., $C_2 + C_4$ condensation. If oxidative organisms exist which are unable to effect a $C_2 + C_4$ condensation, which, if any, is the invariant, the C_2 or the C_4 partner, and what is the other component—a C_1 or a C_3 compound or a longer-chain compound? It follows that, should acetate prove to be the invariant component, the next problem in the intermediate metabolism of acetate flagellates is to identify the fragment which condenses with acetate, and also, the identity of the condensation product.

Clear evidence for the critical importance of acetate comes from studies on *Azotobacter agilis.* An induced mutant was found which grew on ethanol, acetate, and malonate; it was unable, unlike the wild type, to grow on glucose, gluconate, *cis*-aconitate, α-ketoglutarate, succinate, fumarate, malate, lactate, and pyruvate (Karlsson and Barker, 1948a). In the wild-type strain, only the enzyme for oxidation of acetate was constitutive; the other substrates were attacked by adaptive enzymes. Oxalate and formate were not attacked by either strain. Furthermore, the mutant was also unable to oxidize oxaloacetate (which the wild type oxidized energetically), but it was still capable of oxidizing acetate at the normal rate. No succinate or oxaloacetate was formed from tagged acetate, as measured by the isotope dilution method (Karlsson and Barker, 1948b).

2. RESISTANCE OF ACETATE FLAGELLATES TO FREE FATTY ACIDS

Since the utilization of acetate by aerobic protists seems more the rule than the exception, the expression "acetate flagellates" needs careful definition if it is to retain descriptive value. Utilization of acetate within a restricted range of substrates is, by itself, insufficient to define "acetate flagellates." However, they may have additional properties which justify this designation.

Viewed as an ecological and physiological group, the acetate flagellates display an exceptional ability to grow in the presence of high concentrations of free fatty acids; only certain *fungi imperfecti* and a few bacteria appear to rival them in this respect. The growth of several of these flagellates as a function of pH and of concentration of lower aliphatic fatty acids and green and colorless members of the Volvocales, particularly certain species of *Chlamydomonas* (e.g., *C. agloëformis*) and typically in *Polytoma* (the colorless counterpart of *Chlamydomonas*). Among euglenids, *Euglena gracilis* and its colorless counterpart *Astasia chattoni* are outstanding examples. Finally, to emphasize that acetate flagellates are a physiological, not a phylogenetic group, the colorless cryptomonad *Chilomonas paramecium,* already mentioned, is capable of very dense growth on acetate. The resistance of acetate flagellates to customarily toxic concentrations of fatty acids and alcohols may be illustrated by one example from the many listed by Provasoli. *Astasia quartana,* in peptone media at pH 5.3–5.5, grew in (per 100 ml.) 0.05 g.

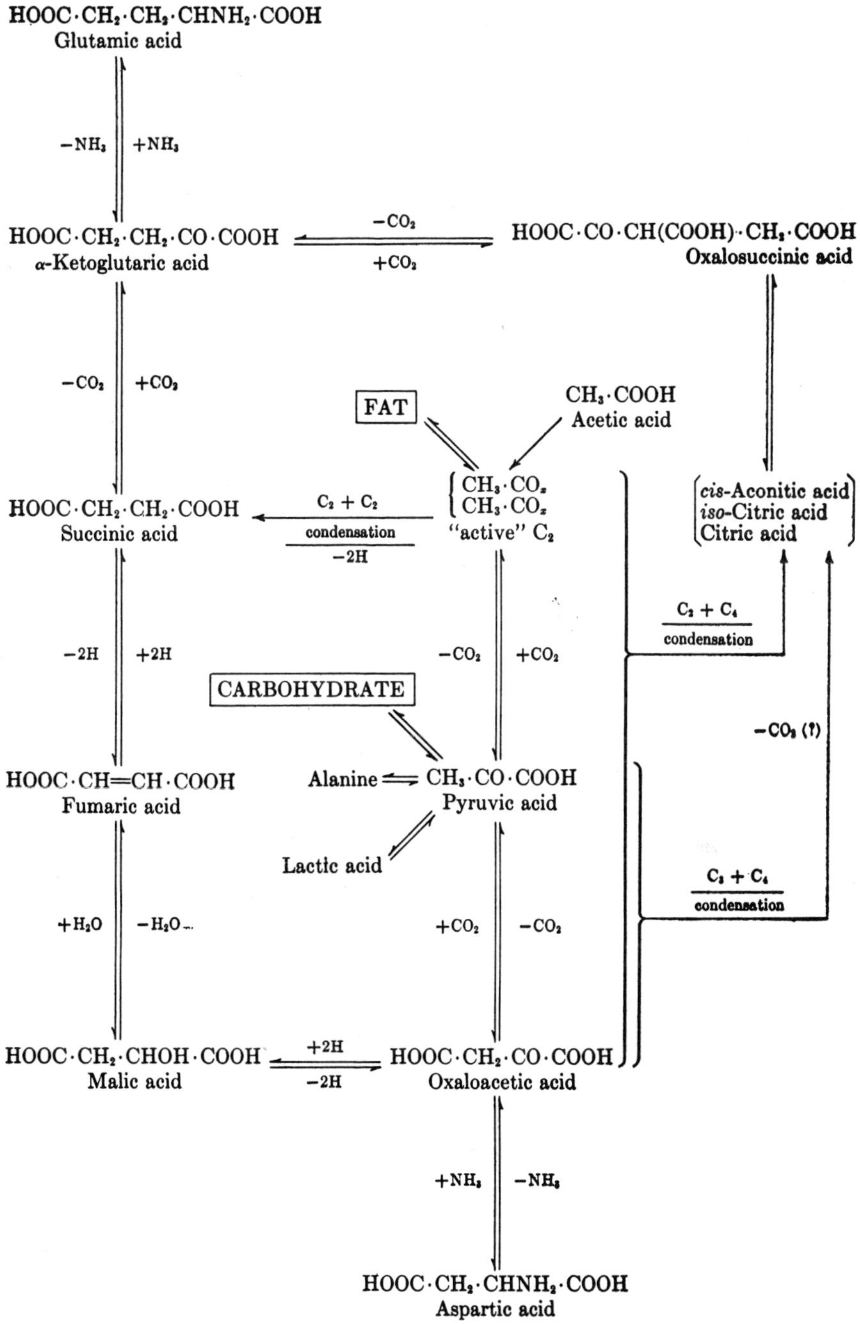

FIG. 2. Aerobic pathways in terminal oxidation.

valeric acid, 0.1 g. butyric or propionic acids, or 0.2 g. of acetic acid. Another striking example of this tolerance is *Polytomella caeca.* In peptone media adjusted to pH 4.5, this colorless volvocine species grew rapidly in 0.2 per cent Na acetate·$3H_2O$ and could be transferred to media adjusted to pH 3.5 containing the same amount of acetate; after heavy growth in this last medium the final pH did not exceed 4.2. In Lwoff's (1941) synthetic medium, which contained half the amount of acetate used by Provasoli, growth took place, after a period of adaptation, in media of pH adjusted to as low as 2.2. Since the pH at which acetic acid in dilute solution is half-ionized is 4.6, and the pH of moderately dilute acetic acid is about 3.0, it is seen that *Polytomella,* and to a lesser extent other acetate flagellates, have an extraordinary resistance to free acetic acid. Purple bacteria utilize lower fatty acids very well, yet are inhibited by 0.1 per cent of the fatty acids when the pH is below 6.0 or so (Van Niel, 1944). Our own experiments with acetate flagellates and purple bacteria have repeatedly shown the distinctly greater resistance to free fatty acids of several of these flagellates.

3. RESISTANCE OF ACETATE FLAGELLATES TO ALCOHOLS AND OTHER COMPOUNDS

The hardiness of the acetate flagellates is also shown by their behavior in alcohol-containing media. Provasoli (1938), showed that *Chilomonas paramecium* and *Euglena gracilis* var. *urophora* grew at neutral pH with (g./100 ml.) 0.2 g. of ethanol, 0.1 g. of butanol, or 0.025 g. of hexanol; other examples from Provasoli (*loc. cit.*) could be cited.

The outstanding resistance to free fatty acids (including acids approaching the soap-formers as the hydrocarbon chain lengthens) and to alcohols (also showing surface activity in the long-chain members) may extend to other classes of compounds. Crude enrichment cultures containing healthy flagellates sometimes had a strong odor of *p*-cresol. Another incidental observation was that *E. gracilis* var. *bacillaris* was not inhibited by 0.05 per cent bile salts at pH 6.4.

4. ECOLOGICAL OBSERVATIONS ON ACETATE FLAGELLATES

The success of Pringsheim's (1946a) enrichment method for colorless flagellates (overlaying pieces of cheese or starch with soil and water) may depend on the concurrence of two conditions: liberation by bacteria of relatively high concentrations of lower fatty acids and alcohols, and the production of other substances killing competing organisms. Acetate flagellates may teem in metabolite-rich milieus; they include several common sewage organisms. Pringsheim's (1946b) general experience was that saprophytic flagellates, such as *Astasia, Polytoma,* and *Chilomonas,* throve in media containing large numbers of bacteria, while most holozoic, i.e., colorless phagotrophic flagellates grew best where there was but little dissolved organic material. Acetate flagellates are specialized to cope with an excess, not a dearth of substrate. Another feature of this pattern of adaptation may be an enhanced ability to withstand low oxygen tensions, as noted in the previous discussion of photoreduction and the photosynthetic utilization of H_2S. Still another environmental factor to be evaluated is their ability to grow in very high concentrations of CO_2; the subject of the role of CO_2 for phytoflagellates is treated later (p. 66). It may be mentioned at this point that some of our strains of acetate flagellates were isolated from black water in ditches surrounding manure piles.

5. ACETATE AND PERMEABILITY

In the usual media (pH 5.5–8.0) acetate flagellates are nearly indifferent to several substrates that might be expected to be utilized by most if not all aerobic organisms, in particular the C_4 dicarboxylic acids of the Szent-Györgyi catalytic respiratory cycle. Acetate organisms have a definite ability to grow on succinate, but they utilize malate poorly (Provasoli, 1938). Several of the characteristics of the acetate flagellates—their resistance to non-polar or to soap-like substrates, and their utilization of a limited number of substrates—may all be consequences of the unusually low permeability of their plasma membrane. It is noteworthy that the substrates which they use are those known to penetrate cells easily. The factors determining permeability are far from clear (Höber, 1945). It is accepted, however, that for predicting the permeation of cells by moderately concentrated solutions, two general rules hold: (1) small molecules penetrate more rapidly than large molecules, and (2) lipid-soluble molecules penetrate more rapidly than water-soluble molecules. Acetate admirably fulfils both conditions. If organisms can oxidize other molecules possessing the physical requisites for penetration, these molecules may constitute superlative substrates: good examples are ethanol, butanol, and butyrate. This unified hypothesis allows an orderly arrangement of much information. Any process which results in an enhancement of lipid-solubility (more precisely definable as a decrease in the dielectric constant of the compound) will increase penetrability too. Acids penetrate more rapidly when they are less ionized, as when the pH of the medium is lowered; bases penetrate more rapidly when the dissociation of hydroxyl ion is repressed by increase in pH; zwitterionic substances such as glycine should penetrate most readily (if slowly) in the region of the isoelectric point.

An inspection of the distribution coefficient *K* between water and ethyl acetate of several important acidic substrates (Marvel and Richards, 1949) brings out the respective differences between acetic and butyric acids and the C_4 dicarboxylic acids in respect to penetrability (*K* is defined as the ratio of the concentration by weight in the water phase to that in the organic phase). *K* for acetic acid is 1.5 and for butyric acid 0.18. *K* is much greater for dicarboxylic acids: for succinic acid 2.7–4.3 (varying with concentration of acid); fumaric acid, 0.59; citric acid, 10 or more. The enormous difference between acetic and succinic acid is evident if one considers that not only is the distribution coefficient of succinic acid roughly twice that of acetic acid, but also that the succinic acid molecule is nearly twice as bulky. The same reasoning applies, with greater force, to citric acid.

It may therefore be conjectured that if an organism is nearly impermeable to substrates and cannot utilize secondary substrates of high penetration ability such as ethanol and butyrate, acetate may be the sole assimilable external substrate. *Chlorogonium euchlorum* and its colorless twin *Hyalogonium klebsii* come close to exemplifying this condition: succinate was reported to be the only substrate utilized besides acetate (Pringsheim, 1937; Provasoli, 1938); fumarate was not tried. One logical extreme of this permeability series would be an organism permeable only to CO_2 as a source of carbon. It would be indifferent to substrates in light, and hence it would be incapable of growth in darkness even in rich media; in other words, it would be a "strict photoautotroph." Failure of certain flagellates to grow in darkness might therefore be attributed to their very low permeability. It may be significant that all the green acetate flagellates tested so far grow in darkness; certain of them (*Euglena gracilis* var. *bacillaris*) very well indeed. Among the obligate photoautotrophs are the following species of *Euglena: deses, pisciformis, klebsii, anabaena, viridis,* and *stellata* (Dusi, 1933, 1941, 1944).

Low permeability cannot be invoked to account for all failures to grow in darkness. As brought out in studies of *Chlamydomonas moewusii* and several autotrophic bacteria (p. 87), an internal metabolic block is more likely to be the immediate cause of this failure. An acetate flagellate incapable of growth in darkness would be of exceptional theoretical interest. It would be premature at this point to speculate in which evolutionary direction the permeability series from acetate flagellate to obligate autotroph is to be read. It is conceivable that some groups may have pursued one path, and other groups the opposite.

Another noteworthy aspect of acetate utilization in acetate flagellates should not be overlooked: the wide range of concentration and pH in which acetate utilization proceeds vigorously. It might be supposed that the acetate flagellates, because of their seeming impermeability, might have difficulty in utilizing ionic acetate. The limited information available does not support this: if media are used which permit growth in the alkaline range (peptone media, or better, synthetic media containing agents preventing the alkaline precipitation of essential metal ions), the upper limit to growth is not determined by decreasing availability of acidic substrates. Values of pH above 9.0 are commonly found after vigorous growth on acetate, even when the initial pH was above 8.0. One is led to infer that acetate flagellates have in addition to the protective permeability barrier an exceptionally efficient internal means of disposing of acetate, or else they might be expected to be overwhelmed by the influx of acetate. Many aquatic protists seem so susceptible to acetate toxicity that the suitability of acetate as a substrate may be missed altogether. A study (Whiffen, 1945) of the nutrition of water molds belonging to the Saprolegniaceae was carried out at pH 6.0. No growth was obtained with 0.01 per cent Na acetate·$3H_2O$—presumably, this low level of acetate was necessary to avoid toxicity. With the aid of synthetic media allowing good growth above pH 6.0, acetate emerged as a good substrate (H. S. Reischer, in press). Organisms may utilize acetate present in relatively high concentrations in neutral or alkaline media and may yet be inhibited by unexpectedly small concentrations of acetate on the acid side of neutrality. The simple assumptions of differential permeability account fairly well for the penetration of *acetic acid,* but the rapid passage of *acetate* across the cell membrane from neutral or alkaline media is probably an activated transport; perhaps a phosphorylation of acetate at the plasma membrane, in the manner now accepted for glucose, maintains the necessary diffusion gradient. The new histochemical techniques for the localization of phosphatase have not been applied to flagellates.

C. Do Acetate Flagellates Have a Tricarboxylic Acid Cycle?

The functioning of acetate in the metabolic pool may be examined in the light of the working hypothesis that acetate participates in a cyclic process resembling (at least in principle) the tricarboxylic acid cycle.

The first steps in acetate metabolism, as mentioned, may lead either to (1) its inclusion into reactions generating energy or (2) transformations preparatory to the utilization in synthesis of one or both of its carbon atoms. Hence the precise characterization of this master reaction is a central problem of phytoflagellate metabolism. Does the tricarboxylic acid scheme—the Krebs cycle—hold for acetate flagellates, or does their metabolism resemble that in *Azotobacter,* where the condensation of acetate with oxaloacetate seems to be unessential even if it may prove to be present? There is enough evidence of the use of C_3, C_4 and C_5 compounds by acetate flagellates (see p. 65 *et seq.*) to support the opinion that their main aerobic cycle will not differ radically from that on which information is gradually being pieced together from a miscellany of other protists. The fragmentary and disjointed data on substrate utilization by acetate flagellates might be regarded as a series of guideposts to a more penetrating study of enzyme systems needed for testing the validity of metabolic cycles.

The failure to utilize one or another exogenous substrate does not rule out the participation in metabolism of the substrate or of closely related compounds. First, the substrate may not have penetrated; second, granted penetration, non-phosphorylated substrates may prove not to be utilized in this case—they are degraded intermediates and the organism may lack the special phosphorylases needed to assimilate them.

1. $C_2 + C_1$ ADDITIONS; GENERAL PRINCIPLES

An increase in the carbon chain, starting from acetate, may be achieved by reductive α-carboxylation, yielding pyruvate:

$$CH_3 \cdot COO^- + CO_2 + H_2 + \sim ph \rightarrow CH_3 \cdot CO \cdot COO^- + HO\text{-}ph - \Delta F_0 \ 5.6 \text{ kg.-cal.}$$

(Lipmann, 1946)

Like other simple carboxylations of this type, this reaction is endergonic, as is seen by the appearance of a negative free energy term on the right-hand side of the equation (for a discussion of thermodynamic notation, see Rabinowitch, 1945, p. 50). Following the convention established by Lipmann, the source of energy, the high-energy phosphate bond, is represented by $\sim$*ph,* and the ordinary phosphate link, left by the draining of energy from the high-energy bond, *HO-ph.* The reversal of this reaction, starting with exogenous pyruvate, may generate a high-energy phosphate bond, along with acetate, a utilizable substrate. If pyruvate can be utilized it may be an important intermediate, although not necessarily an indispensable one. Pyruvate may figure in several additional condensation reactions of the general type just described.

There are two difficulties in work with pyruvate: there is always the risk of spontaneous decarboxylation, and its keto group may be active in forming toxic polymers. Pyruvate reactions, therefore, are better studied in brief respirometer experiments since it is unsafe to draw conclusions from growth experiments extending over several hours or days. The utilization of pyruvate and lactate by *Chilomonas* was noted by Lwoff and Dusi (1934). Some growth results on pyruvate utilization were also recorded by Provasoli (1938) ; they are presented as representative of the type of information needing corroboration by other kinds of experiment. Pyruvate did not support growth of *Hyalogonium klebsii;* it supported slight growth of *Astasia* (2 strains), *Polytomella caeca, Polytoma uvella,* and *Chlorogonium euchlorum;* good growth of *Polytoma ocellatum* and *Chilomonas paramecium;* and dense growth of 2 strains of *Euglena gracilis.*

The study of alanine metabolism contributes to the knowledge of pyruvate metabolism. Alanine may be formed from pyruvate by reductive amination (Fig. 2). The amino group may come directly from ammonia, or else from aspartate or glutamate, respectively in equilibrium with oxaloacetate and α-ketoglutarate; i.e., alanine may be formed by transamination. The equations for the $C_2 + C_1$ route of formation of pyruvate and alanine may formally be condensed to one to express the close tie between these compounds:

$$CH_3 \cdot COO^- + CO_2 + NH_3 \underset{-2H}{\overset{+2H}{\rightleftharpoons}} CH_3 \cdot CHNH_2 \cdot COO^- + H_2O$$

Such a reductive carboxylation requires less energy than does a simple addition of CO_2, hence the amino derivatives of keto acids of the metabolic pool represent thermodynamic stabilization products as well as gateways to protein synthesis. The value of alanine as a substrate may be obscured by its property, shared with other α-amino acids, of forming complexes with metals. Since many of the synthetic media formerly used had scarcely enough essential trace metals, introduction of alanine was likely to induce unsuspected multiple metal deficiencies. Even though these essential metals, e.g., Fe, Mn, Zn, Cu, and Ca, remained soluble while bound in the coordination complex, they became largely unavailable. With the development of media well buffered in respect to metals (p. 102) it became evident that DL-alanine was a suitable substrate for *Euglena gracilis* (Provasoli *et al.*, 1948) and for several other acetate flagellates. In light, in media allowing negligible growth in the absence of substrate, *Chlorogonium tetras* and *C. elongatum* grew well at pH 4.5–5.0 (the only pH values considered). There was slight growth of *C. tetragamum.* Alanine doubled the growth in light of *Chlamydomonas moewusii.* Differences in utilization of alanine deepen the impression that *E. gracilis* var. *bacillaris* and *Hyalognium klebsii* represent opposite nutritional extremes within the acetate flagellates. In several experiments with varied media, *E. gracilis* var. *bacillaris* grew as well with alanine as it did with acetate, even in darkness. *H. klebsii,* on the other hand, was indifferent to alanine in media supporting excellent growth of *Euglena* and of several strains of *Chlorogonium;* a modicum of liver extract supplied the growth factors needed by *H. klebsii.* This ineffectiveness of alanine was not due to its toxicity, because the further addition of acetate allowed growth as usual.

These exploratory findings imply that pyruvate reactions should be taken into consideration in mapping the condensation cycle of the phytoflagellates. "Condensation cycle" as used here connotes the protistan equivalent of the Krebs tricarboxylic acid cycle; the reactants in the condensation stage may differ from those in the Krebs cycle, and the product may not be a C_6 acid. Still, the *principle* of the Krebs cycle is by present indications an indispensable guide in identifying the master reactions of the terminal oxidation and of the interconversions of the constituents in the metabolic pool.

2. CO_2 AS A GROWTH FACTOR

The CO_2 for the postulated carboxylation of acetate to a C_3-acid obviously could come from respiration. In the absence of photosynthesis, far more CO_2 is sloughed off from respired substrates than is needed for assimilatory carboxylations. If, under heterotrophic conditions, the surfeit of respiratory CO_2 is swept from the medium with a current of CO_2-free air, growth slackens or comes to a standstill. This growth inhibition has appeared in all organisms tested so far, sometimes when CO_2 production was vigorous. Given an appropriate organic supplement, the organism may continue to grow in the absence of CO_2 or at the lowest partial pressures of CO_2 experimentally attainable. Compounds able to substitute for CO_2 may represent the immediate products of CO_2 fixation. Therefore the identification of by-passers of CO_2 provides clues to carboxylation reactions vital to the condensation cycle and to other metabolic processes.

A CO_2 requirement has been demonstrated in several colorless counterparts of chlorophyllous flagellates. On a medium containing 0.5 per cent casein hydrolyzate at pH 6.0, growth of *Chilomonas paramecium* was sharply diminished when CO_2-free air was bubbled through (Jahn, 1936). A similar lowering in growth rate was evident when a synthetic medium containing acetate as substrate (Hutchens *et al.,* 1948). Rahn (1941) showed a CO_2 requirement in *Astasia longa* and *Polytomella caeca* by the delightfully simple expedient of placing a drop of inoculated medium on the underside of the coverglass of a depression slide and KOH at the bottom of the cavity. Rahn's medium consisted of 1.0 per cent of a tryptic digest of casein + 0.05 per cent KH_2PO_4, and therefore, as in Jahn's medium, the pH would not be expected to rise unduly as CO_2 was removed. Because of the high content of amino acids in these two media, it might have been difficult to demonstrate the inhibitory effect of deprivation of CO_2, since certain amino acids have been noted to be somewhat effective in replacing CO_2 for bacteria; apparently, amino acids are not very effective by-passers of CO_2 for these phytoflagellates.

The extension of such studies to photoautotrophs is desirable. If a photoautotroph is capable of growth in darkness, are the compounds effective in replacing CO_2 in darkness also effective in light? Photoautotrophs grown in light in mineral media obviously need CO_2 as the source of carbon; the substrates that support growth in darkness may be considered to be by-passing CO_2 *en masse:* a drastic lowering of the partial pressure of CO_2 is best for studying CO_2 as a *growth factor.* The compounds by-passing CO_2 are not necessarily only those concerned in the well-established mass reversible carboxylations of the metabolic pool. Vital reactions requiring only minute amounts of CO_2 might be overlooked until the CO_2 concentration is very low indeed. A reaction involving CO_2, distinct from the terminal oxidation system, exists in *Streptococcus pyogenes* strain C203S which grew without purines in 1.0 per cent CO_2, and had an absolute purine requirement in the absence of CO_2 (Pappenheimer and Hottle, 1940). When it is made probable by isotope techniques that CO_2 is the immediate source of a carbon atom in a compound, it becomes worth while to test this compound for its ability to aid in the by-passing of CO_2. In neat support of the results with *Streptococcus,* there are indications that CO_2 is the source of one of the carbon atoms of the purine molecule (Buchanan *et al.,* 1948).

The concept of the by-passing growth-factor-like activity of CO_2 was first clearly enunciated by Lwoff and Monod (1947). They found that succinate or glutamate, most active when used together, partially replaced CO_2 for *Escherichia coli.* Since, in later sections of this review, succinate and glutamate are listed as substrates for certain acetate flagellates, it is of interest to know how widespread is this means of replacing CO_2. Succinate and glutamate replaced CO_2 for *E. coli* under anaerobiosis (Ajl and Werkman, 1949). As emphasized by Lwoff and Monod (1949), it is doubtful whether, at least with *E. coli,* a complete replacement of CO_2 has yet been achieved. CO_2 was required for the germination of spores of *Clostridium botulinum;* it was replaceable in part by a mixture of DL-malic, fumaric, and succinic acids, and most readily by an unidentified factor in yeast extract. Germination of the spores of the aerobic *Bacillus mesentericus* may be a combination of adenosine and amino acids (Hills, 1949). In the purple bacterium *Rhodospirillum rubrum* S1, oxidation of acetate by O_2 was virtually absent in darkness in the presence of KOH, but was rapid in the presence of CO_2. CO_2 was here replaceable by trace amounts of oxaloacetate, malate, fumarate, or succinate (Van Niel *et al.,* 1947). The ramifications of this line of investigation are evident from the fact that the parenteral administration to dogs of succinic acid stimulated the respiratory center just as CO_2 does.

3. THIAMINE-CATALYZED CARBOXYLATIONS

The decarboxylation of pyruvate to acetaldehyde and CO_2 is the classical locus of the action of thiamine. That the carboxylation of pyruvate to C_4 acid may also be catalyzed by a thiamine enzyme system has been suggested (Jansen, 1949). Thiamine deficiencies occur in phytoflagellates. The pyrimidine and thiazole portions of thiamine were the first growth factors to be identified for phytoflagellates (Lwoff and Dusi, 1937–1938). No example is yet known among phytoflagellates of the need for the intact thiamine, e.g., the inability to couple the pyrimidine and thiazole portions of the molecule. Lwoff (1947) has tabulated activities in fostering flagellate growth of various analogs of the thiazole and the pyrimidine portions of thiamine. The previous literature was discussed in detail by Lwoff (1943). Thiamine sparing by certain substrates and the by-passing of CO_2 are both phenomena leading to the identification of vital carboxylations and decarboxylations in phytoflagellates.

The thiamine-sparing action of fats, well known in mammals, has its parallel in protists, as shown by experiments with *Prototheca zopfii* (essentially a colorless *Chlorella*), which requires thiamine (Anderson,

1945). Thiamine-deficient organisms oxidized acetate vigorously but could not utilize glucose, a good substrate in the presence of thiamine. Thiamine-deficient cultures grown in glycerol accumulated pyruvate. The symptoms of acute thiamine deficiency in birds and mammals are, interestingly, those of pyruvic acid intoxication. Pyruvate in acid solution, in the presence of thiamine, was used as a substrate by *Prototheca*. The thiamine-sparing action of fats in mammals may have this explanation: the breakdown of fats leads to the formation of a C_2 intermediate, thus avoiding the formation of pyruvate, the intermediate in the metabolism of glucose. It has not been proved that thiamine is an *absolute* necessity for birds or mammals fed a mixture of fat, carbohydrate, and protein so adjusted that (1) there is enough fat for energy, (2) enough pyruvate for formation of essential carbohydrates and other pyruvate-derived metabolites, and (3) enough protein or similar material to supply essential N-free C_4 and C_5 acids as well as essential amino acids.

If thiamine is also needed for the catalysis of a $C_3 + C_1$ reaction, such as the familiar Wood-Werkman reaction,

$$CO_2 + CH_3 \cdot CO \cdot COOH \rightarrow HOOC \cdot CH_2 \cdot CO \cdot COOH$$

(this function of thiamine is not established), the by-passing of thiamine would require the provision of oxaloacetate or its aminated equilibrium product, aspartic acid. It follows from these considerations that the nature of the thiamine requirement in phytoflagellates may help in reconstructing their condensation cycle. Such observations may influence the design of animal experiments. There are not enough data on any thiamine-deficient organism from which to judge whether thiamine is completely replaceable by a simple mixture of C_2, C_3, and C_4 substrates. In *Prototheca,* thiamine exerted a stimulating effect even in the presence of acetate. A *Phycomyes* made more growth in a medium lacking thiamine when the C-source was acetaldehyde, ethanol, or acetic acid instead of sugar (Robbins, 1939).

In suspensions of thiamine-depleted staphylococci, oxaloacetate or fumarate stimulated the anaerobic conversion of pyruvate to lactate, acetate, and CO_2 (Smyth, 1940). As these were short-term manometric experiments, the meaning of these results in terms of growth is problematic.

Certain thiamine-requiring acetate flagellates are able to grow in light in substrate-free media and can grow in darkness with a variety of substrates. They should be excellent experimental objects for elucidating the interactions between thiamine and substrates. It is of interest that the first autotroph for which a vitamin requirement was identified is *Euglena pisciformis.* This obligate photoautotroph, when grown in a medium containing asparagine as sole organic constituent, required thiamine (Dusi, 1939). It is likely that in organisms like *E. pisciformis,* which may not be able to utilize exogenous substrates, the thiamine requirement may be absolute. In thiamine-deficient acetate flagellates the thiamine requirement appears to be responsive to changes in substrate, as indicated by some results with *E. gracilis.* Thiamine was reported to stimulate the growth of *E. gracilis* in the dark (Lwoff and Dusi, 1938). It stimulated *E. gracilis* var. *bacillaris* in light in media containing glutamate and butyrate (Hutner *et al.,* 1949). In experiments with media containing either glutamate or alanine as sole substrate, no stimulation by thiamine was evident. A clue to this puzzling variability is provided by the preliminary observation (Cramer and Myers, unpublished) that this flagellate may have an absolute requirement for thiamine when grown in light in an inorganic medium (except for a trace of vitamin B_{12}) in air + 5 per cent CO_2.

These fragmentary observations obviously cannot yet be integrated into a map of the condensation cycle or other reaction chains. But excellent experimental material is available. Thiamine deficiencies have been reported in several other acetate flagellates: *Chilomonas paramecium, Polytomella caeca,* and *Polytoma ocellatum* and *P. caudatum.* Other potentially useful thiamine-deficient organisms are three species of purple bacteria (Hutner, 1946); each of these species includes strains readily adaptable to growth in darkness. With the discovery of a thiamine-deficient induced mutant of *Chlamydomonas moewusii* (Lewin, unpublished), another obligate photoautotroph can be studied alongside *Euglena pisciformis.*

4. UTILIZATION OF C_4-DICARBOXYLIC ACIDS: $C_2 + C_2$ *vs.* $C_3 + C_1$ ADDITION

CO_2 and thiamine function in reactions, among others, in which C_3 and C_4 compounds are more or less reversibly convertible to smaller molecules. The utilization as substrates of C_4 and larger compounds suggests that at least a few of the syntheses shown in other organisms to involve CO_2, or thiamine, or both, occur in acetate flagellates. The familiar C_4-dicarboxylic acid series may be considered first (the components are arranged in order of progressive oxidation; see also Fig. 2):

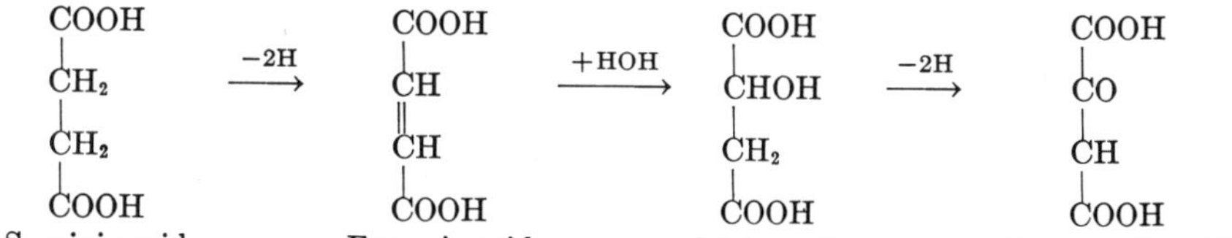

Succinic acid Fumaric acid Malic acid Oxaloacetic acid

The most lipid-like compounds of this series, succinic and fumaric acids, should most readily penetrate the organism. Since all four compounds are strongly ionizing acids, penetration should be better in the acid pH range. Succinate was utilized by certain strains of *Polytoma,* by *Chlorogonium euchlorum, Hyalogonium klebsii, Polytomella caeca, Chilomonas paramecium, Euglena gracilis,* and *Astasia longa* (Pringsheim, 1937b,c). These findings were in the main confirmed by Provasoli, Lwoff, and Dusi (summarized by Provasoli, 1938). *Euglena gracilis* var. *urophora,* Mainx' strain of *E. gracilis,* and *Astasia chattoni* (which corresponds very closely, like *A. longa,* to apochlorotic *E. gracilis*) used succinate hardly at all. Many of the discrepancies in the reports on utilization of succinate might vanish if experiments were conducted at lower pH's, and with higher concentrations of substrate to compensate for the poorer penetrability of this type of compound. Thus, *E. gracilis* var. *bacillaris* used succinate poorly in the region of neutrality, but in acid media (pH 3.0–3.6; media more acidic were not tried) succinate, fumarate, and malate were excellent substrates, almost as good as acetate at higher pH (unpublished results). The favorable influence of low pH was especially striking with glutamate. Similar results were obtained with the metaphyte *Spirodela oligorhiza* (a member of the Lemnaceae). The data on the utilization as substrate of aspartic acid or its monoamide, asparagine, is too incomplete for inclusion in this discussion. It may be noted that asparagine supported the slow growth in darkness of *Chlamydomonas agloëformis* and of *Haematococcus pluvialis* (Lwoff and Lwoff, 1929) and *Chlorogonium euchlorum* (Lwoff and Dusi, 1935b), and is an excellent substrate for *Polytoma uvella* (Rottier, 1936).

Different acetate flagellates utilize the main C_4-dicarboxylic acids at different rates. The limited permeability of the acetate flagellates seems to be an important obstacle to the assimilation of these acids. Are these acids essential parts of the condensation cycle? One must resort to indirect evidence for the most part if one is to understand the carbohydrate metabolism of acetate flagellates.

The usual scheme for the formation of a C_4-dicarboxylic acid from smaller fragments is a $C_3 + C_1$ condensation, such as the Wood-Werkman reaction. The best evidence for this reaction came from the demonstration that propionic acid bacteria took up a mole of CO_2 for each mole of succinic acid formed from glycerol; this was confirmed by isotope methods. Fixation in a compound of isotopic carbon, originally supplied as CO_2, is not by itself proof that the compound acquiring isotopic carbon is a primary reactant in a carboxylation reaction; there must also be proof of a stoichiometric uptake of CO_2. Thus, Foster *et al.* (1949), using C^{14}-labeled CO_2, found radioactivity appearing in the carboxyl group of C_4 acids recovered from *Rhizopus,* probably because of the reversibility of the oxaloacetate decarboxylase; there was no *bulk* fixation of CO_2 under the experimental conditions.

The problem of the role of acetate in the condensation reaction of acetate flagellates takes on a new complexion with the unequivocal proof by Foster *et al.* (1949) that *Rhizopus nigricans* formed C_4-dicarboxylic acids by the long-disputed oxidative methyl-to-methyl condensation of two molecules of acetate (the Thunberg-Wieland condensation). The nearly quantitative yield of labeled succinate + fumarate from labeled ethanol or acetate effectively ruled out the participation of C_6 acids in the condensation, since, had the C_4 acids been formed by way of the C_6 acids, there would have been a loss of some of the radioactivity in the CO_2 sloughed off in passing from C_6 to C_4 acids. When carbinol-labeled ethanol was used, the reaction was formulated as follows:

$$\begin{array}{l} CH_3C^{14}H_2OH \\ CH_3C^{14}H_2OH \end{array} \rightarrow \text{Intermediate I} \rightarrow \text{Intermediate II} \rightarrow \begin{array}{l} \quad\; CH—C^{14}OOH \\ \quad\;\; \| \\ HOOC^{14}—CH \\ \text{Fumaric acid} \end{array}$$

". . . Intermediate I is an 'active' C_2 compound, possibly phosphorylated, and Intermediate II is the primary product of the condensation reaction, presumably an 'active' succinate." It is noteworthy that the intact mycelium did not oxidize succinic acid, whereas preparations of dried mycelium did, an indication that limitations of permeability impeded the use of succinic acid by living mycelium. The dissimilation of the C_4 acids in *Rhizopus* probably proceeds by the usual decarboxylation to pyruvate, followed by the decarboxylation of pyruvate. The metabolism of *Rhizopus nigricans* appears to involve, under anaerobic conditions with glucose as substrate, the usual bulk fixation of CO_2 leading to oxaloacetate (Foster, 1949). It would thus appear that this organism may carry out two types of condensation: the $C_2 + C_2$, and $C_3 + C_1$, the choice depending on the degree of aerobiosis.

The problem of the condensation cycle in acetate flagellates must now be redefined: it is not *what kind* of condensation occurs, but rather *which* of the two types of condensation leading to C_4-dicarboxylic acids is the more important under varying conditions. More data on the by-passing of CO_2 and thiamine might clarify this issue. The same problem arises in deciphering the mode of action of streptomycin on chloroplasts (p. 79). There is no evidence contradicting the utilization of C_2 compounds in mammals via a $C_2 + C_2$ condensation (Lifson *et al.*, 1948).

5. $C_4 + C_1$ ADDITION: GLUTAMATE

Glutamate is in the same class as acetate as a nearly universal substrate for aerobic organisms. In a manner analogous to that accepted for alanine and pyruvic acid, glutamate is in biological equilibrium with α-ketoglutaric acid. This amino acid may be formed by the reductive carboxylation of succinic acid (Ajl and Werkman, 1948):

$$\underset{\text{Succinic acid}}{HOOC\cdot CH_2\cdot CH_2\cdot COOH + CO_2} \underset{-2H}{\overset{+2H}{\rightleftharpoons}} \underset{\alpha\text{-Ketoglutaric acid}}{COOH\cdot CH_2\cdot CH_2\cdot CO\cdot COOH}$$

Glutamate was an excellent substrate for *E. gracilis* (Provasoli *et al.*, 1948; Hutner *et al.*, 1949). A drawback to the use of glutamate and aspartate may be the retention in them of toxic amounts of copper, since these may be purified through the copper salts, a common procedure with amino acids. The use of media heavily buffered in respect to chelating metals minimizes this hazard. In unpublished experiments, glutamate supported excellent growth in light and darkness of those acetate flagellates which were less exacting in respect to utilizable substrates. Glutamate failed to support growth of species of *Chlorogonium* and *Hyalogonium*. In keeping with its extreme water solubility and relative bulkiness, its utilization was most rapid in acidic media, closer to the lower pH limits of many organisms. The importance of glutamate (and, of course, α-ketoglutarate) in many reactions makes it worth while to mention these scanty data. Judging from the efficacy of glutamate as a by-passer of CO_2, one deduces that the carboxylation reaction connecting the C_4- and C_5-dicarboxylic acids may be generally essential. The *Azotobacter* data mentioned previously (p. 57) challenge, however, the idea of the universal indispensability of this reaction.

6. OTHER CARBOXYLATION REACTIONS

The limited information on substrate utilization in acetate flagellates, and the paucity of experiments with enzyme preparations preclude a discussion of the possible existence in acetate flagellates of other reactions closely related to the condensation cycle, such as reversible carboxylations with malate or oxalosuccinate as direct reactants.

It is seen that acetate flagellates may be valuable in studying condensation reactions within the metabolic pool. They may include strains with a high heterotrophic CO_2 requirement, and hence they may prove of value in studies of CO_2 metabolism. For instance, a strain of *Astasia* initially defied isolation: single flagellates did not survive washing with sterile medium. But when the washing was carried out in a current of CO_2, the flagellates survived (Provasoli, unpublished). This strain has since been maintained without difficulty by mass transfer in peptone media; it may be a flagellate counterpart of bacteria like *Brucella,* which need a high concentration of CO_2 (*ca.* 10 per cent) on isolation and then grow without CO_2 enrichment in artificial media (McCullough and Dick, 1942).

D. Glucose Utilization and Phosphorylations

THE GLUCOSE BLOCK

One of the distinctions of the acetate flagellates, as already noted, is their inability to utilize exogenous carbohydrates. An element of paradox long attended this phenomenon, since their reserves are mostly carbohydrate in nature (p. 47). The recognition that the equilibrium between glucose and starch (or glycogen) depends on phosphorolysis made this failure to oxidize glucose less mysterious. The ready utilization of glucose in nature does not alter the belief that glucose in the non-phosphorylated state is a degenerate substrate. *Astasia klebsii* showed an uptake of O_2 when supplied with fructose diphosphate, but not with glucose (Von Dach, 1942). Meanwhile it had been found that several bacteria which were unable to utilize glucose could oxidize glucose monophosphate and fructose diphosphate (Barron and Friedemann, 1941); the addition of adenosine triphosphate did not overcome the block in glucose utilization. It was therefore concluded that the phosphorylating enzyme for glucose (hexokinase) was missing, not the coenzyme. The existence of a similar block in an acetate flagellate was thoroughly documented by Lwoff, Ionesco, and Gutmann (1949), who succeeded in demonstrating a phosphorylase in *Polytomella caeca* which converted starch to glucose-1-phosphate. This phosphorylase was associated with granules, probably mitochondrial. Glucose-1-phosphate was isolated from the cells. Hexokinase and phosphoglucomutase could not be demonstrated, and thus free glucose could not serve as a metabolite for the organism.

The phosphate esters of the glycolytic and phosphate transfer systems were difficult to extract from *Chlorella* by the mild methods successful with muscle and autolyzed yeast and the different phosphate fractions had an unusual distribution of components (Emerson *et al.,* 1944). The respiration of this *Chlorella* was more stimulated by a mixture of hexose monophosphates than by glucose.

Results similar to those with *Polytomella* and, to some extent, with *Chlorella,* were obtained in *Euglena gracilis* var. *bacillaris* (Albaum *et al.,* 1950). Acid-soluble phosphate (P) was extracted and fractionated as barium salts according to the procedures of the Wisconsin group. The following P compounds were identified in the fractions: adenylic acid, adenosine diphosphate (ADP), adenosine triphosphate (ATP), coenzyme I, riboflavin phosphate, glucose-1-phosphate, fructose-6-phosphate, fructose 1,6-diphosphate, phosphoglyceric acid, and a compound which was probably inorganic metaphosphate. The presence of adenylic acid, ADP, and ATP was verified by enzymatic tests as well as by chemical analysis. Since the analyses were incomplete, additional intermediates in the glycolytic cycle were probably present.

These studies show that lack of hexokinase may provide the most reasonable explanation of failure to utilize carbohydrates.

Soluble starch. There has been controversy on the utilization of soluble starch. There are at least two complications: (1) all starch contains some hexose phosphate; and (2) starch tenaciously adsorbs fatty acids. [Starch is commonly added to media for delicate bacteria (*Neisseria, Hemophilus,* etc.) because it adsorbs otherwise toxic fatty acids.] Before drawing conclusions as to the utilization of soluble starch, it might be discreet to investigate the utilization of hexose phosphates.

E. Higher Alcohols and Higher Fatty Acids

Alcohols. Certain acetate flagellates have, it will be recalled, an outstanding ability to withstand concentrations of alcohols inhibitory or toxic to most other organisms. It is generally agreed that higher alcohols, when usable (these are almost exclusively the even-numbered carbon compounds), must first be converted into C_2 fragments, perhaps of the "acetyl phosphate type." Acetate flagellates, in the course of attacking ethanol, may convert it partly to free acetate or acetic acid instead of transforming it directly into acetyl phosphate. Production of acid from ethanol is also found in pseudomonads of the *Ps. fluorescens* group (Stanier, 1947) and in *Acetobacter* (Tosic, 1946). A small amount of C_5-monocarboxylic acid was isolated from mass cultures of *Polytoma ocellatum* supplied with isoamyl alcohol (Provasoli, 1938). Odors of the corresponding fatty acids were detected in cultures containing ethanol, propanol, butanol, or hexanol. The less exacting acetate flagellates showed a rough correspondence between their utilization of longer-chain acids and alcohols.

The utilization of hexanol by acetate flagellates provides a commentary on the limitations of simple methods for studying the turnover of organic compounds in nature. Den Dooren de Jong (1926) made enrichment cultures with hexanol as sole substrate in a simple mineral medium. No hexanol-users appeared. Since higher straight-chain alcohols occur in waxes, the apparent impregnability of hexanol was a puzzle. However, as noted (p. 75) several types of acetate flagellates attack hexanol, among them some photosynthetic forms. Several of the colorless non-euglenid hexanol-utilizers require thiamine, e.g., *Chilomonas paramecium, Polytomella caeca, Polytoma ocellatum;* the term thiamine as used here does not imply that the intact molecule is required: for references to reviews on this subject see section on thiamine metabolism (p. 68). Furthermore, many purple bacteria utilize hexanol (Foster, 1944). The pseudomonads, and to some extent actinomycetes and proactinomycetes, outdo other organisms in attacking most simple dissolved organic compounds. But if appropriate growth factors are present, and if the range of hexanol-utilizers is extended by supplying light, hexanol is restored by acetate flagellates and purple bacteria to the natural cycles of assimilation and dissimilation. It would appear that acetate flagellates and purple bacteria can meet the expected formidable competition for certain substrates offered by the ubiquitous and efficient pseudomonads and other heterotrophic bacteria.

Acetate flagellates which do not utilize ethanol include several species of *Chlorogonium* and *Polytoma, Hyalogonium klebsii,* and *Astasia quartana.* Ethanol, on the other hand, is an excellent substrate for *Polytoma ocellatum, Chilomonas,* and *Euglena gracilis* (Provasoli, 1938). Strains not using ethanol do not grow with the longer-chain alcohols as substrates.

F. Glycine

It has long been known that glycine is an excellent N source for *Polytoma uvella* (E. G. Pringsheim, 1921; Lwoff and Dusi, 1941); *Polytoma uvella* is also able to utilize cysteine, asparagine and glucosamine as N sources (Lwoff, 1929, 1932; Pringsheim, 1937; Lwoff and Dusi, 1941). This amino acid, like alanine, is prone to be inhibitory because of its metal-complexing properties. In metal-buffered media it has proved an excellent substrate for *Euglena gracilis* in light and in darkness. The intermediary metabolism of glycine in phytoflagellates, if it follows the pathway being revealed in other organisms, should present many points of interest. Rats fed N-labeled glycine showed a high isotope concentration in the serine fraction. When serine labeled in the amino and carboxyl groups was fed, almost the same ratio of N:C activity was found in the glycine, indicating the reversibility of the reaction and the loss of the β-carbon atom of serine in the formation of glycine from serine (Rittenberg, 1948).

Glycine, so far, is the only substrate utilized by an acetate flagellate which is neither an immediate derivative of a component of the metabolic pool, nor derivable by simple repetitive clastic reactions from straight-chain acids, alcohols, or aldehydes. For a compound not in the metabolic pool to be utilized as an energy source or to become a building block, it must be transformed by stepwise reactions until it attains the structure of one or another intermediate in the metabolic pool (for excellent examples of such oxidative reaction chains see Stanier and Tsuchida, 1949; Stanier, 1948; Evans, 1947). It follows from this that the ability to oxidize glycine must be accompanied by the ability to oxidize the compounds leading from glycine to the terminal oxidation system. What these compounds are for *Euglena* remains to be seen.

G. Acetate Flagellates and the Origin of the Protozoa

"*Acetate flagellates*" is, then, a convenient shorthand term for chlorophyllous or colorless flagellates capable of flourishing in media rich in acetate and, to a varying extent, in media rich in simple acids, alcohols, and other compounds near the metabolic pool. The acetate flagellates studied thus far have little if any ability to grow with the aid of hexoses and complex substrates, despite the fact that both the chlorophyllous and the colorless forms commonly have carbohydrate reserves.

It is a commonplace observation that many protozoa live in the same substrate-rich environments as those preferred by the acetate flagellates. The hypothesis has been advanced that acetate flagellates may be ancestral to certain protozoa. The amenability of the acetate flagellates to observation and mass culture in the laboratory endows them with practical scientific importance. But in framing comprehensive theories of the evolution of the protozoa, the great host of other free-living flagellates should not be neglected. The features of the acetate flagellates discussed in the previous paragraphs are clearly adjustments to an environment rich in *dissolved* substrates. The more common condition in nature—in the open ocean, in bogs and large lakes, and in non-arable soils—is that of an extremely low concentration of dissolved nutrients. Here the advantage would lie with organisms able to obtain food in concentrated packets—in the form of bacteria and other minute organisms. The specializations for concentrating dilute nutrients are found one step removed in the feeding chain—in the food organisms of the protozoan, not in the protozoan itself. Carnivorous seed plants such as *Drosera, Dionaea,* and *Sarracenia* supplement their diet with trapped animal bodies not because their environment is rich in nutrients but for precisely the opposite reason. (These plants grow in acid bogs or sandy wastes notoriously deficient in nitrogen.) Acquisition of the power to secure nitrogen (and probably other nutrients) directly from animals confers a selective advantage upon them. The orchids, many of which are adapted to a substrate-rich environment and show striking saprophytic tendencies (seeds lacking food reserves, and the reduction or absence of chlorophyll), would seem the ecological counterpart of the acetate flagellates. The pelagic chrysomonads and dinoflagellates, and their colorless phagotrophic, i.e., holozoic, relatives may have sprung from organisms with feeding habits quite unlike those of the acetate flagellates. There appears to be no pressing reason for interpolating a saprophytic stage in the sequence holophyte → colorless phagotroph (or "typical" protozoan) except among forms unmistakably related to acetate flagellates.

Data on the non-sulfur purple bacteria have been cited freely in the present discussion of the physiology and ecology of the acetate flagellates. The parallelism of the two groups is remarkably close. Both are versatile. The majority of the purple bacteria studied to date are readily adapted to growth in darkness, deriving their energy and carbon from ordinary heterotrophic oxidative assimilations. If, by a stretch of the imagination, a greater permeability were allowed the acetate flagellates, their extended range of substrates would probably resemble that of the purple bacteria. Significantly, most of the purple bacteria do not metabolize hexoses. Save for *Rhodomicrobium vannielii* (Duchow and Douglas, 1949) the purple bacteria thus far studied need vitamins. Similarly, it may turn out that one source of highly successful acetate flagellates, the euglenids, may have a phyletic deficiency in thiamine and vitamin B_{12} (p. 95). These vitamin requirements may be taken as further evidence that both groups are attuned to a biologically crowded environment. In both groups the faculty of photosynthesis may be a means of surviving substrate-lean intervals; photosynthesis is not obligate (aerobic strains of purple bacteria can be grown in darkness also). The competitive advantage of both groups would seem to rest, somewhat paradoxically, on their versatility rather than on their specializations. This comparison again shows that the acetate flagellates are a physiological and not a monophyletic group.

V. Streptomycin-Induced Apochlorosis

A. Historical and General

The spontaneous and irreversible change of green into colorless euglenas has been noted only rarely since the original observations of Ternetz (1912). The autonomy of plastids (reviewed by Rhoades, 1946, 1949; Granick, 1949b; and Darlington and Mather, 1949) was underscored when Lwoff and Dusi (1935a) induced apoplastidy in *Euglena mesnili.* Individual flagellates normally have 75–100 chloroplasts. They multiplied slowly in darkness in peptone. After 15 months in darkness most of the individuals had only two, and some only one pale green chloroplast. A few had lost all chloroplasts although retaining the stigma; these individuals no longer formed paramylum granules, which led Lwoff and Dusi to the conclusion that the formation of paramylum necessitated the presence of plastids. Observations on loss of chlorophyll were reviewed by Lwoff (1943); a similar transformation in a chrysomonad is referred to there.

On the basis of observation of two instances of spontaneous green colorless transformations in clone cultures of *Euglena gracilis,* and after a detailed study of various strains of *Astasia* (Pringsheim, 1948b), *Astasia longa* was renamed *Euglena gracilis* forma *hyalina.* How the permanent loss of chlorophyll came about remained hidden. The ease with which *E. gracilis* grows in darkness indicates that apochlorotic forms have a good chance to survive. Like chrysomonads (Pringsheim, 1946b), euglenids normally become colorless in darkness, and their chloroplasts dwindle. When *E. gracilis* is restored to light, fully green chloroplasts may appear in a few hours.

The discovery that streptomycin (SM) readily induced permanent apochlorosis in certain strains of *Euglena gracilis* marked the advent of an exquisitely selective means of probing into chloroplast function. With the sensitive *bacillaris* and *urophora* strains of *E. gracilis*, as little as 40 µg./ml. of SM brought about virtually complete apochlorosis in 15 days; 1.0 µg./ml. caused a permanent average reduction in greenness of individual flagellates to about half the normal value. Concentrations as high as 5 mg./ml. inhibited neither the green forms of these strains when they were grown in darkness nor the induced apochlorotic forms. SM induced apochlorosis equally effectively in light and in darkness, and in non-proliferating and proliferating cultures (Provasoli *et al.*, 1948). More observations on bleaching of *E. gracilis* by SM were reported by Lwoff and Schaeffer (1949a,b). Similar observations were recorded independently by Jírovec (1949): the Mainx strain of *E. gracilis* was rendered colorless by SM. However this strain (in our experience much more resistant to the bleaching action of SM) was inhibited at 500 µg./ml. *Astasia, Polytoma, Polytomella,* and *Chilomonas* were sensitive to SM, and *Chlorogonium elongatum* and *Chlamydomonas agloëformis* were completely inhibited by 10 µg./ml. *Euglena stellata* was also bleached, but grew poorly. The hemoflagellates *Strigomonas oncopelti* and *S. culicidarum* were, on the other hand, not inhibited by 5.0 mg./ml.

One may infer from Jírovec's report of the inhibition of colorless species, and also from the reminder that SM is an antibiotic combatting non-photosynthetic organisms, that the injury by SM may extend beyond chloroplasts. *Euglena gracilis* is exceptional in suffering but negligible damage aside from loss of chloroplasts. Perhaps the commonest effect of SM in phytoflagellates is that represented by *Chlorogonium euchlorum* and *E. mesnili* (Lwoff, 1950). Untreated green individuals grew in the dark on simple media. Colorless organisms were no longer viable—the damage was irreversible. In confirmation of previous reports, colorless *E. gracilis,* obtained by SM treatment, grew.

Chloroplasts of metaphytes are also destroyed by SM. This was observed in the seedlings of gramineous plants (Bracco and von Euler, 1947, 1948). SM changed crown gall tumors of the carrot from green to white and inhibited the growth of sunflower root and tumor tissue (De Ropp, 1948), from which it was concluded that SM is a general inhibitor of embryonic plant tissue.

If the sensitive locus in the chloroplast is the stroma and not the chlorophyll, photosynthetic protists lacking plastids might not show any differential destruction of chlorophyll before the general inhibitory effect of SM was felt. Purple bacteria treated with SM did not show any change in color before complete inhibition set in (Provasoli *et al.,* 1948). These experiments obviously should be extended to blue-green algae. It is tempting to view SM as an important reagent for inducing apochlorosis in nature, especially since SM-producing actinomycetes are common in soils.

B. Specificity of Streptomycin

No other compound has been found to exhibit bleaching activity. All the characteristic effects of SM appear to be properties of the intact molecule. This includes the bleaching phenomenon. The following antibiotics were inactive in inducing apochlorosis (Godfrey *et al.,* 1949):

Actidione
Aureomycin
Borrelidin
Chloromycetin
Clavacin (also tested by Jírovec)
Gliotoxin
Kojic acid
Neostreptomycin
Penicillin (also tested by Jírovec)
Polymyxin
Streptothricin
Tyrothricin
Na usnicate

The following compounds, all having the antibiotic characteristics of SM, were active:

Dihydrostreptomycin A
Dihydrostreptomycin B

The following, all portions of the SM molecule, were inactive:

Streptobiosamine
Streptamine
Guanidine
N-acetylglucosamine

A similar specificity has been shown for SM in its role as a quasi-growth factor for SM-dependent bacteria (Rake, 1948; Miller, 1948).

SM shatters chloroplasts into fragments indistinguishable from mitochondria. Lwoff (1950), in discussing the work of Hovasse (in Pringsheim, 1948b) on the cytological interpretation of the protistan leucoplast, was of the opinion that the normal leucoplast, e.g., the decolorized chloroplast of a *Euglena* maintained in the dark, merges with the mitochondrial system when it is shattered by SM. Since paramylum granules are still laid down, it was assumed that the surviving mitochondrial bodies include the functional equivalent of leucoplasts. It is noteworthy that Pringsheim's apoplastidic *E. gracilis* continued to form paramylum, while the permanently apoplastidic *E. mesnili* obtained by Lwoff and Dusi no longer did so. However they may arise, euglenids with obviously defective photosynthetic apparatus occur in nature. Pringsheim's spontaneously decolorized strains were derived from originally pale green forms. Later they showed a disintegration of chloroplasts similar to that observed under the influence of SM. *Euglena gracilis* var. *urophora* was found colorless in nature and required 2 months in the laboratory to become green (Chadefaud and Provasoli, 1939). It has remained yellowish green since, but otherwise unimpaired in vigor.

C. Mode of Action of Streptomycin

SM differs from penicillin in that resting organisms are inhibited equally with proliferating ones; as noted, the bleaching of *Euglena* followed this rule.

1. THE NUCLEIC ACID THEORY

The basicity of SM, together with the ease with which susceptible organisms acquire resistance, suggested that SM may act by virtue of its electrostatic attraction to nucleic acid, and thus influence the mechanism of heredity. This does not explain why other penetrating bases do not act in the same way as SM. Since it is a base, it is much more active in alkaline media—the usual permeability factors apply. The failure of the basic portions of the SM molecule to exert any effect (antibiosis, bleaching, satisfaction of the SM requirement of SM-dependent strains), and also their inability to influence metabolism in the manner described in the next paragraphs, all speak against this theory. Penetrating bases such as acriflavine and propamidine killed *Euglena gracilis* var. *bacillaris* without affecting greenness (Godfrey *et al.*, unpublished).

2. THE CHLOROPHYLL INTERFERENCE THEORY

Lwoff and Schaeffer (1949a) suggested that SM owes its bleaching properties to a competition, stemming from steric resemblances, between certain dipyrrylmethane derivatives of protoporphyrin and the streptidine portion of the SM molecule. In support, SM was stated to be more effective when applied to cultures in the dark than in light; the slowness of bleaching in light was interpreted as due to protection by chlorophyll. These conclusions differ from those of Provasoli *et al.* (1948), who ascribed the action of light in diminishing the gross bleaching action to its selective favoring of green individuals.

3. THE CONDENSATION-REACTION THEORY OF UMBREIT *et al.*

Umbreit and his co-workers (1949) have made progress toward identifying the site of action of SM. Their results again point to the importance of acetate and the $C_2 + C_2$ condensation. SM *specifically* (only the intact molecule was active) inhibited the concurrent oxidation of pyruvate and oxaloacetate by *Escherichia coli* and by animal tissue preparations. Resistant and SM-dependent variants of *E. coli* survived apparently because they could by-pass the pyruvate-oxaloacetate condensation. The decidedly low susceptibility to SM toxicity exhibited by animals was attributed to permeability barriers at the cell membrane and mitochondria; in tissue preparations where these barriers were disrupted, the pyruvate-oxaloacetate condensation was inhibited.

One difficulty in applying these findings to phytoflagellates is that alternate explanations are provided both for resistance and vulnerability to SM. Organisms might be resistant either to SM because their metabolism does not wholly depend on a sensitive $C_3 + C_4$ condensation, or because the enzymes for this condensation are shielded by impermeable membranes. Sensitive organisms might not have a by-pass for the $C_3 + C_4$ condensation, or else their $C_3 + C_4$ reaction might be all too accessible. Another possible inference from the theory of Umbreit *et al.* is that chloroplasts rely on a condensation reaction which is different either in kind or in relative importance from the condensation reaction of the remainder of the organism.

The sensitivity to SM of the chloroplasts and of the organism as a whole might be expected, on the basis of this theory, to be affected by differences in substrate. The product of the pyruvate-oxaloacetate condensation is assumed to break down immediately to a C_6-tricarboxylic acid. If phytoflagellates could utilize these C_6 acids, the effect of SM might be lessened when they are supplied, by-passing the condensation. The initial difficulty is that of ensuring penetration by these bulky hydrophilic acids. So far the bleaching effect in *Euglena* has not been clearly influenced by wide variations in the substrate component of culture media.

Bleaching has been induced equally well in synthetic and in peptone media (Godfrey *et al.*, unpublished). This uniformity is explicable if the metabolism of chloroplasts is essentially independent of fluctuations in exogenous nutrients. An investigation of the *overall* resistance to SM of a sensitive phytoflagellate as a function of differences in substrate might be a more fruitful line of attack. Respirometric studies of sensitive strains of bacteria have provided corroborative evidence for the condensation-reaction theory: only the oxidation of ethanol was uninhibited; the oxidation of acetate was too slow for determination of its sensitivity to SM (Henry *et al.*, 1948).

4. STREPTOMYCIN DEPENDENCY

A remarkable feature of SM, noted by many workers, is that organisms at first partly inhibited by it may later require it as a growth factor. This dependency, by present indications, cannot be induced unless the organism is at first sensitive and then becomes adapted. As the two strains of *Euglena gracilis* employed by us were, apart from their chloroplasts, insensitive to SM, it was not surprising that they failed to become SM dependent. Dependency might be demonstrated in SM-sensitive flagellates.

5. ANTIGENS OF GREEN *vs.* COLORLESS PHYTOFLAGELLATES

Euglena and other phytoflagellates are not at first agglutinated in the presence of homologous rabbit antisera. Instead, the flagella are damaged and the immobilized organisms may then "clump." This manifestation of an antigen-antibody reaction allows tests of antigen specificity. Rabbit sera containing antibodies to *E. gracilis* did not affect *E. proxima* and *E. polymorpha* (Sauer, 1935). Flagellates of the *Euglena-Astasia-"Khawkinea"* group appeared to have antigens in common. *Chlorogonium* showed no reciprocal immunological reactions with euglenids (Tanzer, 1941). These phenomena may be predominantly the manifestations of surface antigens. The proteins of chloroplasts (Granick, 1949b) may be antigenic. If antisera were to be prepared using disintegrated rather than intact organisms, both green and the derived colorless strains, it might be possible to demonstrate the activity of chloroplast antigens. Chloroplast antigens were probably responsible for the fact that when pigs were sensitized to *E. gracilis* grown in light or in darkness, anaphylactic shock was elicited only by the homologous culture [Elmore (M. E. Sauer), 1928]. The antigenic properties of chloroplasts offer interesting possibilities.

6. SOME THEORETICAL IMPLICATIONS OF STREPTOMYCIN-INDUCED APOCHLOROSIS

The plastid is self-duplicating. Its susceptibility to SM leads to wonder as to whether other autonomous cytoplasmic structures display a comparable sensitivity. This appears to be true of *kappa* in *Paramecium* (J. R. Preer, personal communication; see reviews by Sonneborn). Sensitivity to SM is hardly a criterion of the cytoplasmic nature of a genetic character. e.g., the symbiotic *Chlorella* in *Paramecium bursaria* might be eliminated by SM, but that does not make *Chlorella* a plasmagene. The argument can be inverted: differential sensitivity to SM does not determine whether an intracellular structure is a symbiont or a normal cytoplasmic constituent.

In unpublished experiments, it was found that exposure to elevated temperatures (32°–35° C.) permanently eliminated the chloroplasts in the vars. *bacillaris* and *urophora* of *Euglena gracilis,* giving rise to otherwise apparently healthy strains. Heat treatment was also effective in eliminating *kappa* (Sonneborn, 1946, 1949), and likewise eliminated the factor for CO_2 sensitivity in *Drosophila* (L'Héritier, 1949). The effect of SM on the latter has not been reported. Whatever the ultimate nature of the structures concerned in these observations, inactivation by SM hints that these characters have something biochemical in common besides their extranuclear reproduction.

VI. Growth in Darkness

A. Introduction

The magnitude of the problem of the growth of photosynthesizers in darkness is revealed when stock is taken of failures as well as successes. Actually, nearly all the chlorophyllous phytoflagellates which have been grown in darkness were those already showing pronounced stimulation by substrates in light; most of the organisms investigated belong to the acetate flagellates. One reason for the neglect of this experimental approach to the problem of the immediate products of photosynthesis may have been the success in growing metaphytes and protococcoid algae in darkness on sugars. As sugars—and acetate (for acetate flagellates)—are not today considered very close to the first products of photosynthesis, this lack of substrate specificity deprives these substances of significance in considerations of the first product of photosynthesis.

Experiments on "dark growth" may reveal widely differing physiological types within the same taxonomic group (Lucksch, 1932). Of 7 species of *Chlamydomonas,* 5 grew in darkness on peptone or acetate, and 2 did not. In agreement with results by other workers, Lucksch found that *Haematococcus pluvialis* grew in darkness on acetate. It also grew on gelatin but, curiously, not on peptone. A resemblance to protococcoid nutrition was shown by *Chlamydomonas pseudagloë* and *C. pseudococcum* which grew in darkness on glucose and, in the latter case, on sucrose also.

B. Loss of Chlorophyll

It was previously noted (p. 79) that chlorophyllous euglenids and chrysomonads become colorless in darkness; the pale green chloroplasts retained by *Euglena mesnili* in darkness may be an exceptional condition in euglenids. It has been claimed that the decolorization of the plastids of *Euglena* in the dark depends on the organic nutrition of the organism (Pallares, 1945). In experiments reported in the literature, and in our own experience, variations in media do not influence in a clear manner the rate of disappearance of chlorophyll in *Euglena* in darkness. The rapid reappearance of greenness when cultures are restored to light suggests that the dark block in chlorophyll formation is close to protochlorophyll or to chlorophyll itself in the chain of synthesis. The block must be past heme, a postulated precursor, because heme was identified spectroscopically in *E. gracilis* kept in darkness (Lwoff, 1934). Heme was also observed in *Polytoma.* Uptake of O_2 ordinarily presupposes the presence of a cytochrome system.

C. Nature of the Dark Block

If the elusive "first product" of photosynthesis could, by a stroke of luck, be dissolved in an otherwise complete culture medium, and thence penetrate the organism, *any* photosynthesizer should grow in darkness under those conditions. In other words, the problem of the by-passing of light for an ostensibly obligate phototroph may be posed in terms of the availability of the "first product" of photosynthesis. In the preceding summary of the permeability of the acetate flagellates a series of stages was posited, representing decreases in permeability that would have, at one extreme, organisms permeable only to CO_2, i.e., obligate autotrophs. This general explanation of failures to grow organisms in darkness breaks down when *Chlamydomonas moewusii,* purple bacteria, and *Thiobacillus thioparus* are considered.

1. *Chlamydomonas moewusii*

In our experiments *Chlamydomonas moewusii* failed to grow in darkness with substrates usually satisfactory for other organisms. A puzzling condition was revealed: cultures without substrate—the controls—grew on their being returned to light; cultures which had been supplied with substrates were dead. Similar observations were made independently by Chismore and Lewin (unpublished), and led to a searching study (Chismore, unpublished) of this curious dark-killing: *C. moewusii* failed to grow in the dark on any carbon source tried (Krebs cycle intermediates, sugars, high-energy phosphate compounds, complex proprietary organic media, coconut milk, etc.). No growth was obtained upon the addition of filtrates of light-grown cultures, or when living light-grown organisms were circulated intermittently around cultures in the dark separated from them by a nitrocellulose membrane. Moreover, no dark-produced toxic product, inhibitory to light-grown flagellates, could be demonstrated. A Warburg manometer experiment showed that glucose, acetate, succinate, and pyruvate all caused an increase in O_2 uptake when added to respiring cells in the vessels. Organic carbon compounds were therefore presumably taken up and oxidized in the dark, but the energy from the oxidation was unavailable for growth. It was found that CO_2 was necessary for growth in light: inocula supplied with organic carbon sources, but without CO_2, failed to grow.

2. *Thiobacillus thioparus*

The peculiar disjunction between respiration and growth found by Chismore in *Chlamydomonas moewusii* may be matched by *Thiobacillus,* where once again a permeability barrier does not wholly account for the failure to utilize organic substrates for growth. So far as is known, *Thiobacillus thioöxidans* and *Th. thioparus,* which derive energy by the oxidation of thiosulfate or sulfur, can obtain their carbon only from CO_2 (Stephenson, 1949). Vishniac (1949), working with *Th. thioparus,* pointed out that so large and polar a molecule as tetrathionate had no difficulty in entering the bacteria. The rate of O_2 consumption of a dried-cell preparation was increased by isocitrate, glucose, glycogen, and glucose-1-phosphate (Cori ester). The O_2 uptake with glucose-1-phosphate was about 3 times that required for complete combustion of the substrate; glucose-1-phosphate might thus have acted as a catalyst for the oxidation of cell reserves. As the CO_2 production corresponded to only

15 per cent of that required for complete combustion on the basis of this O_2 uptake, partially oxidized intermediates were sought. An acidic product was isolated which had many of the characteristics of glyceric acid, possibly originating from diphosphoglyceric acid. In spite of the vigorous uptake of O_2 by these dried-cell preparations in the presence of the phosphate ester, growth of intact cells was unaffected by glucose-1-phosphate. Glucose, fructose-1,6-diphosphate, fructose-6-phosphate, 3-phosphoglycerate, pyruvate, and acetaldehyde had much slighter or no effect on O_2 uptake in dried-cell preparations, and a negligible effect on growth of living cells. However, at pH 7.0 in the presence of thiosulfate, succinate increased final growth 30 per cent; succinate could not be used alone, and the presence of at least 2–5 per cent CO_2 in the gas phase was essential.

3. THE ANAEROBIC NARCOTIC IN PHOTOSYNTHESIS

The accumulation of an acidic respiration product like that demonstrated in *Thiobacillus thioparus* might prove fatal to *Chlamydomonas moewusii.* Can phytoflagellates accumulate such products? If metaphytes and *Chlorella* are a reliable indication, phytoflagellates may well do so, at least under anaerobic conditions, according to the information assembled by Franck (1949). It was early noted that metaphytes, after a long period of anaerobiosis, showed a very long induction period before photosynthesis was resumed. Evidently an inhibitor was formed which under these extreme conditions checked photosynthesis completely. This inhibition was removed by admittance of just enough O_2 to replace fermentation by respiration, hence the inhibitor might be a fermentation intermediate, and, as the inhibitor could be removed by alkali, it might be an acid. The inhibitor was also removed by illumination. Increase in concentration of cells increased the poisoning. If the O_2 made photosynthetically was removed in a stream of O_2-free air, the poisoning continued.

4. FAILURE OF DARK ANAEROBIC GROWTH IN PURPLE BACTERIA

Another instance of the inadequacy of permeability to account for failure of growth in darkness is provided by purple bacteria. All attempts to grow them in the dark *anaerobically* have failed. Many of these bacteria grow vigorously in darkness under *aerobic* conditions, with the same substrates as those serving as H-donors in light. In a formal sense there is here an interchangeability in function between light and O_2.

5. THE "FIRST PRODUCT" OF PHOTOSYNTHESIS

The ostensibly disparate phenomena of (1) failure of "dark growth" of *Chlamydomonas moewusii,* (2) the acidic product of glycolysis formed under oxidative conditions by *Thiobacillus thioparus,* (3) the anaerobic narcosis of photosynthesis, and (4) the failure of anaerobic dark growth of purple bacteria, may all be manifestations of a similar incomplete reaction, driven to completion by light, thiosulfate oxidation, O_2, and either O_2 or light (to take these blocks in order). Addition of the immediate product of photosynthesis (and chemosynthesis, since conceivably the same substance is formed) should remove these blocks if it is a molecule which has properties conducive to penetration.

Agreement on the properties of this "first product" must be left to the future, when experimental results become more reproducible. Thus Kamen (1949) and his associates, on the basis of many short-term experiments with isotopic tracers, believed that the CO_2-fixation products in *Chlorella* were not related to any of the commonly known plant or animal constituents. The chemical behavior of the compound indicated a high degree of hydroxylation and carboxylation in what was probably a large molecule. Another group of investigators (Calvin and Benson, 1949; Benson *et al.,* 1949), arrived at a different concept. In very short-term photosynthetic experiments, by far the major portion of labeled C of the newly reduced C^*O_2 was found in the phosphoglyceric acids and triose phosphates. In short-term dark fixation (30 and 90 seconds), considerable activity appeared in alanine, and malic and aspartic acids in addition to the phosphates. In very short-term dark experiments (15 seconds) with *Scenedesmus,* the most concentrated activity was in the phosphoglyceric acid fraction.

The results of the University of Chicago group are closer to those of Kamen *et al.* These workers found that in *Scenedesmus* a water-soluble substance appeared immediately upon illumination and that it contained practically all the labeled carbon taken up from C^*O_2 during the first few moments. It was different from the water-soluble *dark*-fixation products obtained under conditions identical except for the absence of photosynthesis (Brown *et al.,* 1948). They defined the "first product" as the first substance in the photochemical mechanism which, produced wholly or partly in the light from CO_2, was not susceptible to further photochemical reduction. An "intermediate" was any substance able to undergo photochemical reduction, whether it was produced by dark fixation of CO_2, or was a partly reduced product of photochemical action. An *anaerobic* *dark* ("chemosynthetic") fixation of CO_2 was viewed as likely to be the primary CO_2 fixation process. The absence of back reactions at low light intensities suggested that these intermediates were quite stable chemically (Brown *et al.*, 1949). Further chemical data on the first stable intermediate indicated that it was polyhydroxy acid, but *not:*

2-Phosphoglyceric acid	Gluconic acid
3-Phosphoglyceric acid	Dihydroxyacetone
Glyceraldehyde	Glyceric acid

It was not volatile, not any of the familiar Krebs cycle intermediates, nor an amino acid (Fager, 1949).

As a lead to further experiments with substrates in the dark, it may be pointed out that the intermediate in photosynthesis, partly characterized by the Chicago group, may not be an aliphatic compound, but may owe its stability to a limited aromaticity. Two items of circumstantial evidence support this very tentative suggestion. First, the equilibria for the carboxylation of polyhydroxyl aromatic compounds tend to favor carboxylation (Rabinowitch, 1945, p. 185). Secondly, certain pseudomonads oxidized a number of polyhydroxyl cyclic acids, e.g., quinic acid, with such ease (Stanier, 1948) that these compounds may be excellent carbon sources for other organisms as well. An inspection of the list of substrates utilized by pseudomonads (den Dooren de Jong, 1926) gives the impression that the common denominator of these compounds is nearly equivalent to the substrates considered closest to the metabolic pool. It may be said, by way of concluding the discussion of this topic, that work on photosynthetic and chemosynthetic intermediates, and on other compounds presumably close to the metabolic pool, provides indications as to how the block interdicting dark growth of *Chlamydomonas moewusii* may be removed; there appears indeed little alternative, since, as shown, a frontal attack has faltered.

A pertinacious attempt to grow marine dinoflagellates in darkness met with total failure (Barker, 1935; see pp. 53 and 102).

VII. The Nitrogen Requirement and Nitrate as H-Acceptor

A. NO_3^- in Nature

The ability to utilize NO_3^- as a nitrogen source is probably a primitive character in phytoflagellates (Lwoff, 1943). All investigated photosynthetic flagellates utilize nitrate, but this does not mean that all photosynthesizers necessarily have this faculty: there are major phytoflagellate groups yet to be examined. Nitrate reduction is a reaction that seems to be lost readily by phytoflagellates developing strong heterotrophic tendencies. The significance of nitrate will be discussed from the standpoint of van Niel's comprehensive theory of photosynthesis.

It is generally accepted that all the O_2 in the atmosphere came from green plant photosynthesis (p. 38). Nitrate, a labile oxidizing agent, could not exist in significant amounts until organisms developed which could oxidize ammonia to nitrate, taking advantage of the powerful H-acceptor, O_2, newly available. By this reasoning, *all* aerobes are offshoots of anaerobe stocks. Biological organic N compounds contain N at the NH_3 or NH_2 levels of reduction. The only known organic nitro compound of biological origin is an antibiotic, chloromycetin (chloramphenicol). NO_3^- cannot serve as N source for purple bacteria, either aerobically or anaerobically (Hutner, unpublished), hence nitrate reduction to the NH_3 level is not a property of all photosynthetic organisms. The conventional nitrate portion of the nitrogen cycle is a collection of shunts. Ammonia is the significant starting point in biological syntheses; the reduction of nitrate is merely a preparatory step. Loss of the ability to reduce NO_3^- reduction may not seriously diminish an organism's chances of survival. Permanently colorless flagellates seem mostly to have lost the ability to reduce nitrate; *Polytoma ocellatum* is an exception (Lwoff and Dusi, 1938). Certain chlorophyllous phytoflagellates (*Haematococcus pluvialis, Chlamydomonas agloëformis*) were unable to use nitrate as an N source in darkness (Lwoff, 1943, pp. 91–92); Lucksch (1932) stated, however, that *H. pluvialis* grew in darkness on NO_3 and acetate.

B. Nitrate as H-Acceptor

In the dehydrogenation of substrates, certain organisms can use nitrate or O_2 as the H-acceptor. The reduction of nitrate to the level of nitrite and then ammonia—the process making nitrate available as a source of nitrogen—raises the question of whether this reduction, on a larger scale, can serve in *energy-liberating* reactions. Denitrifying bacteria such as the common *Pseudomonas aeruginosa* (*pyocyanea*) can speedily oxidize suitable substrates with nitrate substituting for O_2. Nothing is known directly about this function of nitrate in phytoflagellates, but there are hints. The bulk reduction of nitrate may, as in denitrifying pseudomonads, be indicative of well-developed oxidative abilities.

It was early discovered by Warburg and Negelein that *Chlorella* in darkness oxidized cell materials by means of nitrate. In light, nitrate reduction gave rise to an extra large evolution of O_2, in excess of what could have originated from a stimulation of photosynthesis by the extra CO_2 produced in $NO_3^=$-mediated oxidations. Van Niel (1949b), reviewing these experiments, advanced the hypothesis that the enhanced O_2 production came from a photochemical reaction, with NO_3^- as the H-acceptor instead of CO_2. In any case, *Chlorella* can effect a bulk reduction of nitrate comparable in magnitude to that needed for an appreciable oxidation of substrate. Can nitrate in the dark take the place of O_2 in oxidations needed for growth? Anaerobic conditions would be needed for clear results (Kluyver, 1940). Since the assimilation of glucose in the dark by *Chlorella,* in the presence of nitrate is accompanied by $NO_3^=$ reduction (Myers, 1949), this system might serve for exploratory experiments using phytoflagellates of the sugar-utilizing types described by Lucksch (p. 86). Such experiments might also reveal phytoflagellates growing with the aid of glycolysis.

C. Ammonia and Nitrate as N Sources

Shortcomings in the methods of artificial culture may give a misleading impression that green plants prefer nitrate to ammonia. Most of the synthetic media in which photosynthesizers were grown were acidic, and nitrate, an anion, penetrates best from acid solutions; ammonia, conversely, penetrates best from alkaline media. Plants can store excess nitrate in an inert form (Hoagland, 1948). Excess ammonia may be harder to cope with. Ammonia, furthermore, forms complexes with metals of the transitional series (p. 103), and this may induce trace element deficiencies. As these metal complexes are more stable at higher pH values, the conditions favoring the penetration of ammonia are also conducive to the induction of metal deficiencies. Omitting from consideration the failures of growth clearly due to toxicity, there is no evidence that free-living organisms exist which cannot obtain at least part of their N from ammonia. The uptake of exogenous NH_4^+-nitrogen may be blotted out if amino acids or other prefabricated nitrogenous compounds are supplied, whereas if such nutrients are in slight excess of the specific requirement, deamination or other processes may liberate an excess of ammonia (Algéus, 1948, 1949). By employment of minimal levels of essential amino acids, a net uptake of NH_4^+-nitrogen was demonstrated in so nutritionally exacting an organism as the rat (Lardy and Feldott, 1949). The three amino acids close to the metabolic pool (alanine, glutamic and aspartic acids) are well suited to be organic N sources and C substrates. Compounds may often serve as N sources without being substrates, since a reaction such as deamination, making N available, may leave an unassimilable residue.

VIII. Growth Factors and Chemical Asepsis

The meaning of growth factors has been made amply clear in numerous reviews (see the Introduction by A. Lwoff). The thiamine requirement of phytoflagellates has already been discussed in relation to problems of intermediary metabolism (p. 68). The following discussion centers around certain technical and theoretical problems encountered in studying the vitamin B_{12} requirement of *Euglena* and *Chlamydomonas chlamydogama.* The apparent inconsistencies in the behavior of euglenids in synthetic media have long troubled euglenophiles, and previous reviewers (e.g., Doyle, 1943) have noted the litigious nature of much of the literature on the nutrition of other phytoflagellates as well. The identification of vitamin B_{12} as a growth factor for certain phytoflagellates involved technical difficulties. The partial solution of some of these difficulties may dispel some of the confusion.

Vitamin B_{12}, the anti-pernicious anemia factor of humans, a component also of the "animal protein factor" in poultry, and in the rat and other mammals, is active at exceptionally high dilutions. It is produced in relative abundance by many bacteria. For these reasons that the precautions necessary in compounding vitamin-free basal media, i.e., media supporting negligible growth "blanks" in quantitative assay, are similar to those needed in studying *p*-aminobenzoic acid in its role as a growth factor: PABA is likewise active at very low concentrations and is produced abundantly by many microorganisms.

It was repeatedly observed while studying the PABA requirement in the purple bacterium species *Rhodopseudomonas palustris* (Hutner, 1946) that media compounded from freshly prepared solutions gave satisfactory blanks, but solutions which had been standing for some time frequently did not, though there was no obvious sign of microbial contamination of the solutions. Workers studying PABA as a growth factor for other organisms have also noted vexatious difficulties in obtaining satisfactory blanks (e.g., Lewis, 1942).

These untoward events were gradually recognized as manifestations of a more serious general problem described at length in the medical literature. In the administration of vaccines, blood sera, and fluids in transfusions and other parenteral therapy, these materials frequently gave rise to high fever and chills; routine sterility tests were usually negative. These alarming symptoms were traced to bacterial products, *pyrogens,* produced in particular abundance by gram-negative bacteria (Robinson and Flusser, 1944). A similar, probably closely related substance was found produced by all of a great variety of gram-negative bacteria, including chemo- and photosynthetic types (Zahl *et al.,* 1945), examined. Since many free-living gram-negative bacteria do not grow on ordinary sterility test media, the problem was to eliminate rather than merely to detect these potential vitamin producers. As abundantly shown in the literature on pyrogens, the hazard of biochemical contamination is met not merely with solutions prepared in the laboratory—it attends any manufacturing process involving solutions of non-corrosive intermediates.

A step toward eliminating contaminations was achieved by routine addition to stock solutions of a volatile preservative (Hutner and Bjerknes, 1948), consisting of a mixture (by volume) of 1 part *o*-fluorotoluene, 2 parts *n*-butyl chloride, and 1 part 1,2-dichloroethane. It is hoped that a substitute can be found for the expensive and difficultly available *o*-fluorotoluene; chlorobenzene holds some promise. This preservative passes off on steam sterilization and is virtually inert to nearly all ingredients of culture media. It reacts with free —SH groups—its probable site of action.

A culture technique was developed (Hutner, 1946) in which contact with materials of biological origin was held to a minimum. Cotton is notoriously a significant source of vitamins and even of volatile fatty acids (Robbins and Schmitt, 1945). Cultures were maintained in screw-capped tubes. Beaker-capped 25-ml. flasks containing experimental media were set on Pyrex kitchenware trays and steam sterilized. Such trays of flasks were allowed to cool to room temperature in the unopened autoclave. Slow cooling in essentially a sterile atmosphere was the key to maintenance of asepsis when plugs were eliminated. Pyrex glass wool proved convenient for plugs where rapid cooling was desired. After inoculation, each tray holding flasks was covered with another inverted tray, and the joint between the trays was sealed with adhesive cellulose tape. The flasks were thus in a "micro-greenhouse." Such units could be stacked under continuous illumination (at present a combination of "pink" and "blue" fluorescent lamps is used, but the "daylight" type serves nearly as well) thereby conserving space and light. For routine assay of vitamin B_{12} either small cotton-plugged culture tubes or the seeded agar plate + filter-paper disk method (W. J. Robbins and A. Hervey, personal communication) was suitable. Growth for assay purposes is complete in 4 days or so.

By combining these simple methods with the metal-buffered media developed for other organisms, tracking down the unknown *Euglena gracilis* factor proved easy. This work was accelerated by the use of the vars. *bacillaris* and *urophora* which have a temperature optimum 5° higher than the widely used Mainx strain.

A. *Euglena* AND VITAMIN B_{12}

1. HISTORICAL

Since the announcements in the spring of 1948 of the crystallization of vitamin B_{12}, more than 100 research papers on B_{12} have already been published. This new development in biochemistry compels the re-formulation of several long-debated problems of phytoflagellate physiology.

Several properties of B_{12} were known earlier, because it was a growth factor for *Euglena* (Hutner, 1936). The Mainx strain of *E. gracilis* required a factor tenaciously bound to casein, from which it was removed by repeated isoelectric precipitation. Gelatin contained a small amount of it. The plant proteins edestin and concanavallin A lacked it, and most yeast preparations were poor sources. This is also the distribution of the B_{12} component of the "animal protein factor." The *Euglena* factor was much more soluble in water than in non-polar liquids throughout a wide pH range. Subsequent tests with *E. gracilis* var. *bacillaris* showed that refined liver and alcoholic extracts of casein were highly potent, and the identification was complete when crystalline B_{12} (+ thiamine) proved active.

2. SPECIFICITY OF B_{12}

The pattern of specificity of the B_{12} requirement of *Euglena gracilis,* as known at present, coincides with that of animals and differs sharply from that of lactobacilli. The locus of action of B_{12} is remarkably close to that of folic acid. One aspect of the activity of folic acid is concerned with the synthesis of thymine. Lactobacilli requiring folic acid can be satisfied with a much greater amount of thymine instead. Thymidine or various other desoxyribose derivatives of purines or pyrimidines may similarly replace B_{12} for dependent lactobacilli. The function of B_{12} may be that of attaching desoxyribose to the purines and pyrimidines of desoxyribosenucleic acid (DNA) and may be diagrammed in terms of thymine-thymidine, to select a familiar pair from several which participate in B_{12}-catalyzed reactions:

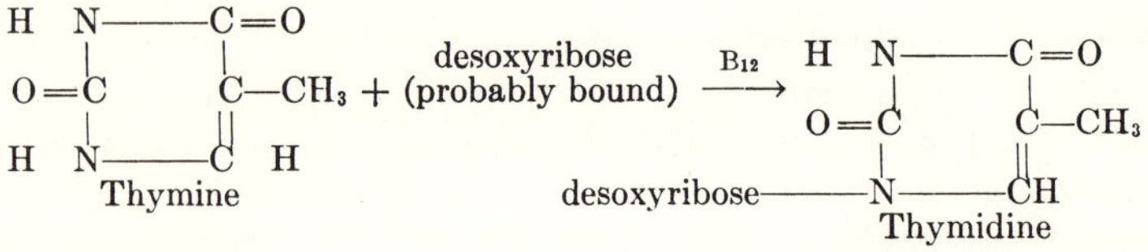

The B_{12} requirement of animals (pernicious anemia in man results from a defect in the gastro-intestinal digestion and absorption of B_{12} and, to a varying extent, folic acid) is not satisfied by thymidine, or by intact or hydrolyzed DNA. Experiments by the authors, in collaboration with the Lederle group, indicated that very high concentrations of hypoxanthine and guanine desoxyribosides, but not thymidine, exhibited a slight and erratic activity that might not be ascribable to contamination of the preparations with B_{12}. Intact DNA and acid or alkaline hydrolyzates of DNA were inactive for *Euglena.*

From experiments performed since the initial description of the B_{12} requirement of *Euglena,* it was calculated that maximum growth of *E. gracilis* (under the particular conditions about 7×10^6 euglenas/ml.) required about 0.15 mμg./ml. of B_{12}. In the linear portion of the growth curve, 0.01 mμg of B_{12} allowed the formation of 880,000 euglenas. Taking 1,400 as the minimal molecular weight of B_{12}, a single organism required 4,900 molecules of B_{12} (Hutner *et al.,* 1950). This extremely high potency is understandable in a compound which, unlike thiamine, riboflavin, or hemin, does not participate in the chief energetic processes of the organism, but, like PABA and folic acid, is concerned in the synthesis of the irreducible core of the organism, its genic complement. The activities of PABA and folic acid clearly implicate them in syntheses not confined to the formation of DNA. PABA and folic acid requirements are by-passable, at least in part, by mixtures of adenine and amino acids in addition to thymine, as indicated by sulfanilamide experiments. In contrast, the activity of B_{12} is channeled narrowly in the direction of the biosynthesis of DNA and, by implication, the gene. The greater potency on a weight and molar basis of B_{12}, despite the fact that it is a larger molecule, thereby becomes perhaps more explicable.

3. ECOLOGICAL ASPECTS OF VITAMIN B_{12}

Unlike the other B vitamins, the distribution of B_{12} in natural materials is not correlated in the main with the original content of protoplasm or of nuclear substance. B_{12} is nearly lacking in metaphytes and, as mentioned, yeast. The very fact that *Euglena* requires B_{12} suggests that in organisms with low B_{12} activity, B_{12} is present but is bound in a nutritionally unavailable form. *Euglena gracilis* has shown little ability to utilize B_{12} in high-molecular combination.

In nature, *Euglena* may satisfy most of its B_{12} requirement from bacterial production. Actinomycetes are common in pond muds (Erikson, 1941) as well as in soils, and certain actinomycetes, among them producers of aureomycin and streptomycin, produce B_{12} in abundance (Stokstad *et al.*, 1949). Certain bacteria belonging to *Flavobacterium*, a genus prominent in the microflora of soils and water, also form appreciable amounts of B_{12} (Petty and Matrishin, 1949).

4. IS A B_{12} REQUIREMENT A PHYLETIC CHARACTER IN EUGLENIDS?

Several euglenids which grew moderately well in chemically defined media all showed a requirement for B_{12}. These included *Euglena stellata* and several strains of *Astasia*. The speculative question of whether this B_{12} requirement is found in all euglenids is broached because the direction of experimentation is involved.

Certain lines of evolution of colorless euglenids lead into saprophytic and holozoic flagellates so simplified in gross morphology that they must be merged for taxonomic purposes with flagellates of other origins. If a B_{12} requirement is a phyletic character of the euglenid stock, a flagellate's failure to exhibit this requirement might exclude it from the euglenids. For phagotrophs like *Peranema* and *Entosiphon*, whose euglenid affinities are unmistakable, this question may acquire practical importance. There is no microorganism suitable for the assay of B_{12} in conjugated form; recourse must be had to the chick or rat, with their batteries of digestive enzymes (Stokstad *et al.*, 1949). With the exception of the assay methods based on Tetrahymena, all the microbiological assay procedures currently in use measure *free* metabolites. Where metabolites are stable to drastic chemical hydrolysis, as are most amino acids and a few vitamins (among them PABA, pyridoxine, and nicotinic acid), dependent organisms of weak digestive powers, such as lactobacilli and yeasts, are inadequate substitutes for animal assays in determining *total* metabolite. When the metabolite is predominantly in high-molecular combination and is inactivated on hydrolysis with acid or alkali, as is the case with folic acid, the only type of microorganism that appears adequate is an organism with a digestive system, like *Tetrahymena,* which can utilize conjugates of folic acid (see the Chapter on ciliates by Kidder and Dewey). There is impressive evidence that in many natural materials B_{12} is present mostly as conjugates. Assays for B_{12} based on dependent micro-animals have not yet been developed: neither insects nor *Tetrahymena* appear to need B_{12}. A suitable micro-assay now lacking, the determination of total B_{12} in natural materials must depend on feeding trials with the chick or rat. If B_{12} is a phyletic requirement of euglenids, a euglenid phagotroph such as *Peranema* might prove to be the urgently needed micro-animal.

The presumably morphologically primitive, i.e., biflagellated, genera *Eutreptia* (green, salt or brackish water) or *Distigma* (colorless, fresh water), should allow, in conjunction with morphologically advanced organisms such as *Peranema*, a good preliminary test of the universality of the B_{12} requirement in euglenids.

5. THE MINIMAL ORGANIC REQUIREMENTS OF EUGLENIDS

Several workers claimed to have grown euglenids in synthetic media without growth factors. This raises the question of the magnitude of the experimental error caused by random chemical contaminations taken as a whole, glassware, chemicals, water, dust, and inoculum. A certain error is inherent in all nutritional investigations, and it becomes proportionately large in experiments where the desideratum is the rigorous determination of the minimal growth requirements. In determining the vitamin requirements of Protozoa, the main reliance should be placed on the serial transfer technique, not upon cultures grown from a single transfer. Hall's recommendation (1943), that "... serial dilution must not be too rapid—in other words, relatively large and long incubations are often preferable to small inocula and short incubation periods" demands analysis. From the practical standpoint, the best procedure may combine the use of small inocula from just-above-minimal media followed by a fairly short growth time. The many observations on the ubiquity of biochemical contaminations do not foster optimism in evaluating identifications of the absolute minimal requirements for growth.

Even all-glass culture methods cannot rule out significant biochemical contaminations. An intimation of this came from Beijerinck and van Delden (1903) who discovered a bacterium that grew in a liquid inorganic medium with nitrate as the N source, so that there was no potential source of energy from NH_4^+. This organism was inhibited by ordinary media and its energy source remained unknown. The pyrogens previously discussed (p. 93) provided further warnings of the dangers of invisible microbial contaminations. It has been often observed that distilled water in Pyrex containers may become intensely pyrogenic. The origin and growth of these microbial communities utilizing oxidizable compounds commonly present in laboratory air have been recently clarified to some extent. Bacteria (pseudomonads?) oxidizing methane and ethane were found to be common in water and soils; certain of these strains also oxidized ammonia to nitrite (Hutton and ZoBell, 1949). These organisms failed to grow on nutrient agar, broth, or gelatin in the absence of gaseous hydrocarbons. The behavior of *Pseudomonas fluorescens* in substrate-free media (Englesberg and Stanier, 1949) indicates more traps for the unwary. In liquid cultures in containers without cotton plugs, this extremely common bacterium made a definite subvisible and subculturable growth. It was calculated that formation of 1–10 organisms required about 2 µg. of substrate, an amount of oxidizable organic matter easily present as a contaminant in ordinary C.P. chemicals. The problem of pyrogens demonstrates that these microbial populations may be large.

There is obviously more opportunity for phytoflagellates to utilize these subtle and disturbing substrates when growth is slow: there is time for autolysis and liberation of the nutrients locked within the microbial bodies killed in sterilization; there is time also for nutrients to dissolve out of the walls of the containers and to enter from the air. For these reasons the evaluation of growth, obtained in minimal media over a prolonged period, is difficult in the extreme. Some of these hard-to-interpret observations may be cited. A strain of *Astasia* grew on acetate as sole organic nutrient, and *Euglena gracilis* grew in light on an entirely inorganic medium (Schoenborn, 1940), or, with acetate, in darkness (Schoenborn, 1942). It was later suggested that *Astasia,* growing without any substrate, might derive its carbon from CO_2 (Schoenborn, 1946, 1949), but the energy source remained unknown. Another case in point is *Lobomonas piriformis* which grew fairly well without added N sources (Osterud, 1946). The dreary history of unsubstantiated N fixations should also counsel caution in interpretations of results obtained with minimal media.

In our unpublished experiments, *Euglena gracilis* var. *bacillaris* showed the typical response curve to B_{12} when grown in a liquid inorganic medium with thiamine and the presumably unassimilable ethylenediamine tetraacetic acid (used as a metal-buffering agent—see p. 106) as sole organic constituents. The plastic screw-capped culture tubes were kept illuminated in a closed refrigerating incubator. The B_{12} cultures grew slowly and steadily over a period of months; there was no growth without B_{12}. These essentially inorganic media are proving useful for maintenance of cultures. Cultures grown with assimilable substrates are in danger of having their photosynthetic apparatus impaired by mutational erosion unless the utilization of CO_2 be made a condition for survival. The healthy growth under these autotrophic conditions probably is limited by the rate at which CO_2 diffuses past loosely fitted caps. Similar but more critical results were obtained by growing the organism in a wholly inorganic medium (save for thiamine) through which was bubbled air + 5 per cent CO_2; the B_{12} requirement then seemed an absolute one, as was also the reciprocal requirement for thiamine (p. 70) (Cramer and Myers, personal communication).

B. *Chlamydomonas chlamydogama:* Vitamin B_{12}, Histidine, and Aspartic Acid Requirements

Both mating types of the heterothallic *Chlamydomonas chlamydogama* isolated from soil grew in mineral media plus soil extract (Bold, 1949) and so poorly without soil extract that growth factor requirements were suspected. These were identified as B_{12}, histidine, and aspartic acid (Pintner *et al.*, unpublished). Bold noted that the organism grew without soil extract on mineral agar. The apparent contamination of agar with significant amounts of amino acids was unexpected; agar does have, however, an appreciable content of B_{12} (W. J. Robbins, personal communication). Specific amino acid requirements have not hitherto been reported in phytoflagellates.

The interrelationships of the histidine requirement in lactobacilli (Broquist and Snell, 1949) illustrate the arbitrariness of simple codifications of nutritional function. *Lactobacillus arabinosus* 17–5 required either histidine or purines (especially xanthine); histidine was not replaceable by purines in the absence of pyridoxine. The histidine requirement of *Streptococcus faecalis* was unaltered by purines or pyridoxine. The histidine requirement of *C. chlamydogama* resembled that of *S. faecalis,* since vitamin mixtures, an alkaline hydrolyzate of yeast nucleic acid, or combinations had no effect.

The existence of these growth substance requirements, especially that for aspartic acid (or asparagine), suggested that the organism might be permeable enough to substrates to grow in the dark. *Chlamydomonas chlamydogama* behaved instead as an obligate photoautotroph: it did not grow in darkness on acetate, glucose, or complex materials such as gelatin hydrolyzate, tryptic digest of casein, and tomato juice; these were nontoxic in light. It has not been determined whether it behaves in darkness like *C. moewusii.* The B_{12} requirement may be of interest for bioassay purposes, since the organism grows well on agar plates and has a higher temperature optimum than have the euglenids now employed for B_{12} assays.

C. Growth Factor Requirements for Other Phytoflagellates

1. growth factors and isolation procedures

Chlorophyllous phytoflagellates needing exogenous organic metabolites may be commoner than completely autotrophic forms. Enrichment and isolation techniques which depend on inorganic media are selective for complete autotrophs. Where other procedures are employed, e.g., the washing until pure of an organism taken directly from the natural habitat or from a crude enrichment culture and then inoculating the washed organism into complex media, many of the phytoflagellates thus obtained in culture require growth factors. Most of the cultured euglenids were isolated by the washing procedures essentially as described. The incorporation of soil extract in nutrient media is thus a means of extending the range of isolated organisms.

2. soil as a source of growth factors

An inkling of the manifold effects that soil extracts may exert was evident from the finding (Lwoff and Lederer, 1935) that for *Polytoma* and *Polytomella* earth extracts furnished not only "humic substance" but also (1) an available source of N and (2) the pyrimidine and thiazole components of thiamine. A mud or soil with an abundant and varied microflora should contain very nearly the gamut of microbial metabolites; a phytoflagellate grown with the aid of soil extracts may have complex nutritional requirements. Vitamins for animals may occur in soils in comparatively high concentration: a recent diet for swine included 5 per cent soil to supply unidentified vitamins (Cunha *et al.*, 1949).

Pringsheim (1946a) emphasized that the initial isolation of phytoflagellates from nature was relatively simple—the more onerous task was to grow them afterward. Important conditions for maintaining cultures await recognition. One underestimated factor may be thermolabile growth factors. Culture media are commonly autoclaved. This may appreciably lessen the range of culturable phytoflagellates. There is evidence for a thermolabile growth factor for free-living algae: an organism needing complex media (supplied as unheated blood) was isolated, curiously, from a hot H_2S spring (Dyar and Ordal, 1946).

Pringsheim (1949) considered it to be *Spirulina albida,* a colorless blue-green alga "occurring in the surface film of waters where the bottom is covered with black mud." He had not succeeded in culturing it free from bacteria. The growth factor for this blue-green may be identical with one required by another blue-green alga, *Gloeotrichia echinulata* (Rodhe, 1948).

a. Hyalogonium klebsii. The authors, in collaboration with T. J. Starr, found that *H. klebsii* did not grow without a factor supplied conveniently in refined liver extract. The usual mixtures of vitamins (including B_{12}), amino acids, and nucleic acid components did not support growth.

b. Cryptomonas and Synura. Rodhe (1948) was unable to cultivate *Cryptomonas ovata* in a medium proved suitable for many fresh-water planktonic diatoms and green algae, until an extract of soil or of lake sediment was added. Ashing destroyed the activity. In our experiments, the growth of a *Cryptomonas* proceeded only in the presence of soil extract, peptone, and the like. Mainx (1929) observed that *Synura uvella* would grow with soil extract. Pringsheim (1946a) found beef extract also suitable.

The marine dinoflagellates *Prorocentrum micans* and a species of *Peridinium* did not grow in mineral-enriched natural or artificial sea water unless soil extract was added. Hoagland's "A–Z" mineral supplement did not substitute for the soil extract (Barker, 1935).

The work of Ondratschek should be accepted with reserve if not rejected entirely. As pointed out by Lwoff (1947), many of his claims are unverifiable. His experiments on inorganic nutrition, which include an account of species variations in arsenic requirements for growth, are as much a strain on credulity as are his other reports.

IX. Inorganic Nutrition

A. Trace Elements

CHELATING AGENTS AS METAL-ION BUFFERS

Several essential metal ions, unless very dilute, precipitate as phosphates, carbonates, or hydroxides at physiological pH. In mineral culture media this limited solubility may limit growth, or restrict it to certain limits of pH. The metals, a shorthand expression for their ionized and non-ionized states, may be present in complex natural media in concentrations far above those which would lead to precipitation if the metals were by themselves. Natural materials contain solubilizing agents hindering this precipitation. Most of these agents act by forming chelate complexes with metals, and in so doing they reduce the concentration of free metal ions below the point at which the solubility-product relations predict precipitation. These solubilizing compounds have one feature in common: they contain two groups capable of donating or accepting electrons. These groups are so spaced that, by the inclusion of a metal ion, 5- or 6-membered rings may be formed, stabilized by what are loosely called resonance forces. Citrate is a familiar example of a biological solubilizing chelating agent. Other such compounds are α-amino acids, α,β-glycols (e.g., glycerol), and aromatic compounds with *ortho* substituents (e.g., anthranilic acid). Ammonia and substituted ammonias form Werner complexes which similarly reduce the concentration of free metal ion. For a treatment of the chemistry of chelation the reader is referred to J. R. Johnson (1943). As such metallo-organic compounds hold the key to the designing of reproducible media allowing heavy growth, and chelation controls the activity of many enzymes, some pertinent aspects of chelation will be described. Essentially the same ground was covered in an outline of the procedures followed in culturing a marine diatom in synthetic substitutes for sea water (Hutner, 1946). The development of culture media containing non-metabolizable chelating agents has been described *in extenso* (Hutner *et al.*, 1950).

The structural formulas for citric acid as ordinarily conventionalized (e.g., as on the left-hand side of the following equation) do not adequately convey its capacity—allowed by the bond angles of its C and O atoms—to participate in ring formation by inclusion of an appropriate metal. The reaction of citrate with Ca may be formulated as follows; other arrangements are possible.

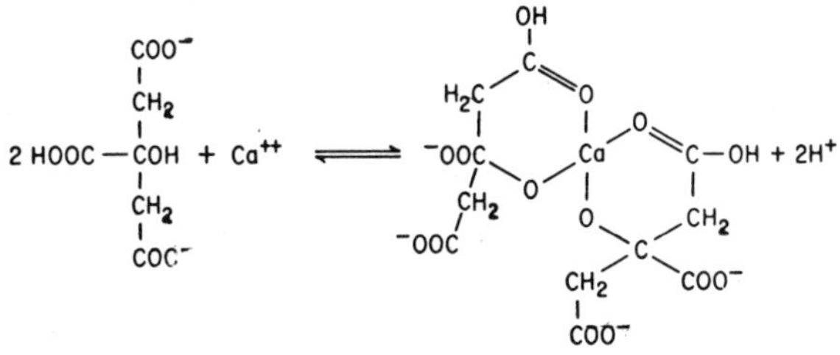

a. Effect of pH and Organic Structure on Metal Binding. It is difficult to assign definite formulas to inner complexes of the Ca-citrate type. The multiple acidic groups of citric acid provide the surplus of hydrophilic groups needed for solubilization of the complex as a whole. The H^+ that is displaced by the chelated Ca comes from the aliphatic hydroxyl group. The forces binding Ca in the ring may be seen, from even a cursory inspection of the formula, to bear at least a formal resemblance to that responsible for the extra stability ("aromaticity") of the benzene ring. In instances where resonance is weak, as in complexes with glycols such as glycerol, the metal binding is rather weak although definite. Where a system of double bonds is present, is in many analytical reagents for metals, the complex is very stable. Finally, where a conjugated double-bond system is highly developed, as in porphyrins, appropriate metals may be bound with exceeding firmness. These considerations must be kept in mind when designing experiments with chelating substrates such as α-hydroxy carboxylic acids and α-amino acids. Since the H^+ is highly mobile and acts to displace metal ions from the ring, a lowering of pH immediately releases Ca^{++} from the chelate ring. In other words, the lower the pH the greater the dissociation of the Ca-citrate complex ion, and the greater the proportion of free Ca^{++} in the solution. The facility with which the composition of such complexes varies with pH has defeated efforts to isolate definite molecular species.

b. Stability of Metal Binding, and Solubility. The solubility of a complex metal ion does not provide a direct measure of its stability. Dyes offer a comparable situation: the chromophore groups may be distinct from the solubilizing groups. Judging from the ease with which metal deficiencies are induced by citrate as compared with oxalate (chemical data on this point seems to be lacking), metals are bound more tightly by citrate than they are by oxalate. Citrate has a surplus of solubilizing (hydrophilic) groups; oxalate has not.

c. Specificity of Metal Binding. If each different complex-former bound metals with different relative tightness, then for each type of metal binder the order of stability of bound metals would have to be determined anew. This appears to be unnecessary: the order of stability of complexes was found to be independent of the chelating agent and followed the sequence Cu > Co > Zn > Fe > Mn > Mg (Irving and Williams, 1948). To what extent the steric rigidities introduced by the presence of double bonds influence this sequence has not been worked out; a box-like molecule like a porphyrin might not admit the larger metal ions. Inspection of the order of stability indicates that metals of the so-called transitional series of the periodic table, roughly those that have exceptionally loose electrons because of incomplete inner shells, and consequently prone to form colored compounds, are bound more tightly than the others. Therefore the presence of a chelating agent may make it necessary to add more of an element such as Zn, which is tightly bound, than an element like Fe or Mg, even though the organism's actual need for Zn is less than that for the others. Zn is exceptional in being colorless and tightly bound. If an organism itself produces chelating agents, as do many fermentative fungi, this will cause an increase in the apparent metal requirements without there being any fundamental difference in its metal metabolism. It is not surprising therefore that the basal media developed to take advantage of the properties of chelating agents should have a distinct resemblance to those used for the culturing of an organism such as *Aspergillus niger*.

d. Chelating Agents in Culture Media for Phytoflagellates. Chelating agents, by lowering the effective concentrations of metal ions, may exert two opposing effects: (1) concentrations of metal ions otherwise toxic are rendered innocuous; and (2) metal deficiencies may be induced. Cu and Ag are the traditional bug-bears in growing algae. Glass-distilled water is generally specified. The growth of the alga is still dependent on the absence of toxic metals from the other ingredients of the medium. Difficulties of this sort disappear when purified chelating agents are supplied: Cu and Ag (also the dangerous Hg) are strongly bound. In fact, the writers have found it necessary to add appreciable amounts of Cu to their culture media, although an absolute requirement has not been demonstrated.

The chelating agent may induce simultaneous deficiencies for several metals. By addition of the metals in question in suitable proportion, this difficulty may be overcome. Such media are, in effect, buffered in respect to these metals. As the growing organisms adsorb metals (presumably because their surface is provided with chelating groups which can compete with the exogenous complex former), more metal ions dissociate from the complex in compliance with the law of mass action. By this means dense growth may be obtained whereas, if the chelating agent were not present, the metal concentrations needed to secure this growth would be such that precipitation or toxicity might inhibit growth altogether. Metal-buffered media of this type with citrate as chelating agent were instrumental in making practical the *Euglena* assay for vitamin B_{12} (Hutner *et al.*, 1949). The failure to recognize that the addition of a chelating agent might induce *multiple* metal deficiencies may have been responsible for the general lack of exploitation of this principle in nutritional investigations, although citrate and similar compounds have been used for many years to inhibit the precipitation of Fe. Per 100 ml., a milligram or so of a compound such as ferric ammonium citrate or tartrate (such preparations are variable in composition) is insufficient for massive growth; much greater amounts of citrate are needed. It is better to supply the necessary minor elements as such instead of relying on their being adequately present as impurities of iron salts and the major constituents of the medium.

The need for metal-buffered culture media may become distressingly apparent when a growth factor is under investigation. The crude preparations of natural materials generally used in early stages of such investigations are likely to be rich in essential elements. At the same time, the presence of chelating agents blots out metal toxicities. As more refined preparations are used, both metal deficiencies and toxicities may appear to bewilder the investigator. There are described in the literature a number of instances where what was thought to be a vitamin turned out to be an essential metal. The use of metal-buffered media from the outset, even when the need for such media is not immediately demonstrable, represents an insurance against contretemps of this sort.

e. The Calcium Requirement. An estimate of the order of magnitude of a calcium requirement may be obtained by successively increasing the concentration of citrate and then determining the increments of Ca needed to restore growth. This procedure presupposes that simultaneous deficiencies in other metals have been avoided. One may then plot concentration of citrate *vs.* Ca. By extrapolating the curve to zero citrate, the intercept with the citrate axis allows an estimate of the Ca requirement. For a marine diatom this has been (very approximately) established at 0.5 mg./100 ml. culture medium. Unpublished experiments with *Euglena gracilis* and *Chlorella* yielded nearly the same value. This amount of Ca is well within the amount that might be expected to occur as an impurity in ordinary reagent grade chemicals. This low value for the Ca requirement appears to explain many previous failures to demonstrate a Ca requirement in algae, e.g., those of Pringsheim (1926) and Trelease and Selsam (1939). The present results have a wider significance in that certain enzymes transferring phosphate appear to require Ca as a coenzyme, and absence of a Ca requirement would have implied that phytoflagellates had a different metabolism. The same extrapolation procedure can of course be applied to other essential metals which form chelate complexes.

f. Use of Non-Metabolizable Chelating Agents. In studies of substrate utilization it may be desirable to use a non-metabolizable chelating agent. Citrate and other metal binders of biological origin may introduce a disturbing factor. Ethylenediamine tetraacetic acid ["EDTA"; indexed in *Chemical Abstracts* as ethylene-bis (iminodiacetic) acid] has proved very useful for the purpose (Hutner *et al.*, 1950; also many unpublished experiments).

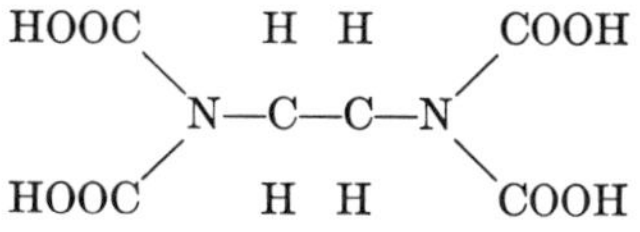

Ethylenediamine tetraacetic acid (EDTA)

A typical basal medium for massive growth of many phytoflagellates has the following composition per 100 ml.:

EDTA	0.05 g.	Boron	2.0 mg. (as H_3BO_3)
K_2HPO_4	0.02 g.	Mn	2.0 mg. (as $MnSO_4 \cdot H_2O$)
$MgSO_4 \cdot 7H_2O$	0.08 g.	Fe	0.8 mg. (as $FeSO_4 \cdot 7H_2O$)
NH_4Cl	0.02 g.	Mo	0.6 mg. (as $Na_2MoO_4 \cdot 2H_2O$)
Zn	5.0 mg.	Cu	0.4 mg. (as $CuSO_4 \cdot 5H_2O$)
Ca	2.0 mg.	Co	0.4 mg. (as $CoSO_4 \cdot 7H_2O$)

PH adjusted to 6.5–6.9 with KOH. A double-strength basal solution will not form a precipitate. *This "base" is much too concentrated for many delicate organisms; it may have to be diluted and modified considerably for them.* Citrate may be substituted for the EDTA. The medium as given represents only a first approximation toward a thoroughly satisfactory mineral base; it was, for instance, satisfactory for *Euglena gracilis* var. *bacillaris, Chlamydomonas moewusii* and *C. chlamydogama,* whereas a closely similar medium, higher in Mn, was suitable for eliciting zoospore formation and for growing mycelium of several species of *Saprolegniaceae* (H. S. Reischer, unpublished).

g. Cobalt. Since each molecule of vitamin B_{12} contains one atom of cobalt, formation of one individual *Euglena* was calculated to require 4,900 atoms of cobalt (see p. 96). It has been calculated (Hutner *et al.*, 1950) that a sample of Fe (purified for spectroscopic purposes but still contaminated with Co to the extent of 0.001%) when used at the customary concentration in the assay medium, may supply at least 30 times as much Co as that corresponding to the B_{12} requirement. This illustrates the need for purer chemicals than are yet available if certain trace element requirements are to be demonstrated.

B. Nutrition of Plankton Phytoflagellates

Exceedingly low concentrations of certain inorganic nutrients may affect plankton phytoflagellates in any of three ways: (1) they may be insufficient to supply the mineral requirements for growth; (2) they may suffice to support good growth; or (3) they may be inhibitory. Certain marine dinoflagellates in sea-water media furnish a striking example of both effects (Barker, 1935). *Prorocentrum micans* and a *Peridinium* were inhibited by 0.001 per cent NH_4Cl. Phosphate was only limiting when it was below 0.005 mg. per cent. Nitrate was much less toxic than NH_4^+.

The fresh-water chrysomonads *Dinobryon divergens* and *Uroglena americana* were killed by PO_4 concentrations in excess of 2.0 µg. per cent (Rodhe, 1948). In keeping with this sensitivity, these species are not found in polluted water. Rodhe calculated that at times [PO_4] in the lake where the flagellates lived might rise to inhibitory levels, and that fluctuations in [PO_4] might be an important factor in determining the periodicity of their abundance. Rodhe's medium "VIII" was suitable for the hardier planktonic algae. It had the following composition (mg./100 ml. final medium):

		Milligrams salt/100 ml.
Ca	0.147 as $Ca(NO_3)_2$	6.0
Mg	0.1 as $MgSO_4$	0.5
Na	0.75 as Na_2SiO_3	2.0
K	0.22 as K_2HPO_4	0.5
SO_4	0.4 as $MgSO_4$	
N	1.02 as $Ca(NO_3)_2$	
P	0.089 as K_2HPO_4	
Fe	0.018 as Ferric citrate + citric acid	0.1 + 0.1
Mn	0.001 as $MnSO_4$	0.003
Si	0.46 as Na_2SiO_3	

pH 7.0–7.5

As is frequently recommended for media of this type, Fe and phosphate were autoclaved and added separately. Two species of *Chlamydomonas* and the coenobial volvocines *Eudorina elegans, Gonium pectorale,* and *Pandorina morum* were grown in this medium, also many non-flagellate species. A similar, if simpler, medium was used by Dusi (1940) for *Eudorina*. There was no growth in the presence of NH_4 salts. Oddly enough, Dusi's strain was isolated by washing in 0.4 per cent peptone, which might be expected to have an appreciable amount of free NH_4^+.

Does the toxicity of NH_4^+ in synthetic media arise from secondary metal-binding effects, as reported for the basidiomycete *Coprinus* (Fries, 1945)? In our experiments, *Chlamydomonas moewusii* tolerated high concentrations of NH_4^+ (supplied as 0.1 per cent NH_4Cl) at pH 8.0 when complex-forming metals were augmented (Zn, Mn, Cu, Fe).

The toxicity of media containing the high concentrations of mineral nutrients necessary for massive growth is an insufficiently appreciated obstacle to the growing of plankton organisms in synthetic media. If, to avoid toxic concentrations, the major salts be reduced, then essential trace elements hitherto supplied in the form of contaminants of these salts may be reduced to suboptimal levels. Deficiencies in knowledge of inorganic nutrition are nowhere more obvious than where the task is to assemble reproducible synthetic media which will allow good growth of delicate plankton organisms such as those just mentioned.

C. The Absorption of Phosphate

Organisms vary greatly in respect to phosphate requirements. The ability of plankton organisms to utilize very small concentrations of PO_4 implies the existence of an extremely effective trapping mechanism for PO_4. An appealing activated-transport hypothesis was advanced to account for the accumulation of PO_4 by yeast (Nickerson, 1949). It postulates that the diffusion gradient is maintained by the polymerization into metaphosphate of intracellular PO_4, and the polymerization of glucose-1-phosphate is regarded as the immediate source of the metaphosphate-P:

$$\text{Extracellular inorganic } PO_4 \rightarrow \text{glucose-1-phosphate} \rightarrow \\ \text{glycogen} + \text{polymetaphosphate}$$

A general mechanism by which osmotic work may be done against a gradient was outlined by Dixon (1949, p. 52). If the cell membrane is permeable to a substance B, but not to its phosphorylated derivative, phosphorylation of B at the membrane would lead to the trapping of B. It would be interesting to see whether the simultaneous supply of assimilable substrates to PO_4-sensitive phytoflagellates might diminish the toxicity of PO_4.

It is probable that in phytoflagellates many of the cytoplasmic inclusions which are strained by basic dyes and described as *volutin* are in fact metaphosphate. *Haematococcus pluvialis, Chlamydomonas,* and *Polytoma* were rich in volutin, and its abundance varied with supply of PO_4: in its absence volutin disappeared quickly, especially from dividing organisms (Reichenow, 1928). Volutin in yeast behaves similarly and was there identified as metaphosphate (Wiame, 1949). The probable occurrence of metaphosphate in *Euglena gracilis* was noted previously (p. 75). A volutin-rich mutant of *Chlamydomonas moewusii* has been obtained (R. A. Lewin, unpublished) as determined by the abundance of intracellular granules stained metachromatically with methylene blue. It shows promise of being a useful cytoplasmic marker. When the mutant was crossed with the wild type, tetrad analysis of the germinated zygotes showed a 1:1 segregation for this character.

D. Acid-Tolerant Phytoflagellates

Waters draining mines may be very acid because of the H_2SO_4 originating from the oxidation of iron pyrites and other sulfides. Lackey (1938) observed that *Euglena mutabilis* formed a brilliant dark green coating on submerged objects in such acid-polluted streams. Densities of 11,000,000 organisms per milliliter were not uncommon. A *Chromuliña,* and a *Chlamydomonas,* and, on one occasion, *Cryptomonas erosa,* accompanied *E. mutabilis.* The pH range for a pure culture of *E. mutabilis* in a casein digest medium was 2.1–7.7 (Von Dach, 1943); at pH 1.4 some euglenas remained alive as long as 12 days.

One may wonder whether the resistance to acid, shown by these organisms, is purchased at the price of an extremely low permeability. The Mainx strain of *E. klebsii* [identical, according to Pringsheim (1948a), with *E. mutabilis*], could not be grown in darkness (Dusi, 1933). Its growth in light never was very vigorous even in the presence of the peptones and acetate so favorable to *E. gracilis.*

E. Inorganic Requirements of Marine Phytoflagellates

Animal life in the ocean depends on the activities of the diatoms, dinoflagellates, chrysomonads, and cryptomonads if the plankton. Chlorophyceae, so common in fresh-water plankton, are nearly lacking in the open ocean; their place is taken by members of the brown-pigmented groups (Pascher, 1917) where, as noted, there has been an elaborate development of phagotrophic lines. Furthermore, investigation of the inorganic requirements of marine flagellates, particularly that for NaCl, offers some provocative parallels to the electrolyte requirements of animals. A renaissance in marine protistology, an advance beyond statistical faunistic and floristic studies, has too long been delayed.

SUBSTITUTES FOR SEA WATER

In one respect sea water is an unsatisfactory base for culture media: it is likely to be saturated in respect to a number of precipitate-forming salts, since sea water is in equilibrium with the shore and bottom. Enrichment of sea water with nutrients such as phosphates or iron is likely to lead to precipitates. The inorganic nutrients in sea water, as factors determining the multiplication of plankton organisms, were outlined by Harvey (1935). Many more or less simplified substitutes for sea water have been proposed. One for the culture of *Ulva* (Levring, 1946) has the following composition (g./100 ml.):

NaCl	2.36	$MgSO_4 \cdot 7H_2O$	0.594
KCl	0.064	$NaNO_3$	0.01
$MgCl_2 \cdot 6H_2O$	0.453	$Na_2HPO_4 \cdot 12H_2O$	0.002
$CaCl_2$	0.098	$NaHCO_3$	0.0192

The nitrate and phosphate represent enrichments. The pH was adjusted to 8.2 with NaOH.

With the possible exception of the often-cited pioneering study by Barker (1935) (it is not clear whether his organisms were pelagic or neritic), there is no record of pure cultures of any pelagic phytoflagellates. The ocean is a comparatively constant environment, and pelagic organisms may be so delicately attuned to their environment that maintenance of the necessary constancy of laboratory conditions might prove very difficult; the fragility of pelagic organisms is well known. Life in the littoral and neritic zones is subject to much greater, sometimes drastic, alterations in environment, and might therefore offer material more suitable for laboratory study. The common diatom of marine investigations, *Nitzschia closterium* var. *minutissima,* common on mud flats, is conspicuously euryhaline; pelagic organisms are likely to be stenohaline, i.e., restricted to a very narrow range of sea-water concentrations.

Comparative studies with *Nitzschia closterium* and a marine *Chlamydomonas,* very kindly sent by H. W. Harvey of the Plymouth Laboratory of the Marine Biological Association of the United Kingdom, showed that satisfactory media more flexible than simulated sea water were not difficult to prepare. A medium developed for study of electrolyte requirements (Hutner *et al.,* 1950) had the following composition per 100 milliliters:

K_2HPO_4	0.02 g.	Zn	3.0 mg.
$MgSO_4 \cdot 7H_2O$	0.25 g.	Ca (as NO_3^-)	2.5 mg.
Glycine	0.25 g.	Mn	1.0 mg.
K acetate	0.2 g.	Mo	1.0 mg.
NaCl	0.25–4.0 g.	Fe	0.6 mg.
		Cu	0.5 mg.

Ethylenediamine tetraacetic acid 0.05 per cent, or K_3 citrate$\cdot H_2O$ 0.15 per cent, were used as the metal carriers. The pH was brought to 7.0 with KOH. A comparison of this medium with Levring's medium is instructive in showing the wide range of permissible variation in substitutes for sea water. The Mg and Ca constituents must be present in relative abundance for good growth. Two impure cultures of chrysomonads and a red cryptomonad, all marine, required more Mn for good growth than was present in sea water from the entrance to Plymouth Sound. The addition of as little 0.01 µg. per cent Mn was definitely stimulatory and larger amounts greatly increased growth (Harvey, 1947).

The inorganic requirements of obligately marine or halophilic organisms are the sum of two separate requirements: ionic balance and minimum osmotic pressure. The marine *Chlamydomonas* grew in a medium in which the total salt concentration was about one-tenth that of sea water, provided the proper relationships of the various ions were maintained (Vishniac *et al.,* unpublished). Luminous bacteria, on the contrary, had a pronounced osmotic requirement, as has also been found by others, e.g., Hill (1929). In the experiments of Vishniac *et al.,* a species of *Photobacterium* had a salt requirement no greater than that of the marine *Chlamydomonas,* provided the osmotic pressure was restored with an indifferent substance such as mannitol.

It was further observed that certain quaternary ammonium compounds, notably tetraethylammonium (Et_4N) bromide partly replaced Na for the marine *Chlamydomonas.* In the aforementioned glycine medium, NaCl 0.1 per cent allowed fair growth. Growth was quadrupled by the addition of 0.3 per cent Et_4NBr. At a lower concentration of NaCl (0.05 per cent) no growth occurred unless 0.3 per cent Et_4NBr was added. That this was not merely an osmotic effect was shown by the relative inefficacy of other compounds (KCl, $MgSO_4$, NH_4Cl); moreover, tetramethylammonium bromide (Me_4NBr) was not as effective as Et_4NBr. The action of Et_4NBr in replacing Na in Ringer's solution is remarkably similar to that shown in these growth experiments: Lorente de Nó (1949) noted that nerves of slow conduction maintained their conductivity in Na-free Ringer's solution if Et_4NBr was supplied. Me_4NBr

and choline chloride were much less effective. The effect was non-specific: other quaternary compounds had similar activity, among them phenyltriethyl- and *n*-propyltriethylamine. Na-free extract of ox brain acted like a quaternary base in restoring conductivity to nerve in Na-free Ringer's solution.

In an effort to make more sensitive the conditions for replaceability of Na, it was found that the Na requirement was lowered in the presence of Mg and raised by K, and, more markedly, by NH_4^+ salts. These results appear to be in harmony with the increase in the K requirement for growth of lactobacilli in the presence of Na and NH_4 salts (MacLeod and Snell, 1948).

An extension of these findings to other organisms promises similar results. *Dunaliella viridis* was benefited by Et_4NBr at low concentrations of NaCl (Vishniac *et al.*), but the results with this halophilic organism were not as striking as those obtained for the *Chlamydomonas*.

These Na-replacement relations have far-reaching implications for the theory of protoplasmic conductivity. Osterhout and Hill (cited in Höber, 1945, pp. 352–353) early observed that *Nitella* cells immersed in distilled water lost the ability to respond to an electrical stimulus. Their irritability was restored by water in which similar cells had been standing for some time. The restorative effect was produced by guanidine, adrenaline, Et_4NCl, NH_4Cl; or by additions of blood plasma, saliva, urine, white of egg, or milk. It was supposed that an organic substance continuously diffused from the protoplasm and had to be newly produced for maintenance of irritability. Evidence was adduced that the leaching of this substance from the cell markedly affected the properties of its plasma membrane (Osterhout, 1949). One can suppose that marine organisms may derive this substance, presumably an electrolyte, from the environment, and that a physiological equivalent is produced endogenously by fresh-water organisms. Et_4N^+ salts and one set of related compounds are becoming important in the therapy of hypertension (Arnold *et al.*, 1949), and related compounds of another class have a curare-like activity (Castillo *et al.*, 1949). A fuller discussion of this subject is beyond the compass of this review; so prosaic-seeming a topic as the electrolyte requirements of phytoflagellates makes an unforeseen and urgent demand on the store of fundamental knowledge of the attributes of protoplasm and of mammalian tissue.

The halophilic genus *Dunaliella* has been the subject of studies of ionic balance (Baas-Becking, 1930, 1931). Since the cultures were impure, and very incomplete media were used, it is difficult to evaluate the experiments described, particularly those reporting the removal of Mg toxicity by Ca.

X. Sexuality in Chlamydomonas

A. Background of the Problem

The vegetative individuals and gametes of *Chlamydomonas* and related flagellates are haploid and therefore one may expect all genes to be expressed phenotypically. In several heterothallic species there is little or no obvious morphological difference between the mating types, and there is no clear differentiation between a vegetative organism and a gamete. This may be regarded as the minimum of phenotypically expressed sexuality, and a study of chlamydomonads promises an unusually direct approach to the chemical factors determining gametogenesis, copulation, and the maturation and germination of zygotes. Unfortunately, one cannot rely on *Neurospora* for the genetics of factors affecting the photosynthetic apparatus; here sexual phytoflagellates have a special value.

Starting in 1933, Moewus, at times with associates, published about 25 papers on the phenomenon of sexuality in algae. The assertion made for his principal experimental material, *Chlamydomonas eugametos,* was that gametes secreted a diffusable sex-specific substance, which could elicit mating activity in potential gametes rendered sexually inactive by prolonged stay in darkness.

The history of investigations on *Chlamydomonas moewusii* is interwoven with that of *C. eugametos.* Gerloff (1940), in reviewing the genus *Chlamydomonas,* considered that *C. eugametos* (*sensu* Moewus) was identical with a species previously described by Pascher as *C. sphagnophila.* Gerloff did not receive cultures from Moewus. In the foreword of Gerloff's paper, Moewus was thanked for a letter regarding this species; Gerloff used the cultures of *C. eugametos* isolated by Czurda (1935) from an impure culture sent to Czurda by Moewus. Gerloff found that Czurda's isolates belonged to a new species, *C. moewusii,* differing from *C. sphagnophila* (= *C. eugametos sensu* Moewus) in having a distinct membranous papilla lacking in the other species. Gerloff confirmed Czurda's finding that the papilla was constant despite wide variations in cultural conditions; Moewus on the other hand stated that this was a variable character for his species. Gerloff then discovered *C. moewusii* in many locations in Germany and elsewhere and was able to mate these strains with Czurda's strains.

One of us (L.P.), from material kindly forwarded by Dr. P. R. Burkholder, isolated mating strains of a *Chlamydomonas* morphologically identical with *C. moewusii,* and succeeded in mating them with the Czurda strains kindly sent us by Dr. S. Prát of the Institute of Plant Physiology of the Charles University in Prague. Our strains of *C. moewusii* formed gametes more readily than did the Czurda strains and therefore the new isolates are the *C. moewusii* strains mentioned in recent American literature.

Moewus' cultures have not been available to other investigators. Lwoff (1947) recounted some attempts to obtain *Polytoma* cultures from him; Pringsheim and Ondraschek (1939) also tried unsuccessfully. Until there is independent verification of Moewus' claims, his papers might well rest in limbo. The requirements for experiments to obtain similar results were outlined by Smith (1946). Moewus' theories have been reviewed many times, e.g., by Sonneborn (1941).

B. Gametogenesis

Moewus reported that gametes of *Chlamydomonas eugametos,* after being kept in the dark, copulated only in light at the blue end of the visible spectrum (= 435.8 and 496.1 mμ). *Chlamydomonas moewusii,* which like *C. eugametos* does not copulate in darkness, copulated in red or in blue light, but only very slowly in the essentially monochromatic light of a sodium lamp (Provasoli *et al.,* unpublished).

G. M. Smith (1948), continuing his earlier (1946) studies with heterothallic species of *Chlamydomonas,* described the behavior of *C. reinhardi,* which could be grown in darkness on acetate media. When grown in white light and stored for a week in darkness, then flooded with water, cultures produced motile and sexually functional cells. When grown in darkness and flooded in darkness, motile cells not sexually functional, appeared. Finally, when grown in blue (435.7 mμ) or in red (615–690 mμ) light, cells became motile and sexually functional immediately.

Experiments on diffusibility of a sex substance. These experiments were carried out with *C. moewusii* (Provasoli *et al.,* unpublished). All attempts to demonstrate the diffusion of a sex substance between active cultures of the two sexes failed—extracts from one mating type induced no clumping in suspension of opposite-type cells. In certain experiments cultures of only one sex were illuminated, but still the supernatant culture fluid of one sex did not influence the other. Artificial substitutes for diffusible sex substances, including infusions of saffron, were inert. Cultures were also grown on solid media, and portions of washed agar on which organisms had grown were introduced into cultures of the opposite sex. All experiments were absolutely negative.

Critical observations of cell behavior in pairing. Continuous observation of the behavior of many mating pairs in cultures of low density strongly suggested that the initial clustering depended on surface phenomena through contact of individual organisms. Contact began with flagella. There was no evidence of attraction over a distance. Initial contacts appeared to be random, but, once made, stimulation of flagellar motion and adherence ensued.

Antigenic sexual factors. No antigenic distinction could be demonstrated between sexes when rabbit antisera prepared against each mating type were tested. Serological differences between *C. moewusii, C. chlamydogama,* and *C. reinhardi* were readily detected by these methods.

Pairing experiments with organisms killed by various means. Individuals of one mating type of *C. moewusii* were killed with formalin, heat, ultraviolet light, streptomycin, or combinations of these agents, marked clumping ensue between the killed organisms of one mating type and living organisms of the complementary mating type. These experiments strongly suggest the reaction of surface-active rather than diffusible substances in the initial clumping—a conclusion substantiated by all other work, including the morphological observations by Gerloff (1940) on pairing, and strongly suggestive of those of Metz (1948) with *Paramecium.*

According to Lewin (1949a), sexually active individuals of *C. moewusii* were readily obtained by flooding young cultures which had been grown on the surface of agar for 2–5 days. These gametes, typically formed in packets of 8, were paler and smaller than motile vegetative cells; they could transform into these if mating did not take place. A minimum of 14 hours of illumination (at 470 footcandles = 5060 lux) was needed for the completion of gamete fusion.

C. Maturation and Germination of Zygotes

1. *Chlamydomonas moewusii*

Preliminary observations on the germination of the zygotes of this species (Lewin, 1949a) indicate that a precise control of illumination was necessary if zygotes were to germinate with regularity. In light, the zygotes enlarged, and within 48 hours acquired a thick wall. Such fully matured zygospores gave only a low percentage of germination even after 1–2 months. It was found (Lewin, 1949b) that if the mixture of mating organisms was transferred to the surface of agar, illuminated for 24 hours at 25° C., and subsequently transferred to darkness, the walls of the zygote remained thin and in 5–6 days the organism began to divide. At the end of a further 5–6 days virtually all the zygospores had produced 4 to 8 zoospores. Zygospores allowed to mature in continuous light for 3 days or longer could not be germinated. The light-mature zygospores of *Chlamydomonas moewusii* were remarkably impermeable: many retained viability after storage for a week in pure acetone (Lewin, personal communication). Similar observations were made for zygospores of other species of *Chlamydomonas,* e.g., Kater (1929) found that they resisted fixation by Schaudinn's fluid and alcohols.

2. *Chlamydomonas chlamydogama*

The brief literature on the germination of the zygospores of Volvocales was reviewed by Starr (1949). Zygospores of *Chlamydomonas chlamydogama* of various ages gave 90 per cent germination within 48–96 hours after a 48-hour exposure to a temperature of 37° C. A *Chlorococcum*-like alga gave similar results.

General remarks. Much painstaking effort has been expended to make *Chlamydomonas* a useful genetical tool, and much additional effort will be needed. But the outlook is encouraging, and, before long, tetrad analysis in chlamydomonads may become quite simple. There is need for a heterothallic strain that will grow vigorously in darkness. *Chlamydomonas reinhardi,* in our hands, has not grown as rapidly in darkness as some of our non-mating strains. In focusing attention on the photoactivation of gametic behavior, one should not ignore the fact that most strains of the colorless genus *Polytoma* are markedly sexual; unfortunately heterothallic strains are unknown, and the determinants of sexuality and zygote germination are unclear (Pringsheim and Ondratschek, 1939). A vigorous culture of *Polytoma uvella* or *P. obtusum* may in a few days form a pellicle consisting mainly of copulating forms and may then accumulate a sediment of orange-red zygotes. These zygotes germinate in fresh culture media without special attention.

XI. Summary

This introductory section of this review, on the general characteristics of the brown and the green flagellate stocks, indicated that there were several, perhaps many, phytoflagellate → protozoan transitions, and therefore many points of evolutionary departure. The morphological picture indicates that several distinct although interconnected evolutionary trends, some clearly exhibited in present-day groups, shaped these transitions. Only the tendency to loss of chlorophyll has been studied, and at that only in two groups: the chlamydomonads and the euglenids. The photophagotroph, "phytozoan," pattern is still, biochemically, a closed subject.

Postulation of the biochemical make-up of the photoautotrophic archetypal flagellate is complicated by the multiplicity of bacterial photosyntheses. Which is the photosynthesis most nearly resembling that in primitive phytoflagellates? Which bacterial photosynthesis is the least primitive? There is some likelihood that the purple bacteria, as the most aerobic, may be less primitive than the others. The issue of the primitiveness in bacteria of chemoautotrophism *vs.* photoautotrophism is also under discussion (Van Niel, 1949b). These basic uncertainties imply that the conventional metaphytan photosynthesis, characterized by utilization of NO_3 and production of O_2, may not necessarily represent the most primitive form of photosynthesis in phytoflagellates. The existence here of reactions common to bacterial photosynthesis, e.g., photoreduction, and, possibly, a dark anaerobic fixation of CO_2 as the primary step in its photosynthetic reduction, may represent primitive characters that must be taken into consideration in setting up starting points in phytoflagellates. It is puzzling that the ability to liberate oxygen in photosynthesis is shared by groups so unlike the phytoflagellates as the blue-green and red algae.

The euglenids are rich in phagotrophic species, and because the investigation of the nutrition of a few pigmented and unpigmented saprophytic euglenids is well under way, work with this group promises immediate progress toward understanding the biochemical origins of phagotrophy and the related phenomenon of the loss of chlorophyll. A priori, there is little objection to postulating a close link between euglenid saprophytes and phagotrophs, but the nutritional resemblances between the *Euglena gracilis* group and the phagotrophic species remain to be determined. *Euglena gracilis* may be induced, experimentally, to lose its chloroplasts without obvious damage to its other systems. A more difficult problem is not how may photosynthesis be lost in a saprophyte of this type, but rather, how are the obligately photosynthetic species related to the photosaprophytes, and why is photoautotrophy obligate in certain species? The availability of pure cultures of obligately autotrophic euglenids enhances the value of this group for the study of the evolution of protozoa. The demonstration of a B_{12} requirement in a few euglenids may be an intimation that this is a phyletic character and therefore useful in tracing affinities. The existence of a B_{12} requirement in the quite unrelated *Chlamydomonas chlamydogama* is noteworthy in this connection.

The chlamydomonads may be of value in developing the concept of the archetypal phytoflagellate, but, as they almost entirely lack green or colorless phagotrophic relatives (there are a few rhizopodial forms), their value is indirect: it is mainly that of providing a standard of comparison, though they present a bewildering variety of physiological types. They may contribute to clarifying the nature of euglenid photosynthesis, especially if the blocks to growth in darkness, which are found in the chlamydomonads, prove to resemble those in euglenids. Indeed, their tendencies toward saprophytism may hardly be more pertinent to the problem of the evolution of the phagotrophs than are the saprophytic tendencies exhibited by certain metaphytes.

The common saprophytic and parasitic angiosperm genera provide fascinating material for comparative studies of the biochemical temptations to abandon phytosynthesis. Some, like dodder (*Cuscuta*) and mistletoe (*Phoradendron* and *Viscum*) contain chlorophyll in certain stages of their life cycles—does this mean that their parasitism is a newly acquired habit? Others are wholly parasitic (*Epifagus, Orobanche,* etc.). Others are highly developed saprophytes, such as the nearly colorless orchids (*Corallorrhiza,* etc.) and colorless Ericaceae (*Monotropa,* etc.). The comparison of the nutritional loss mutations in these groups with those in flagellates would be of extraordinary interest. The first stage of nutritional investigation of these plants is necessarily that of germinating the seeds aseptically, but very little work has been done in this direction.

Ignorance of the dynamic biochemistry of the various brown-pigmented algal groups is nearly total. This is an especially serious gap since phagotrophy has an abundant development in the chrysomonads and dinoflagellates. Little is known of the organic nutrition of the saprophytic or phagotrophic dinoflagellates. Investigation of this essentially marine series is crippled by ignorance of the inorganic requirements of marine organisms in general. It would be hard to overestimate the potential theoretical value of pure cultures of colorless dinoflagellates such as *Oxyrrhis marina* or *Noctiluca.* Energetic efforts should be made to culture forms like *Oodinium limneticum,* a parasite readily maintained as an infection of common fresh-water aquarium fish (Jacobs, 1946). It has a motile *Gymnodinium*-like stage and well-developed chromoplasts which may be photosynthetically functional; starch is present. Similarly, in the chrysomonad series, cultures of the pigmented phagotrophs *Ochromonas, Chromulina,* and the like are badly needed.

The results obtained with *Tetrahymena* show how complicated the minimum diet of a phagotroph may be. A tangle of growth factors awaits unraveling in the study of phagotrophic phytoflagellates. The isolation and identification of an unknown growth factor is likely to be a long-drawn-out and very expensive undertaking. Thousands of kilograms of material may have to be refined to yield a few micrograms or milligrams of crystalline vitamin, a feat which may be beyond the resources of academic laboratories. It would seem the practical course, in identifying new growth factors for phytoflagellates, to be alert to the possibilities of linking phytoflagellate factors with those being investigated in animals. Several fairly well-defined animal vitamins remain to be identified (e.g., the "animal protein factor"), and in this field phagotrophic phytoflagellates may prove valuable as assay organisms.

Every newly recognized growth factor represents the identification of a new cog in the cell machine. For reasons presented earlier, it is likely that thermolabile growth factors will assume a prominent position in general biochemistry, since they are very likely to be dynamic constituents of the organism that might be overlooked by the customary enzyme study techniques. The phagotrophic phytoflagellates may provide opportunity to study such factors without our having to depend, as at present, mainly on pathogenic bacteria or very exacting protozoa as assay organisms.

An attempt has been made to show that a satisfactory knowledge of the Protozoa rests on an understanding of the phytoflagellates. This increase in understanding cannot be attained without constant support from other disciplines. A smörgåsbord of phytoflagellate experiments has been laid out; if it is appetizing enough to attract workers, the object of this review will have been achieved.

Acknowledgments

The writers are indebted to the workers whose generous disclosures of unpublished data and fruitful ideas have enriched this effort to collate the scattered clues to the nature of the phytoflagellates. Special thanks are due Ralph A. Lewin, Helen S. Reischer, and Dr. Wolf Vishniac for critical readings of the manuscript; sins of omission and commission are our own.

This review is dedicated to the memory of Professor H. D. Reed of Cornell University. A master of morphological zoology, in a tradition stemming from Agassiz, he became convinced that many principles of evolution could best be revealed by investigating protists with the tools of chemistry combined with those of morphology. In 1932 he asked the senior author to report on lines of future experimentation that might allow insight into evolutionary processes, especially those concerned in the origin of protozoa and metozoa. The present review represents an attempt to fulfil that assignment. The junior author is likewise indebted to his former chief, Dr. André Lwoff, a transmitter of a parallel tradition established by Édouard Chatton, for the leitmotif which has guided his work in this field.

REFERENCES

Ajl, S. J., and Werkman, C. H. (1948). *Proc. Natl. Acad. Sci. U. S.* **34**, 491.

Ajl, S. J., and Werkman, C. H. (1949). *Proc. Soc. Exptl. Biol. Med.* **70**, 522.

Albaum, H. G., Schatz, A., Hutner, S. H., and Hirshfeld, A. (1950). *Arch. Biochem.* **29**, 210.

Algéus, S. (1946). *Botan. Notiser.* 278 pp.

Algéus, S. (1948). *Physiol. Plant.* **1**, 382.

Algéus, S. (1949). *Physiol. Plant.* **2**, 266.

Anderson, E. H. (1945). *J. Gen. Physiol.* **28**, 297.

Arnold, P., Goetz, R. H., and Rosenheim, M. L. (1949). *Lancet* **II**, 408.

Baas-Becking, L. G. M. (1930). *Stanford Univ. Contrib. Marine Biol.* 102; (1931). *J. Gen. Physiol.* **14**, 764.

Baas-Becking, L. G. M., and Ross, P. A. (1926). *J. Gen. Physiol.* **9**, 111.

Baldi, E. (1941). Richerche Idrobiologiche sul Lago di Tovel. *Mem. Museo Storia Naturale Venezia Tridentina* **6**, 1–296 (pp. 244–283); *Studi Trentini Scienze Naturali* **19**, fasc. 2, 1.

Baldwin, E. (1947). Dynamic Aspects of Biochemistry. Macmillan, New York, 457 pp.

Baldwin, E. (1948). An Introduction to Comparative Biochemistry. 3rd ed., University Press, Cambridge. 164 pp.

Barcroft, J. (1943). *Proc. Nutrition Soc. (Engl. and Scot.)* **3**, 247.

Barker, H. A. (1935). *Arch. Mikrobiol.* **6**, 157.

Barron, E. S. G., and Friedemann, T. E. (1941). *J. Biol. Chem.* **137**, 593.

Beijerinck, M. W., and van Delden, A. (1903). *Zentr. Bakt. Parasitenk. Abt. II*, **10**, 33.

Bell, D. J. (1948). Introduction to Carbohydrate Biochemistry. [Cambridge] University Tutorial Press, London, 107 pp.

Benson, A. A., Calvin, M., Haas, V. A., Aronoff, S., Hall, A. G., Bassham, J. A., and Weigl, J. W. (1949). Chap. 19 *in* Franck, J., and Loomis, W. E. (eds.). Photosynthesis in Plants. Iowa State College Press, Ames.

Bessey, E. A. (1942). *Mycologia* **34**, 355.

Bold, H. C. (1949). *Bull. Torrey Botan. Club* **76**, 101.

Braarud, T., and Fagerland, E. (1946). *Avhandl. Norske Videnskaps—Akad. Oslo I. Mat.—Naturv. Klasse* **No. 2**, 1.
Bracco, M., and von Euler, H. (1947). *Kem. Arb. II,* **10**, 4 pp.; von Euler, H. (1948). *Arkiv Kemi, Mineral. Geol.* **A25**, no. 8, 1–9; Bracco, M., and von Euler, H. (1948). *Kem. Arb. II,* **10**, 4 pp.
Bréchot, P. (1937). *Compt. rend. soc. biol.* **126**, 555.
Broquist, H. P., and Snell, E. E. (1949). *J. Biol. Chem.* **180**, 59.
Brown, A. H. (1948). *Plant Physiol.* **23**, 331.
Brown, A. H., Fager, E. W., and Gaffron, H. (1948). *Arch. Biochem.* **19**, 407.
Brown, A. H., Fager, E. W., and Gaffron, H. (1949). Chap. 20 *in* Franck, J., and Loomis, W. E. (eds.). Photosynthesis in Plants. Iowa State College Press, Ames.
Brown, H. P. (1945). *Ohio J. Sci.* **45**, 247.
Buchanan, J. M., Hastings, A. B., and Nesbett, F. B. (1943). *J. Biol. Chem.* **150**, 413.
Buchanan, J. M., Sonne, J. C., and Delluva, A. M. (1948). *J. Biol. Chem.* **173**, 81.
Calvin, M., and Benson, A. A. (1949). *Science* **109**, 140.
Castillo, J. C., Phillips, A. P., and De Beer, E. J. (1949). *J. Pharmacol. Exptl. Therap.* **97**, 150.
Chadefaud, M., and Provasoli, L. (1939). *Arch. zool. exptl. et gén.* **80**, 55.
Chismore, J. Unpublished.
Chismore, J., and Lewin, R. A. Unpublished.
Cook, A. H. (1945). *Biol. Revs. Cambridge Phil. Soc.* **20**, 115.
Copeland, H. (1947). *Am. Naturalist* **81**, 340.
Cramer, M. L., and Myers, J. Unpublished.
Cunha, T. J., Burnside, J. E., Buschman, D. M., Glasscock, R. S., Pearson, A. M., and Shealy, A. L. (1949). *Arch. Biochem.* **23**, 324.
Czurda, V. (1935). *Botan. Centr. Beihefte* **A53**, 133.
Darlington, C. D., and Mather, K. (1949). Chap. 8, The Elements of Genetics. Macmillan, New York.
Deflandre, G. (1937). *Bull. soc. franç. microscop.* **6**, 109.
den Dooren de Jong, L. E. (1926). Bijdrage tot de Kennis van het Mineralisatieproces. Dissertation, Delft.
De Ropp, R. S. (1948). *Nature* **162**, 459.
Dixon, M. (1949). Multi-Enzyme Systems. Cambridge University Press, 102 pp.
Dobell, C. C. (1911). *Arch. Protistenk.* **23**, 269.
Doyle, W. L. (1943). *Biol. Revs. Cambridge Phil. Soc.* **18**, 119.
Duchow, E., and Douglas, H. C. (1949). *J. Bact.* **58**, 409.
Dusi, H. (1933). *Ann. inst. Pasteur* **50**, 550, 840.
Dusi, H. (1939). *Compt. rend. soc. biol.* **130**, 419.
Dusi, H. (1940). *Ann. inst. Pasteur* **64**, 340.
Dusi, H. (1941). *Ann. inst. Pasteur* **66**, 159.
Dusi, H. (1944). *Ann. inst. Pasteur* **70**, 311.
Dyar, M. T., and Ordal, E. J. (1946). *J. Bact.* **51**, 149.
Elliott, A. M. (1939). *Trans. Am. Miscroscop. Soc.* **58**, 42.
Elmore, M. E. (1928). *J. Immunol.* **15**, 21, 33.
Emerson, R. L., Stauffer, J. F., and Umbreit, W. W. (1944). *Am. J. Botany* **31**, 107.
Englesberg, E., and Stanier, R. Y. (1949). *J. Bact.* **58**, 171.
Erikson, D. (1941). *J. Bact.* **41**, 277.

Evans, W. C. (1947). *Biochem. J.* **41**, 373.
Fager, E. W. (1949). Chap. 21 *in* Franck, J., and Loomis, W. E. (eds.). Photosynthesis in Plants. Iowa State College Press, Ames.
Florkin, M. (1949). Biochemical Evolution. Edited, translated [from the French], and augmented by S. Morgulis. Academic Press, New York, 157 pp.
Foster, J. W. (1944). *J. Bact.* **47**, 355.
Foster, J. W. (1949). Chap. 11, Chemical Activities of Fungi. Academic Press, New York.
Foster, J. W., Carson, S. F., Anthony, D. S., Davis, J. B., Jefferson, W. E., and Long, M. V. (1949). *Proc. Natl. Acad. Sci. U. S.* **35**, 663.
Franck, J. (1949). Chap. 16 *in* Franck, J., and Loomis, W. E. (eds.). Photosynthesis in Plants. Iowa State College Press, Ames.
French, C. S. (1946). *Ann. Rev. Biochem.* **15**, 397.
Frenkel, A. (1949). *Biol. Bull.* **97**, 261.
Frenkel, A., Gaffron, H., and Battley, E. H. (1949). *Biol. Bull.* **97**, 269.
Fries, L. (1945). *Arkiv Bot.* **32A**, No. 10, 1.
Fritsch, F. E. (1935). The Structure and Reproduction of the Algae. I. Macmillan, New York, 791 pp.
Gerloff, J. (1940). *Arch. Protistenk.* **94**, 311.
Godfrey, R. R., Schatz, A., and Hutner, S. H. (1949). Unpublished.
Goodwin, T. W., and Srisukh, S. (1949). *Biochem. J.* **45**, 263.
Gottlieb, J. (1850). *Ann.* **75**, 51.
Granick, S. (1948a). *J. Biol. Chem.* **172**, 717.
Granick, S. (1948b). *J. Biol. Chem.* **175**, 333.
Granick, S. (1949a). *J. Biol. Chem.* **179**, 505.
Granick, S. (1949b). Chap. 5 *in* Franck, J., and Loomis, W. E. (eds.). Photosynthesis in Plants. Iowa State College Press, Ames.
Granick, S., and Gilder, H. (1947). *Advances in Enzymol.* **7**, 305.
Gross, F. (1937). *J. Marine Biol. Assoc. United Kingdom* **21**, 753.
Gunter, G., Williams, R. H., Davis, C. C., and Smith, F. G. W. (1948). *Ecol. Monographs* **18**, 309.
Habermann, J. (1874). *Ann.* **172**, 11.
Hall, R. P. (1941). Chap. 9 *in* Calkins, G. N., and Summers, F. M. (eds.). Protozoa in Biological Research. Columbia University Press, New York.
Hall, R. P. (1943). *Vitamins and Hormones* **1**, 249.
Hardin, G. (1942). *Physiol. Zoöl.* **15**, 466.
Harvey, H. W. (1935). Recent Advances in the Chemistry and Biology of Sea Water. University Press, Cambridge, 164 pp.
Harvey, H. W. (1947). *J. Marine Biol. Assoc. United Kingdom* **26**, 562.
Heilbron, I. M. (1942). *J. Chem. Soc.* 79.
Hendey, N. I. (1946). *Nature* **158**, 588.
Henry, J., Henry, R. J., Housewright, R. D., and Berkman, S. (1948). *J. Bact.* **56**, 527.
Hill, S. E. (1929). *J. Gen. Physiol.* **12**, 863.
Hills, G. M. (1949). *Biochem. J.* **45**, 353, 363.
Hirst, E. L., and Jones, J. K. N. (1946). *Advances in Carbohydrate Chem.* **2**, 161.
Hoagland, D. R. (1948). Lectures on the Inorganic Nutrition of Plants, p. 136. Chronica Botanica, Waltham, Mass.

Höber, R. (1945). Physical Chemistry of Cells and Tissues. Blakiston, Philadelphia, Pennsylvania.

Horowitz, N. H. (1945). *Proc. Natl. Acad. Sci. U. S.* **31**, 153.

Hovasse, R. (1934). *Compt. rend.* **198**, 402.

Hutchens, N. L., Podolsky, B., and Morales, M. F. (1948). *J. Cellular Comp. Physiol.* **32**, 117.

Hutner, S. H. (1936). *Arch. Protistenk.* **88**, 93.

Hutner, S. H. (1946). *J. Bact.* **52**, 213.

Hutner, S. H. (1948). *Trans. N. Y. Acad. Sci.* **10**, 136.

Hutner, S. H., and Bjerknes, C. A. (1948). *Proc. Soc. Exptl. Biol. Med.* **67**, 393.

Hutner, S. H., and Hockett, R. C. (1936). Unpublished.

Hutner, S. H., Provasoli, L., Schatz, A., and Haskins, C. P. (1950). *Proc. Am. Phil. Soc.* **94**, 152.

Hutner, S. H., Provasoli, L., Stokstad, E. L. R., Hoffmann, C. E., Belt, M., Franklin, A. L., and Jukes, T. H. (1949). *Proc. Soc. Exptl. Biol. Med.* **70**, 118.

Hutton, W. E., and ZoBell, C. E. (1949). *J. Bact.* **58**, 463.

Hyman, Libbie H. (1940). The Invertebrates: Protozoa through Ctenophora. McGraw-Hill, New York, 726 pp.

Irving, H., and Williams, R. J. P. (1948). *Nature* **162**, 746.

Jacobs, D. L. (1946). *Trans. Am. Microscop. Soc.* **65**, 1.

Jahn, T. L. (1936). *Proc. Soc. Exptl. Biol. Med.* **33**, 494.

Jahn, T. L. (1941). Chap. 6 *in* Calkins, G. N., and Summers, F. M. (eds.). Protozoa in Biological Research. Columbia University Press, New York.

Jahn, T. L. (1946). *Quart. Rev. Biol.* **21**, 246.

Jansen, B. C. P. (1949). *Vitamins and Hormones* **7**, 83.

Jírovec, O. (1949). *Experientia* **5**, 74.

Johnson, J. R. (1943). *In* Chap. 25, p. 1868, *in* Gilman, H. (ed.). Organic Chemistry: an Advanced Treatise, Vol. II, Wiley, New York.

Johnson, L. P. (1939). *Trans. Am. Microscop. Soc.* **58**, 42.

Johnson, L. P., and Jahn, T. L. (1942). *Physiol. Zoöl.* **15**, 466.

Kamen, M. D. (1949). Chap. 18 *in* Franck, J., and Loomis, W. E. (eds.). Photosynthesis in Plants. Iowa State College Press, Ames.

Karlsson, J. L., and Barker, H. A. (1948a). *J. Bact.* **56**, 670.

Karlsson, J. L., and Barker, H. A. (1948b). *J. Biol. Chem.* **175**, 913.

Kater, J. M. (1929). *Univ. Calif. (Berkeley) Pubs. Zoöl.* **33**, 125.

Kluyver, A. J. (1940). *Intern. Congr. Microbiol. Rept. Proc. 3rd Congr. N. Y. 1939*, 73.

Kofoid, C. A., and Swezy, O. (1921). The Free-Living Unarmored Dinoflagellata. *Mem. Univ. Calif.* **5**, 1.

Knight, B. C. J. G. (1936). Bacterial Nutrition. Med. Research Council (Brit.), Special Rept. Scr. No. 210. H. M. Stationery Office, London, 182 pp.

Kutscher, F. (1898). *Z. physiol. Chem.* **24**, 360.

Lackey, J. B. (1938). *U. S. Pub. Health Service Repts.* **53**, 1499.

Lackey, J. B. (1944). Chap. 7 *in* Phelps, E. B. Stream Sanitation. Wiley, New York.

Lackey, J. B., and Smith, R. S. (1940). *U. S. Pub. Health Service Repts.* **55**, 268.

Lardy, H. A. (ed.) (1949). Respiratory Enzymes. Burgess Publishing Co., Minneapolis, 290 pp.

Lardy, H. A., and Feldott, G. (1949). *J. Biol. Chem.* **179**, 509.
Levring, T. (1946). *K. Fysiograf. Sällskap. Lund Förh.* **16**, no. 7, 12 pp.
Lewin, R. A. (1949a). *Biol. Bull.* **97**, 243.
Lewin, R. A. (1949b). *Nature* **164**, 543.
Lewin, R. A. Unpublished.
Lewis, J. C. (1942). *J. Biol. Chem.* **146**, 441.
L'Héritier, P. (1949). *In* Lwoff, A. (ed.). Unités biologiques douées de continuité génétique. Centre National d.l. Recherche Scientifique, pp. 113–122.
Lifson, N., Lorber, V., Sakami, W., and Wood, H. G. (1948). *J. Biol. Chem.* **176**, 1263.
Lindemann, R. L. (1942). *Ecology* **23**, 1.
Lipmann, F. (1946). *Advances in Enzymol.* **6**, 231.
Lipmann, F. (1948). *Cold Spring Harbor Symposia Quant. Biol.* **13**, 127.
Lorente de Nó, R. (1949). *J. Cellular Comp. Physiol.* **33**, 1.
Lucksch, I. (1932). *Botan. Centr. Beihefte* **A50**, 64.
Lwoff, A. (1932). Recherches biochimiques sur la nutrition des Protozoaires. Masson, Paris, 158 pp.
Lwoff, A. (1934). *Zentr. Bakt. Parasitenk. Abt.* I Orig. **130**, 497.
Lwoff, A. (1938). *Arch. Protistenk.* **90**, 194.
Lwoff, A. (1941). *Ann. inst. Pasteur* **66**, 407.
Lwoff, A. (1943). L'Évolution physiologique. Étude des pertes de fonctions chez les microorganismes. Actualités scientifique et industrielles 970. Hermann, Paris, 308 pp.
Lwoff, A. (1947). *Ann. Rev. Microbiol.* **1**, 101.
Lwoff, A. (1950). *New Phytologist* **49**, 72.
Lwoff, A., and Dusi, H. (1934). *Ann. inst. Pasteur* **53**, 641.
Lwoff, A., and Dusi, H. (1935a). *Compt. rend. soc. biol.* **119**, 1092.
Lwoff, A., and Dusi, H. (1935b). *Compt. rend. soc. biol.* **119**, 1260.
Lwoff, A., and Dusi, H. (1937). *Compt. rend.* **205**, 630, 756, 882.
Lwoff, A., and Dusi, H. (1938a). *Compt. rend. soc. biol.* **127**, 53.
Lwoff, A., and Dusi, H. (1938b). *Trav. station zool. Wimereux* **13**, 431.
Lwoff, A., and Dusi, H. (1941). *Ann. inst Pasteur* **67**, 229.
Lwoff, A., Ionesco, H., and Gutmann, A. (1949). *Compt. rend.* **228**, 342.
Lwoff, A., and Lederer, E. (1935). *Compt. rend. soc. biol.* **119**, 971.
Lwoff, A., and Monod, J. (1947). *Ann. inst. Pasteur* **73**, 323.
Lwoff, A., and Monod, J. (1949). *Arch. Biochem.* **22**, 482.
Lwoff, A., and Schaeffer, P. (1949a). *Compt. rend.* **228**, 511.
Lwoff, A., and Schaeffer, P. (1949b). *Compt. rend.* **228**, 779.
Lwoff, M., and Lwoff, A. (1929). *Compt. rend. soc. biol.* **102**, 569.
McCullough, N. B., and Dick, L. A. (1942). *J. Infectious Diseases* **71**, 198.
MacLeod, R. A., and Snell, E. E. (1948). *J. Biol. Chem.* **176**, 39.
Mainx, F. (1929). *Tabul. Biol.* **5**, 1.
Manten, A. (1948). Phototaxis, Phototropism, and Photosynthesis in Purple Bacteria and Blue-Green Algae. Dissertation, Utrecht.
Marvel, C. S., and Richards, J. C. (1949). *Anal. Chem.* **21**, 1480.
Metz, C. B. (1948). *Am. Naturalist* **82**, 85.
Meyerhof, O. (1925). *Biochem. Z.* **162**, 42.
Miller, C. P. (1948). *Ann. Internal Med.* **29**, 765.

Myers, J. (1949). Chap. 17 *in* Franck, J., and Loomis, W. E. (eds.). Photosynthesis in Plants. Iowa State College Press, Ames.

Nakamura, H. (1937). *Acta Phytochim. (Japan)* **9**, 189; 1938. **10**, 259; Yamagata, S., and Nakamura, H. (1938). **10**, 297. (Cited by Rabinowitch, p. 128.)

Nickerson, W. J., (1949). *Experientia* **5**, 202.

Oparin, A. I. (1938). The Origin of Life. Translated by S. Morgulis, Macmillan, New York, 270 pp.

Osterhout, W. J. V. (1949). *Proc. Natl. Acad. Sci. U. S.* **35**, 548.

Osterhout, W. J. V., and Hill, S. E. (1936). *Physiol. Revs.* **16**, 216. (Cited by Höber, p. 352.)

Osterud, K. L. (1946). *Physiol. Zoöl.* **19**, 19.

Pallares, E. S., Barrenecheas, Jr., M., and Villalba (1945). *Química (Mex.)* **3**, 5. (cited by Strain, 1949).

Pappenheimer, A. M., Jr., and Hottle, G. A. (1940). *Proc. Soc. Exptl. Biol. Med.* **44**, 645.

Pascher, A. (1917). *Biol. Zentr.* **37**, 312.

Pascher, A. (1927). Die Süsswasserflora Deutschlands, Österreichs und der Schweiz. Heft. 4: Volvocales—Phytomonadinae. G. Fischer, Jena, (a) pp. 114–116; (b) pp. 52–58.

Pascher, A. (1930). *Biol. Zentr.* **50**, 1.

Pascher, A. (1931). *Botan. Centr. Beihefte A.* **48**, 481.

Pascher, A. (1932). *Botan. Centr. Beihefte A.* **48**, 675.

Pascher, A. (1940). *Arch. Protistenk.* **93**, 331.

Petty, M. A., and Matrishin, M. (1949). *Proc. Soc. Am. Bact.* 49th Gen. meeting, p. 47.

Pintner, I. J., Provasoli, L., and Hutner, S. H. Unpublished.

Pringsheim, E. G. (1921). *Beitr. allgem. Botan.* **2**, 88.

Pringsheim, E. G. (1926). *Planta* **2**, 555.

Pringsheim, E. G. (1937a). *Nature* **139**, 196.

Pringsheim, E. G. (1937b). *Planta* **26**, 631, 665.

Pringsheim, E. G. (1937c). *Planta* **27**, 61.

Pringsheim, E. G. (1946a). Pure Cultures of Algae. Cambridge University Press, 119 pp.

Pringsheim, E. G. (1946b). *Trans. Roy. Soc. (London) B.* **232**, 311.

Pringsheim, E. G. (1948a). *Biol. Revs. Cambridge Phil. Soc.* **23**, 46.

Pringsheim, E. G. (1948b). *New Phytologist* **47**, 52; with addendum by Hovasse, R. (pp. 68–79).

Pringsheim, E. G. (1949). *Bact. Revs.* **13**, 47.

Pringsheim, E. G., and Ondratschek, K. (1939). *Botan. Centr. Beihefte A.* **59**, 117.

Provasoli, L. (1938). *Boll. zool. agrar. e. bachicolt. univ. Milano* **9**, 1.

Provasoli, L., Hutner, S. H., and Schatz, A. (1948). *Proc. Soc. Exptl. Biol. Med.* **69**, 279.

Provasoli, L., Pintner, I. J., and Haskins, C. P. Unpublished.

Rabinowitch, E. I. (1945). Photosynthesis and Related Processes. I. (a) Chap. 6, pp. 128–149. Interscience, New York, 599 pp.

Rahn, O. (1941). *Growth* **5**, 197.

Rake, G. (1948). *Proc. Soc. Exptl. Biol. Med.* **67**, 249.

Reichenow, E. (1928). *Arch. Protistenk.* **61**, 144.

Reischer, H. S. (1951). *Mycologia.* In press.
Rhoades, M. M. (1946). *Cold Spring Harbor Symposia Quant. Biol.* **11**, 202.
Rhoades, M. M. (1949). *In* Lwoff, A. (ed.). Unités biologiques donées de continuité génétique. Centre National d.l. Recherche Scientifique, pp. 37–44.
Riegel, B., Stanger, D. W., Wikholm, D. M., Mold, J. D., and Sommer, H. (1949). *J. Biol. Chem.* **177**, 7.
Rieke, F. F. (1949). Chap. 22 *in* Franck, J., and Loomis, W. E. (eds.). Photosynthesis in Plants. Iowa State College Press, Ames.
Rittenberg, D. (1948). *Cold Spring Harbor Symposia Quant. Biol.* **13**, 173.
Robbins, W. J. (1939). *Science* **89**, 303.
Robbins, W. J., and Schmitt, M. B. (1945). *Bull. Torrey Botan. Club* **72**, 76.
Robinson, E. S., and Flusser, B. A. (1944). *J. Biol. Chem.* **153**, 529.
Rodhe, W. (1948). *Symbolae Botan. Upsalienses* **10**, 1.
Rottier, P.-B. (1936). *Compt. rend. soc. biol.* **122**, 65, 776.
Sauer, M. E. (1935). *J. Immunol.* **29**, 157.
Schoenborn, H. W. (1940). *Ann. N. Y. Acad. Sci.* **40**, 1.
Schoenborn, H. W. (1942). *Physiol. Zoöl.* **15**, 325.
Schoenborn, H. W. (1946). *Physiol. Zoöl.* **19**, 430.
Schoenborn, H. W. (1949). *J. Exptl. Zoöl.* **111**, 437.
Schopfer, W. H. (1943). Plants and Vitamins, pp. 215–218. Chronica Botanica Co., Waltham, Mass.
Skoog, F. K., and Lindegren, C. C. (1947). *J. Bact.* **53**, 729.
Smith, G. M. (1938). Cryptogamic Botany. Vol. I. Algae and Fungi. McGraw-Hill, New York, 545 pp.
Smith, G. M. (1946). *Am. J. Botany* **33**, 625.
Smith, G. M. (1948). *Science* **108**, 680.
Smith, J. H. C. (1949). Chap. 3 *in* Franck, J., and Loomis, W. E. (eds.). Photosynthesis in Plants. Iowa State College Press, Ames.
Smyth, D. H. (1940). *Biochem. J.* **34**, 1598.
Sonneborn, T. M. (1941). Chap. 14 *in* Calkins, G. N., and Summers, F. M. (eds.). Protozoa in Biological Research. Columbia University Press, New York.
Sonneborn, T. M. (1946). *Cold Spring Harbor Symposia Quant. Biol.* **11**, 236.
Sonneborn, T. M. (1949). *Am. Scientist* **37**, 33.
Stacey, M. (1946). *Advances in Carbohydrate Chem.* **2**, 161.
Stanier, R. Y. (1947). *J. Bact.* **54**, 191.
Stanier, R. Y. (1948). *J. Bact.* **55**, 477.
Stanier, R. Y., and Tsuchida, M. (1949). *J. Bact.* **58**, 45.
Stanier, R. Y., and Van Niel, C. B. (1941). *J. Bact.* **42**, 437.
Starr, R. C. (1949). *Proc. Natl. Acad. Sci. U. S.* **35**, 453.
Stephenson, M. (1949). Bacterial Metabolism, 3rd. ed., pp. 272–276. Longmans, Green, New York.
Stokstad, E. L. R., Jukes, T. H., Pierce, J., Page, A. C., Jr., and Franklin, A. L. (1949). *J. Biol. Chem.* **180**, 647.
Strain, H. H. (1949). Chap. 6 *in* Franck, J., and Loomis, W. E. (eds.). Photosynthesis in Plants. Iowa State College Press, Ames.
Tanzer, C. (1941). *J. Immunol.* **42**, 291.
Ternetz, C. (1912). *Jahrb. wiss. Botan.* **51**, 435.
Tischer, J. (1944). *Z. physiol. Chem.* **281**, 143.

Tosic, J. (1946). *Biochem. J.* **40**, 209.

Trager, W. (1941). *Physiol. Revs.* **21**, 1.

Trelease, S. F., and Selsam, M. E. (1939). *Am. J. Botany* **26**, 339.

Umbreit, W. W. (1949). *J. Biol. Chem.* **277**, 703; Oginsky, E. L., Smith, P. H., and Umbreit, W. W. (1949). *J. Bact.* **58**, 747; Smith, P. H., Oginsky, E. L., and Umbreit, W. W., 761; Umbreit, W. W., and Tonhazy, N. E., 769.

Van Niel, C. B. (1944). *Bact. Revs.* **8**, 1.

Van Niel, C. B. (1946). *Cold Spring Harbor Symposia Quant. Biol.* **11**, 285.

Van Niel, C. B. (1949a). *Bact. Revs.* **13**, 161.

Van Niel, C. B. (1949b). Chap. 22 *in* Franck, J., and Loomis, W. E. (eds.). Photosynthesis in Plants. Iowa State College Press, Ames.

Van Niel, C. B., Volcani, B., Elsden, S., and Vishniac, W. (1947). Unpublished.

Vishniac, W. (1949). On the metabolism of the chemolithoautotrophic bacterium *Thiobacillus thioparus* Beijerinck. Dissertation, Stanford.

Vishniac, W., Hutner, S. H., Kisliuk, R., and Storm, J. Unpublished.

Von Dach, H. (1940). *Ohio J. Sci.* **40**, 37.

Von Dach, H. (1942). *Biol. Bull.* **82**, 356.

Von Dach, H. (1943). *Ohio J. Sci.* **43**, 47.

Wald, G. (1943). *Vitamins and Hormones* **1**, 195.

Whiffen, A. J. (1945). *J. Elisha Mitchell Sci. Soc.* **61**, 114.

Wiame, J. M. (1949). *J. Biol. Chem.* **178**, 919.

Wilkinson, I. A. Unpublished.

Work, T. S., and Work, E. (1949). The Basis of Chemotherapy. Interscience, New York, 435 pp.

Wynne, E. S., and Foster, J. W. (1948). *J. Bact.* **55**, 331.

Yonge, C. M. (1945). *Biol. Revs. Cambridge Phil. Soc.* **19**, 68.

Zahl, P. A., Starr, M. P., and Hutner, S. H. (1945). *Am. J. Hyg.* **41**, 41.

The Nutrition of Parasitic Flagellates (Trypanosomidae, Trichomonadinae)

MARGUERITE LWOFF

Pasteur Institute, Paris

CONTENTS

I. Introduction

Two large groups of parasitic flagellates which have long held the interest of parasitologists have also furnished the biochemist with material for study: these are the Trypanosomidae and the Trichomonadinae.

Both groups comprise very numerous species, almost all of them parasitic. Some of these are pathogens capable of causing severe infections (human and animal trypanosomiasis, etc.) ; others appear to have merely a commensal action and are perfectly tolerated by the host. Some are easily cultured *in vitro;* others multiply poorly outside their natural host. In brief, all their biological characteristics make them both extremely varied and interesting as material for research. Furthermore, the comparison of their growth factor requirements or, in other words, their power of synthesis with those of free-living protozoa constitutes a subject of the greatest interest which is far from being developed in the manner it deserves. A few unusual growth factors have been defined for flagellates belonging to one or the other family. They will be discussed for the Trypanosomidae and Trichomonadinae successively in the following pages.

II. Trypanosomidae

Growth Factors for the Trypanosomidae

The Trypanosomidae, as known, are a very homogenous group of flagellates, all parasites to a greater or less degree. Some live exclusively in invertebrate hosts, mostly insects; others are found in vertebrates, and still others make use of two hosts, vertebrate and invertebrate, and are transmitted by one or the other. In the insect they parasitize the digestive tract and its annexes; in the vertebrate they sometimes live in the bloodstream and sometimes in the cells of various tissues. Certain characteristics are common to many representatives of the group; others are restricted to a few species or even to a few strains. Thus, the hematin requirement, which is one of the best known physiological requirements of the group, is not found in species parasitic in the digestive tract of Hemiptera (*Strigomonas oncopelti*) or of Diptera (*Strigomonas culicidarum* var. *culicis*).

Likewise, the requirement for ascorbic acid, whatever its significance, is only possessed by certain highly evolved forms, *Leishmania* and *Trypanosoma.* The requirement for constituents of blood serum exists only in representatives of these two genera. After having studied the nature of the growth factors identified for these flagellates, we would be in a position to define some factors in the biochemical evolution of the family Trypanosomidae which parallel their biological evolution.

1. HEMATIN

One of the oldest cultural characteristics recognized in trypanosomes is their requirement for blood. From 1903 on Novy and MacNeal used blood agar, a well-known medium ever since the work of Pfeiffer on *Hemophilus influenzae* (1892, 1893), for the rat trypanosome (*Trypanosoma lewisi*) and for the agent of nagana (*T. brucei*). It was with this same medium that Charles Nicolle, in 1908, for the first time grew the flagellate causing oriental sore (*Leishmania tropica*) and kala-azar (*L. donovani*). Since then many species of Trypanosomidae have been cultured on various media, solid or liquid.* Many require blood, a characteristic which has rapidly drawn the interest of investigators. To Zotta (1923) we owe the first experiments on the ''stimulating'' action of blood. Realizing the great activity of catalase obtained from calves' liver, which could replace whole blood for the growth of a *Leptomonas* of Hemiptera, *L. pyrrhocoris,* Zotta concluded that blood had a ''catalytic'' action. Although this was later proved inexact, the work of Zotta is striking for its comprehension of the problems involved.

Of the indispensable constituents found in blood, a few are now known. One of the most interesting of these is hemin. Hemin was known as a growth factor for bacteria since the work of Olsen (1921), Thjötta (1921), Thjötta and Avery (1921) Fildes (1922) on the Pfeiffer bacillus, *Hemophilus influenzae.* It was found to play the same role in the culture of a parasitic trypanosome of the digestive tract of the mosquito, *Culex pipiens: Strigomonas fasciculata.* Easily cultured in liquid medium, it multiplied rapidly in peptone water with a very small amount of blood added (as little as 1 part per million or less); this blood could be replaced by protohematin (M. Lwoff, 1933a). Therefore the only factor that exists in blood necessary for the growth of *Strigomonas fasciculata* is hematin. This fact made particularly easy the further study of hematin as a growth factor—a study interesting for the following reasons.

Hematin, an iron salt of protoporphyrin or 1,3,5,8-tetramethyl-2,4-divinyl-6,7-propanoic porphyrin is in fact an absolutely constant constituent of animal and plant cells, as well as of aerobic microorganisms. It is perhaps also present in certain anaerobes. The great majority of living organisms synthesize protohematin. The only exceptions are certain rare bacteria, protozoan or insects, which require an exogenous source of hematin. The only insect thus far known as incapable of synthesizing

* See bibliography in M. Lwoff, 1940.

hematin is a reduviid blood sucker *Triatoma infestans,* one of the vectors of Chagas disease, which is caused by *Trypanosoma cruzi* (M. Lwoff and Nicolle, 1945, 1947a,b). Hematin is an essential metabolite for these organisms since they are incapable of synthesizing it; in other words, it is a growth factor or a vitamin.

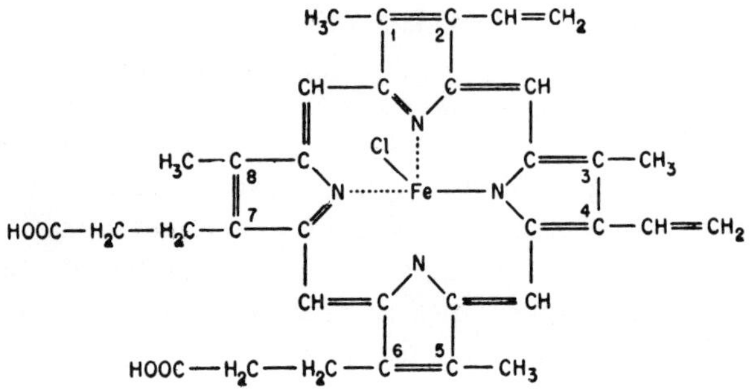

How is hematin concerned in the metabolism of *Strigomonas fasciculata?* Because of a lack of convincing evidence, one could hardly say that blood and particularly hematin played a "catalytic" role, exerting a catalase-like or a peroxidase-like action (see Zotta, 1923). But many substances which have peroxidase activity are without effect on the multiplication of the flagellate. Thus neither deuterohematin nor hematohematin, nor cytochrome C, nor other hemins—all substances which have peroxidase activity—can replace protohematin (Table I); whereas protoporphyrin, a substance devoid of peroxidase activity, is just as active in promoting the multiplication of the flagellates as protohematin (M. Lwoff, 1933, 1940). And, as will be seen further in this discussion, the effect on growth parallels that on respiration.

TABLE I. Effect of Substitutions in 2,4 position on heme activity toward *Strigomonas fasciculata*

	2,4	Activity
Protoheme	$—CH=CH_2$	+
Hematoheme	$—CH(OH)\cdot CH_3$	0
Mesoheme	$—CH_2\cdot CH_3$	0
Deuteroheme	—H	0

The specificity of the structure of protohematin, which is replaceable by protoporphyrin, is thus shown, but it is effective not because of any peroxidase activity; its activity is related to its structure. Any change destroys the activity of the porphyrin, particularly the suppression of the vinyl group in the 2,4 position.

The physiological role of hemin must therefore be sought elsewhere. A quantitative study of the action of blood (A. Lwoff, 1933, 1934) has shown the site of activity.

Development of the flagellate is abundant at concentrations varying between 1 part/100 and 1 part/10,000, but there is no relation between the concentration of blood and the number of flagellates. The development

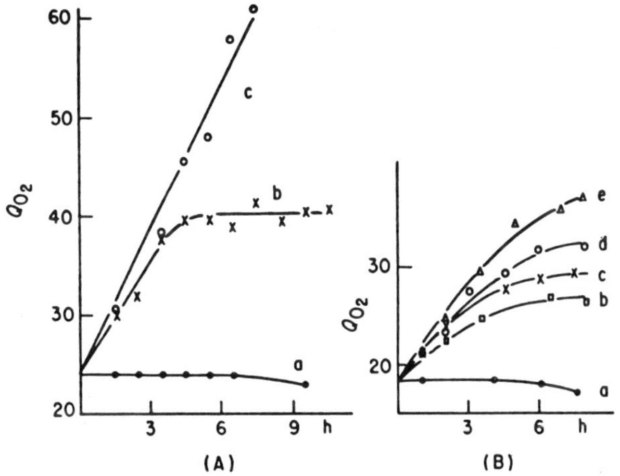

FIG. 1. Effect of blood on the respiration of *Strigomonas fasciculata*. (After A. Lwoff 1933–1934.)

A—Flagellates, 450 μg. (dry weight)

a, control; b and c, blood, 2.5 and 10 μg., respectively

B—Flagellates, 708 μg. (dry weight)

a, control; b, c, d, and e, blood, 2, 2.5, 3.3, and 5 μg., respectively

is limited by another factor. In contrast, at concentrations of blood between 2×10^{-5} and 1×10^{-6}, the number of flagellates is proportional to the blood concentration; 10^{-9} g. of blood allows the growth of 6,200 flagellates. Therefore, 0.16×10^{-12} g. of blood is needed for the development of one flagellate, or approximately 7,500 molecules of hematin. As there are 1.5×10^{11} organisms in 1 g. of flagellates (dry weight), hematin represents 1/12,000 of the dry weight of the flagellate.

What could be the function of hematin as a growth factor? Ninety per cent of the respiration is inhibited by KCN and CO. The inhibition due to CO is considerably lessened by illumination. The respiration is therefore catalyzed by iron. This has also been shown by comparing the respiration of normal and hematin-deficient organisms. The hematin-deficient flagellates are obtained by growing the organisms in a medium in which the quantity of hematin is the only limiting factor for growth. Addition of a small amount of hemin to this medium elicits an intense multiplication of the hematin-deficient flagellates which had no longer been dividing. This very simple technique is actually in general use for the quantitative study of the action of a growth factor in respect to a given organism.

The QO_2 of normal flagellates grown in the presence of excess blood is 55 at 28°, while that of the deficient organisms is 22. The respiration of deficient flagellates remains constant. But if blood or hematin is added to the Warburg vessel, the respiration will increase without an appreciable change in the weight of the flagellates. Once the blood is exhausted, respiration remains constant. Manifestly hemin is a part of the catalytic respiratory system. One milligram of blood will increase the QO_2 of 1 mg. (dry weight) of flagellates by 3 units. The amount of blood and the number of hematin molecules necessary to produce a QO_2 of 55 can be calculated. The number found was 520,000 (A. Lwoff, 1933, 1934).

By measuring multiplication as a function of added blood—an entirely different method—one arrives at the figure 750,000. The agreement of these two figures calculated by different methods is satisfactory considering the many steps involved.

The stimulation of growth by protohematin is strictly parallel to the stimulation of respiration in so far as it is brought about by protoporphyrin, but not by the porphyrins corresponding to the inactive hematins. (Table II, p. 134.)

TABLE II. Effect of Different Hemes and Porphines on Respiration and Growth of *Strigomonas fasciculata*
(After A. and M. Lwoff)

	Peroxidases	Respiration	Growth
Blood	+	+	+
Protoheme	+	+	+
Deutero-, hemato-, meso-, pyrro-, heme	+	0	0
Cytochrome c	+	0	0
a and b Pheophorbid-hemes	+	0	0
Artificial peroxidase and Baudisch "active iron"	+	0	0
Protoporphine	0	+	+
Deutero-, hemato-, meso-, aetio-, pyrro-porphins	0	0	0
a-Pheophorbid, rhodin	0	0	

The hematin requirement therefore seems to be linked with the inability of the flagellate to synthesize protoporphyrin. It can insert iron into the protoporphyrin and it can utilize protohematin for the construction of its catalytic respiratory system.

What does the protoporphyrin requirement signify? All free-living protozoa as well as certain Trypanosomidae of insects (see below and Table V, p. 145) can multiply in the absence of protoporphyrin. Like all strict aerobes they contain heme pigments; they synthesize protoporphyrin. The ancestor of *Strigomonas fasciculata,* which one can imagine as a free-living flagellate resembling fairly closely the present form and living, for example, in bodies of water, could certainly synthesize protoporphyrin. Protohematin is therefore an essential metabolite, at least for aerobic organisms, and it would appear that the protoporphyrin requirement results from loss of ability to synthesize it.

Thus we know that when hematin acts as a growth factor it has no "catalytic" action, but that on the contrary its effect is quantitative, that each flagellate requires approximately 600,000 molecules of hematin, and that this hematin is used in the synthesis of the respiratory system. We also know that this activity is associated not with iron and peroxidase function, but rather with a specific structure. Finally, we know that this vitamin requirement is a result of a loss of the ability to synthesize this vitamin. It is recognized today that the properties formerly attributed to hematin, in its role of a growth factor, are also those of a great many essential metabolites acting as vitamins for microorganisms as well as for higher organisms.

Hematin is necessary for many, although not for all, Trypanosomidae. Certain among them, such as *Strigomonas oncopelti,* a parasite of Hemiptera, and *S. parva, S. media,* parasites of flies, multiply perfectly well in peptone water without the addition of blood or hemin. For others, such as *S. fasciculata* which has been used as an example in the preceding paragraphs, the only factor brought in by blood is hematin. This is found to be the case principally in parasites of insects such as, among others, *Strigomonas culicidarum* var. *culicis* (mosquitoes, M. Lwoff, 1935) and *S. muscidarum* (flies, M. Lwoff, 1936). It is also so for many leishmanias (M. Lwoff, 1939, 1940) and trypanosomes (M. Lwoff 1938a,b, 1940) (see Table IV, p. 138) ; but in these, hematin is no longer the only factor brought by blood; for certain species or strains, blood is also the source of other factors to be discussed later.

It may be noted at this point that no leishmania or trypanosome is known that does not require exogenous hematin. No one has given proof that hematin is not required by leishmanias and trypanosomes. All who have sought to define the indispensable constituents found in blood and who have used, correctly, hemoglobin-free basal media (Salle and Schmidt, 1928; Ray, 1932; Berrebi, 1936; Saïto, 1937; M. Lwoff) have noted the impossibility of growing these flagellates in the complete absence of hemoglobin or of the formed components of blood—the erythrocytes—which bear it. M. Lwoff has shown (1931) that hematin was the active factor in hemoglobin and one of the active factors in erythrocytes. The results of Adler (1934), or Senekjie (1943), and of Senekjie and Lewis (1945), who grew various leishmanias and *Trypanosoma cruzi* on a peptone medium enriched only with rabbit serum, lead one to believe that the serum was not entirely free of sufficient traces of hemoglobin to allow the multiplication of the flagellates in an otherwise complete medium.

The requirement for hematin is not an exclusive property of protozoa, and it does not take us too far afield to show its role in bacterial metabolism.

For example, the addition of hematin to hematin-deficient cells of *Hemophilus influenzae* produces in a few minutes an increased respiration proportional to the quantity of hematin. There can be no doubt that in this case, as in the case of *S. fasciculata,* hematin is used in the construction of the catalytic respiratory system (A. and M. Lwoff, 1937).

Hemophilus influenzae can use protoporphyrin as well as ferroporphyrin, and is therefore capable of inserting the iron in protoporphyrin, thus synthesizing a ferro-protoporphyrin (Granick and Gilder, 1946). But *Hemophilus* differs from *Strigomonas* in respect to the specificity of the hematin requirement and in the role played by various groups in the porphyrin nucleus. We cannot go very far into this interesting subject, but we must mention the existence of certain important differences among cultures of *Hemophilus,* notably the use of the vinyl groups as shown by the inability to utilize meso-, hemato-, or deuterohematin, and the role of the vinyl groups as direct or indirect controllers of the insertion of iron

TABLE III. Protoheme in the anaerobic growth of *Hemophilus influenzae* (After Granick and Gilder, 1946)

Anaerobic transfers	h*: 0	h*: 0.01	h*: 0.001
1	0.087E†	0.088E†	0.094E†
2	0.00	0.054	0.075
3	0.00	0.055	0.045

* h = protohematin μg./ml.

† E = extinction under 1 cm. at 380 mμ.

The first anaerobic transfer was made from an aerobic strain grown in presence of 0.02 μg./ml. hematin; the second and third transfers from anaerobic transfers with 1.1 μg./ml. hematin.

in the molecule as well as their participation in the reduction of nitrate or of the synthesis of the system carrying out this reduction, and, finally, the role played by the propanoic groups (Table III, p. 136).

To conclude this study of hematin as a growth factor, one may ask how the need for hematin in the trypanosomes can be interpreted. The Trypanosomidae are aerobic, and offer no exception to the general rule: they contain hematin. Those among them which can multiply in the absence of hemin (*Strigomonas oncopelti, S. parva, S. media*) can therefore synthesize hematin. As to the others, they do not possess this power. For one reason or another, by some mechanism of which we are not sure, they have lost this ability. All these Trypanosomidae are parasitic whether or not they are capable of synthesizing hematin, and one might suppose that loss of this power of synthesis must have been established earlier than the parasitic state. If this were so, what relation exists between the nature of the habitat and the hematin requirement? This habitat, whatever it may be—digestive tract of an insect, blood or tissue of vertebrates—is always rich in hematin. There is, then, no apparent relationship between these two conditions: content of hematin in the environment and ability to synthesize it. There seems to be no doubt that the Trypanosomidae have a tendency toward the loss of this synthetic ability and that their parasitic way of life allows species to survive which have lost this ability.

Numerous parasitic microorganisms possess an extremely limited power of synthesis. In complex media when a certain metabolite is constantly present, natural selection would not tend to eliminate the organisms which no longer synthesize this metabolite. This is true of hematin. Besides, it is likely that the mutations which control this synthesis are rare. *Strigomonas oncopelti* was cultured for several years in a medium containing blood; the ability to synthesize hematin was not lost. All attempts to obtain strains which synthesize hematin from strains which do not have been futile. It would seem probable that the biosynthesis of hematin is dependent upon several genes. The stability of the mutant "hematinless" is probably due to the high improbability of the reverse mutations of several genes.

2. ASCORBIC ACID

For a very long time, the multiplication of many Trypanosomidae from the genera *Leishmania* and *Trypanosoma* was obtained only in the presence of whole blood. Analysis of the function of whole blood has shown that indispensable substances were present in both serum and erythrocytes. One of the factors, as just seen, found in red blood cells, is hematin. But erythrocytes also contain another factor, ascorbic acid.

Certain leishmanias or trypanosomes can be maintained indefinitely in a peptone medium containing a small amount of agar (1 to 2 parts/1,000) and enriched with horse serum (1/10) and hematin (*L. agamae, L. ceramodactyli,* certain strains of *L. tropica*) ; other strains cannot (*L. donovani,* certain of *L. tropica, L. brasiliensis, Trypanosoma cruzi*). Multiplication of the latter strains will take place in a peptone + serum + hematin medium if ascorbic acid is added under suitable experimental conditions (M. Lwoff, 1938a, 1939, 1940; Goat and Mora, 1947). If Senekjie (1943) found no favorable action of ascorbic acid on *Trypanosoma cruzi,* it was doubtless because his experiments were done in media devoid of serum and hematin.

Referring to Table IV, which summarizes the growth factor requirements of leishmanias and trypanosomes in peptone media, one can see that all the organisms cited require serum and hematin, and only certain ones require ascorbic acid.

TABLE IV. Growth Factors for *Leishmania* and *Trypanosoma* (Basal Medium : Peptone Agar Solution)

	Hematin	Ascorbic acid	Blood serum
Leishmania tropica			
strains: Tunis	+	+	+
Baghdad	+	+	+
Greece	+	+	+
H'ra	+	−	+
Dim	+	−	+
L. donovani			
strains: Tunis	+	+	+
Khar	+	+	+
L. brasiliensis	+	+	+
L. ceramodactyli	+	−	+
L. agamae	+	−	+
Trypanosoma cruzi			
strains: Schiz.	+	+	+
Cruz.	+	+	+

The study of the ascorbic acid requirement holds interest because it gives rise to various interpretations. Its action, well known to be that of a reducing agent in bacterial cultures, immediately poses the problem of its mode of action in cultures of Trypanosomidae. If ascorbic acid acts essentially only to lower the oxidation-reduction potential of the culture media or to destroy certain toxic oxidation products formed in the media, it is quite evident that here its action would not be specific, and other reducing agents could replace it. This is known to be true in various bacterial cultures where glutathione, cysteine, etc. can completely replace ascorbic acid.

But this is not the case with Trypanosomidae. Neither cysteine nor sodium sulfide nor other reducers ("reductose") were shown to be able to promote growth of leishmanias or trypanosomes in the presence of peptone, hematin, and serum. Yet these reducers have no toxic or inhibitory action for these flagellates (M. Lwoff, unpublished); they are inert.

Furthermore, certain analogs of ascorbic acid, D-isoascorbic acid, D-glucoascorbic acid, and 3-methylascorbic acid are inactive in replacing ascorbic acid (M. Lwoff, unpublished). One or two transfers are possible but seldom more. This favors the hypothesis that L-ascorbic acid acts as a growth factor.

Besides, it is well known that Trypanosomidae are strict aerobes (see p. 177), as shown by the work of von Brand and more recently by that of Chang (1948). For example, in a culture of *T. cruzi*, an increase of E_h (310 to 330 mv.) is followed by an increased multiplication of the flagellates; a decrease (280 mv.) decreases the rate of multiplication. The flagellates tend to disappear from cultures contaminated with certain bacteria which lower the E_h of the medium, while they can live in mixed cultures of bacteria which do not produce this effect. Hence the presence of ascorbic acid or other reducers should be rather unfavorable. However, ascorbic acid does favor growth; reducers are without action, no doubt because Trypanosomidae adjust themselves fairly readily to a reduced oxygen tension.

It will be seen that ascorbic acid probably acts as a reducer for *Trichomonas* (p. 168). But what holds for *Trichomonas* does not necessarily hold for all anaerobic microorganisms. Thus a spirochete, *Spirochaeta gallinarum,* which grows in poorly oxygenated media still requires ascorbic acid for growth, even in media where the oxygen tension is lowered. Ascorbic acid here cannot be replaced by other reducers (M. Lwoff and Chorine, 1943). The objection can be raised that the media used for spirochetes contain serum and that the ascorbic acid detoxifies oxidation products formed as the culture grows; other reducers might not possess this property. As for this, it may be mentioned that only fresh serum is suitable for the cultivation of *Spirochaeta gallinarum* and that the addition of ascorbic acid is not enough to restore the activity of an aged serum.

An analogous problem posed in lactobacilli appears to have been solved. Under certain conditions, ascorbic acid stimulates the growth of certain of these bacteria, an effect which might not be due only to its reducing properties, since, on the one hand, other reducers have the same effect and, on the other hand, the growth of lactobacilli may not be the same in media having the same E_h (Rose and Peterson, 1949) ; the action of ascorbic acid is not specific: glucoascorbic and thioglycolic acids, glutathione, and Na formaldehyde-sulfoxylate can replace it for *Lactobacillus leichmannii* in a medium based on amino acids. Here where its stimulating action was observed, it presumably acts as a reducer of the oxidation products of vitamin B_{12}, which are formed only under certain conditions, e.g., upon autoclaving.

To return to the Trypanosomidae, other arguments in favor of the direct participation of ascorbic acid in the metabolism of certain Trypanosomidae are furnished by considerations of cytochemistry and pathogenicity, biological arguments to be added to those already enumerated. It is possible to detect the presence of ascorbic acid in the cytoplasm of trypanosomes by treatment with silver nitrate (Giroud's technique). In this manner the presence of vitamin C in *Trypanosoma brucei* (experimental infection in the guinea pig) is demonstrated by the formation of blackish granules of variable size, number, and shape (Roskin and Nastiukowa, 1941). Systematic observations by the Russian authors on the variation in the ascorbic acid content of the trypanosomes during the course of infection in the guinea pig have shown that the concentration of vitamin C in the cytoplasm varies with the severity of the infection. In guinea pigs on a normal diet containing sufficient ascorbic acid, the concentration of vitamin C in the trypanosomes varies from day to day. When the trypanosomic infection is started in an already C deficient guinea pig, no vitamin C is detectable in trypanosomes. But if the animal is put on a C deficient diet at the very start of an infection, trypanosomes are found to contain some ascorbic acid. It would seem that the parasite can concentrate the free vitamin of the host. If a deficient guinea pig with trypanosomes containing no detectable vitamin receives subcutaneous injection of 0.01 g. of ascorbic acid, a few days later the trypanosomes will be found to contain large quantities of vitamin C. The amount of intracytoplasmic vitamin then diminishes, the granules become fewer and smaller. The vitamin can be made to reappear by the same procedure.

Furthermore, the ascorbic acid concentration in the liver of the guinea pig is found to diminish during the course of an experimental infection with *Trypanosoma brucei*. In rats infected with *T. hippicum*, this decrease is found in the kidneys, the adrenals, and the liver, while a net increase is found in the plasma; there must therefore be transport of the vitamin from some sites to others (Nyden, 1948).

Finally, phenomena of a similar nature have been found in relation to vitamin C and *Plasmodium.* In human malaria (tertiary malignant) a progressive vitamin C deficiency of the host develops (Krishman, 1938; Lotze, 1938; Mohr, 1941) which makes necessary the administration of ascorbic acid. The daily requirement can reach, according to Mohr, 47 to 60 mg. Experiments on *Macacca mulatta* infected with *Plasmodium knowlesi* have shown (McKee and Geiman, 1946) that the ascorbic acid concentration of blood and plasma is lower in infected monkeys than in normal monkeys; in either spontaneous or experimentally produced deficiencies in monkeys, the infection follows an aberrant course: lowering of the rate of increase in the number of parasites, and instead of the usual rapid increase in the number of parasites, there is a slow increase and gradual spontaneous cure. Administration of pure ascorbic acid to these deficient animals restores the usual characteristics of this disease, namely, a significant increase in the number of *Plasmodium* in the bloodstream and death within the usual period; furthermore, if this treatment is continued long enough, it permits the appearance of immunity, with consequent effective treatment of the infection.

Although McKee and Geiman did not find direct evidence for the necessity of ascorbic acid *in vitro* for the multiplication of *P. knowlesi,* they nevertheless found that in its absence "the usual protoplasmic mass is not achieved," and that the slowing of growth and degenerate aspect of the *Plasmodium* in the deficient host were striking.

Without considering these arguments as irrefutable proof of the necessity of ascorbic acid as a growth factor for certain parasitic protozoa, this accumulation of converging evidence is clearly one which undoubtedly deserved to be taken up again and extended.

3. SERUM

Hematin and ascorbic acid are not the only factors in blood necessary for *Leishmania* and *Trypanosoma.* Whatever the composition of the basal medium, serum must be present (see Table IV). Sera from extremely diverse animals is active, that from the rabbit probably being best. Horse serum is very convenient for experiments because it can be obtained in large quantities and free of hemoglobin.

Agreement has not been reached on the active constituents of serum. Adler and Ashbel (1934) found some activity in the albumin fraction, but

Senekjie and Lewis (1945) found no activity in any serum proteins: albumins and eu- and pseudoglobulins. These authors thought that in their experiments the active substances must have been lost in the course of fractionating the serum proteins by dialysis. The principal characteristic of these substances seems to be the ability to withstand heating at 70° C. for 30 minutes but to be destroyed by heating at 100° C. for the same length of time. All this is even in greater need of confirmation than the experiments on *T. cruzi* in which growth was obtained in the presence of serum fractions and in the absence of hematin, a substance which, as seen, is indispensable for this trypanosome. Besides, is the role attributed to serum really that of a growth factor? Keeping in mind the example of *Trichomonas columbae* (p. 162), it might be supposed that the serum supplement might correspond to a cholesterol requirement. A fatty acid requirement (linoleic acid and *Trichomonas vaginalis*, p. 169) might also be invoked. Even the possibility of a simple neutralization of toxicity cannot be eliminated, i.e., the detoxification of unsaturated fatty acids, analogous to what has been observed in certain bacteria and perhaps also in enteric amebae (p. 248).

4. THE B VITAMINS

It is seen that the blood required by certain Trypanosomidae contains several indispensable factors which have to be added either together or separately in a basal medium of peptone water. But peptone water is itself a highly complex medium which contains many growth factors besides these indispensable factors. Work on these growth factors was begun by the study of a *Strigomonas* parasitizing Hemiptera, *S. oncopelti*, which can be transferred indefinitely in a medium consisting solely of peptone water (p. 130).

a. Thiamine Requirement. Widely different peptones allow growth of *Strigomonas oncopelti:* more or less degraded muscle peptones of peanut peptone; as well as yeast autolyzate. Others cannot be used, such as certain alkaline hydrolyzates of silk. The addition of thiamine to such hydrolyzates, in the presence of glucose as carbon source, makes them entirely suitable for the culture of *S. oncopelti* (M. Lwoff, 1937, 1940). The action of thiamine is quantitative. The lower limit of activity is about 1 part in 1.4×10^{12}, a concentration which allows a density of 15 to 50,000 flagellates/ml. at the point of maximum growth.

The need for thiamine can also be demonstrated by using peptones permitting growth, then subjecting them to heating (120° C.) in alkali (pH 9.8), thus destroying the vitamin. The addition of thiamine restores the original activity of these media. The whole thiamine molecule is necessary. Neither thiazole alone (4-methyl, 5-β-hydroxyethylthiazole), nor pyrimidine alone (2-methyl,4-amino,5-methylpyrimidine), nor a mixture of the two is capable of replacing thiamine.

Strigomonas oncopelti is therefore incapable of synthesizing the thiamine molecule and of joining the two parts of this molecule: 4-amino, 5-β-hydroxyethylthiazole and 2-methyl,4-amino,5-methylpyrimidine.

How specific is thiamine? The study of six derivatives with various modifications have shown the following results.

Changes can be made either in the thiazole moiety (three derivatives studied) or in the pyrimidine moiety (two derivatives studied).

1. 4-Methyl, 5-β-hydroxypropyl-N-[(2-methyl, 4-aminopyrimidyl [5]) methyl] thiazole. The change has been made in the thiazole moiety by replacing the β-hydroxyethyl group (in position 5) by a β-hydroxypropyl group.

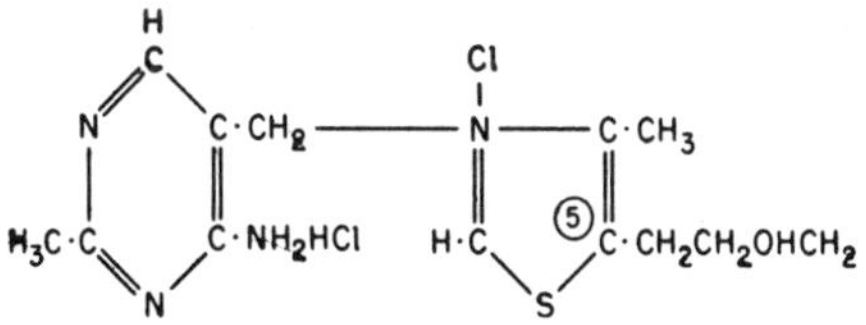

This is active.

2. 4-Methyl, 5-γ-hydroxypropyl-N-[(2-methyl, 5-aminopyrimidyl[5]) methyl] thiazole. The change is again in the thiazole moiety. As in the last derivative, there is a hydroxypropyl group in position 5, instead of the hydroxyethyl group of thiamine, but the hydroxyl is in the γ position.

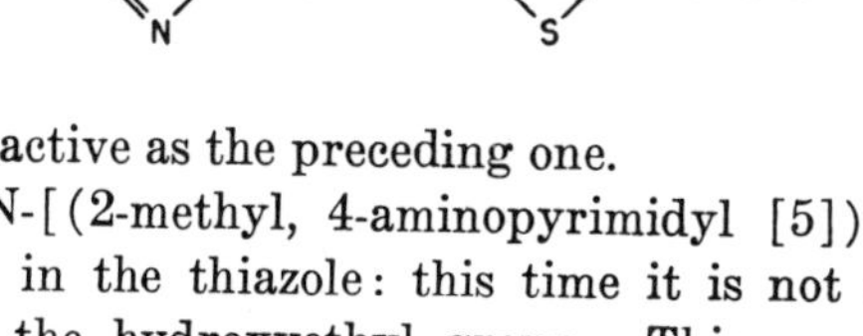

This molecule is active as the preceding one.

3. 4-Methyl-N-[(2-methyl, 4-aminopyrimidyl [5]) methyl] thiazole. Again a change in the thiazole: this time it is not a substitution but the omission of the hydroxyethyl group. This compound is inactive. Whereas fairly important changes in the structure of the hydroxyethyl group in position 5 do not affect the activity, the omission of this group destroys activity.

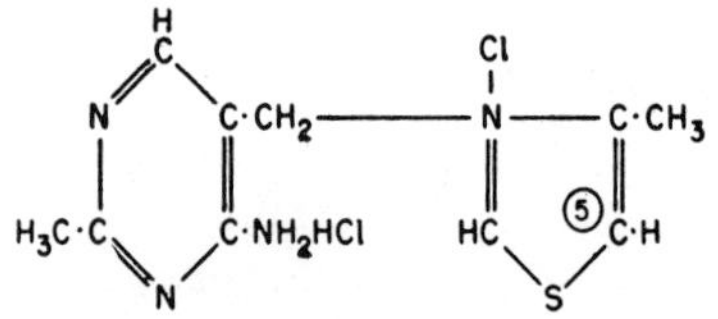

In the two compounds that follow the changes are made in the pyrimidine moiety.

4. 4-Methyl, 5-β-hydroxyethyl-N-[(6-methyl, 4-aminopyrimidyl [5]) methyl] thiazole, a compound also known as "isovitamin B_1." Here the CH_3 usually in position 2 of the pyrimidine is in position 6: Isovitamin

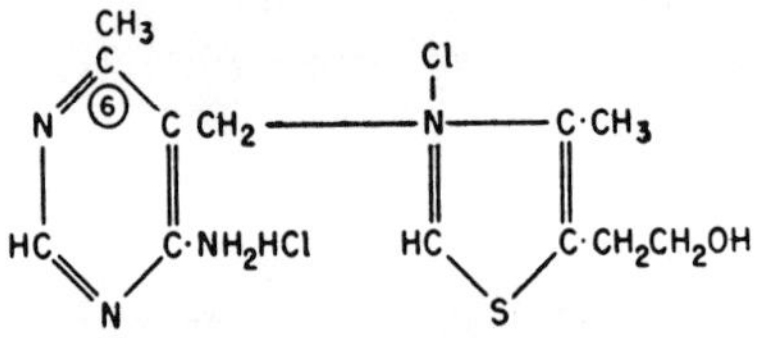

B_1 is inactive, which demonstrates the importance of the position of the methyl group of the pyrimidine for the growth of *Strigomonas*.

5. 4-Methyl, 5-β-hydroxyethyl-N-[(2-ethyl, 5-aminopyrimidyl [5]) methyl] thiazole. In this molecule ("ethylthiamine"), the methyl group in the 2 position on the pyrimidine is replaced by an ethyl group.

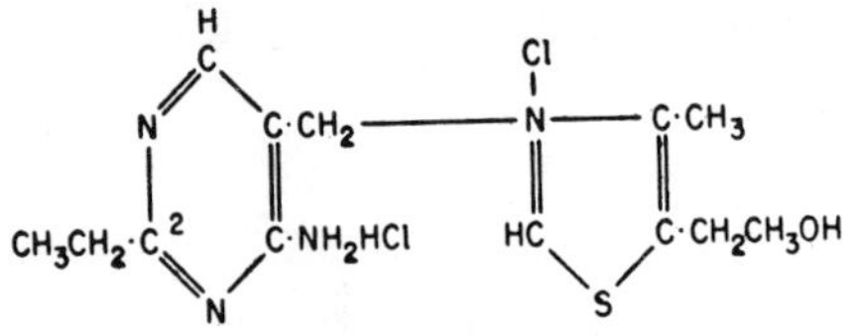

Ethylthiamine is about as active for *Strigomonas oncopelti* as thiamine.

6. Thiochrome, which has the following structure, is inactive for *Strigomonas oncopelti*.

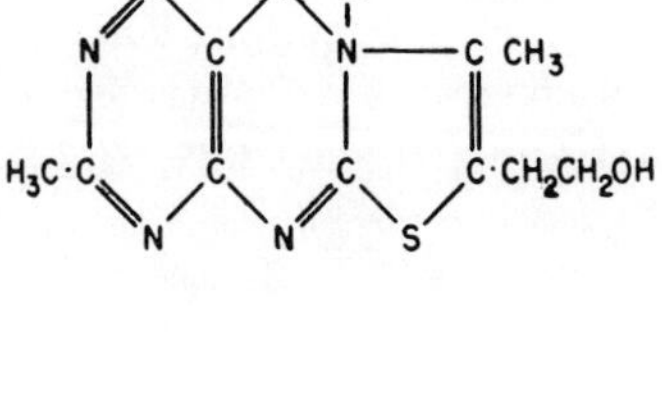

To summarize:

1. The substitution of a β- or a γ-hydroxypropyl group for the hydroxyethyl group in position 5 of the thiazole moiety does not destroy the activity of the thiamine molecule for *Strigomonas oncopelti.* However, the absence of this group in position 5 inactivates the molecule.

2. Substitution of an ethyl group for the methyl group in position 2 on the pyrimidine does not suppress the activity of the molecule, but a methyl group in position 6 instead of 2 yields a completely inactive compound.

3. Finally, a change in the manner of coupling of the molecules of pyrimidine and the thiazole also inactivates the molecule.

The role of thiamine in the metabolism of protozoa has not been studied, but there is every reason to suppose that they utilize the vitamin in the same way that other organisms do.

TABLE V. Growth Factors for Trypanosomidae

	Hematin	Thiamine	Ascorbic acid	Serum
Strigomonas oncopelti	0	+*	0	0
Strigomonas parva	0		0	0
Strigomonas media	0		0	0
Strigomonas muscidarum	+		0	0
Strigomonas culicidarum culicis	0		0	0
Strigomonas culicidarum anophelis	+	+	0	0
Strigomonas fasciculata	+	+	0	0
Leptomonas ctenocephali	+		0	?
Leishmania ceramodactyli	+		0	+
Leishmania agamae	+		0	+
Leishmania donovani	+		+	+
Leishmania tropica	+		+ or 0	+
Leishmania brasiliensis	+		+	+
Trypanosoma cruzi	+		+	+
Trypanosoma rotatorium	+		?	+

* And probably also for the other strains and species.

Other Trypanosomidae, like *Strigomonas oncopelti,* are incapable of synthesizing thiamine, e.g., *S. fasciculata, S. culicidarum.* Thus far the thiamine requirement of other species of the family has not been established. It is highly likely that a requirement will be found upon systematic investigation. While on this subject, it is interesting to recall the *in vivo* observations of Reiner and Paton (1932). These authors established a relation, in the rat, between vitamin B_1 avitaminosis and experimental infection with *Trypanosoma equiperdum,* the organism which causes dourine; the greater the vitamin deficiency in the rats the fewer the organisms in the bloodstream. This is explained perfectly if multiplication of the trypanosomes is limited by the available thiamine. Facts of the same kind are known for other vitamins, namely of the B group, and for other Protozoa such as *Plasmodium* (see Clark, 1950).

The need for the entire thiamine molecule has been found only among ciliates *Leucophrys piriformis* (*Tetrahymena geleii*) (A. and M. Lwoff, 1938) while the phytomonad flagellates only require either one of the constituents of the molecule (pyrimidine or thiazole) or both together. It is well known that a requirement for thiamine or its constituents is of wide occurrence in bacteria and fungi.

b. Other B Vitamins. It is not possible to grow *Strigomonas oncopelti* as well as related *Strigomonas* strains indefinitely on a medium consisting of hydrolyzed casein (vitamin-free) with an added carbon source, such as glucose. The addition of a very small amount of yeast extract renders this medium entirely suitable for the multiplication of the flagellates. The yeast extract can be replaced by a mixture of vitamins, with slightly inferior results (M. Lwoff, unpublished). This vitamin mixture includes, naturally, thiamine (an absolute requirement), nicotinic acid, *p*-aminobenzoic acid, and pantothenic acid. In this medium (vitamin-free casein hydrolyzate plus thiamine, nicotinic acid, *p*-aminobenzoic acid and calcium pantothenate), results are irregular and are sometimes aided by the addition of pyridoxine or biotin. The casein hydrolyzate cannot evidently be considered as free of traces of vitamins. But there is more. According to recent results of M. M. Weber, J. Cowperthwaite and S. H. Hutner (unpublished), an unknown factor, present in liver, appears necessary for *Strigomonas*. These authors have experimented with a flagellate from the mosquito, *S. culicidarum,* and have used a basal medium having the following composition:

K_2HPO_4	0.02 g.	DL-Tryptophan	2.0 mg.
Ethylenediamine tetraacetic acid	0.06	Hematin	2.5
$MgSo_4 \cdot 7H_2O$	0.04	Thiamine	0.1
NH_4Cl	0.02	Riboflavin	0.02
L-Arginine HCl	0.02	Pyridoxamine·2HCl	0.02
DL-Lysine HCl	0.02	Zn (as $ZnSO_4 \cdot 7H_2O$)	2.0
L-Histidine HCl	0.02	Mn (as $MnSo_4 \cdot H_2O$)	1.0
DL-Leucine	0.005	Mo (as $Na_2MoO_4 \cdot 2H_2O$)	0.4
DL-Isoleucine	0.005	Boron (as H_3BO_3)	1.0
L-Tyrosine	0.005	Co (as $CoSO_4 \cdot 7H_2O$)	0.2
DL-Valine	0.005	Cu (as $CuSo_4 \cdot 5H_2O$)	0.02
DL-Methionine	0.005	Fe (as $FeSO_4 \cdot 7H_2O$)	0.2
Sorbitol	1.0	Ca (as $CaCO_3$ + HCl)	1.0
Triethanolamine	0.4		

Note. Concentrations in weights per 100 ml. of final medium; the pH is brought to 8.5 with KOH.

In this sorbitol constitutes the principal energy source, triethanolamine the pH buffer for the alkaline region, and ethylenediamine tetraacetic acid the metal carrier. It may be noted that the medium contains thiamine and hematin. All the above amino acids are indispensable with the possible exception of tyrosine and methionine. In this medium addition of a certain quantity of a preparation of liver used for the treatment of pernicious anemia (0.2 ml./100 ml. of commercial "15 U.S.P. units" refined liver) permits an abundant growth of *Strigomonas.* The unknown factor in liver has not yet been separated from vitamin B_{12}; however, vitamin B_{12} by itself is completely inactive as a replacement for the liver factor. Additional required known factors are probably present in the liver fraction. The unknown factor is very low in yeast and in tryptic digests of casein. It appears different from protogen (see p. 324), the *Leuconostoc citrovorum* factor, and the *Lactobacillus bulgaricus* factor. It should be pointed out here that yeast, as an autolyzate, a hydrolyzate or an aqueous extract, with or without blood, is an excellent culture medium for the strigomonads. That is because either this new factor is present in yeast or the yeast preparations contain essential metabolites synthesized by virtue of the presence of this factor.

In 1943, however, Senekjie noted that thiamine, nicotinic acid, and pyridoxine had "temporary growth stimulating" properties for *Trypanosoma cruzi in vitro,* and that riboflavin and pantothenic acid had a certain toxicity. But the conditions under which these experiments were done (a basal medium consisting of macerated beef plus peptone and therefore already well stocked with these vitamins) make it difficult to ascribe significance to the results. At present, only the thiamine requirement has been well established but other factors will be rapidly identified if experiments of the type done by Weber *et al.* cited above proceed fruitfully.

A remark finally, on the subject of *p*-aminobenzoic acid: this compound will relieve the inhibitory action of *p*-aminophenylsulfamide on *Strigomonas oncopelti* (M. and A. Lwoff, 1945). A concentration of $M/72$ *p*-aminophenylsulfamide inhibits the division of the flagellates without killing them, as this inhibition is relieved, even after 53 days in contact with the sulfonamide, with $M/1.9 \times 10^7$ *p*-aminobenzoic acid. The sensitivity of *S. oncopelti* to *p*-aminobenzoic acid is very great: the flagellates develop quite normally in the presence of a concentration of PAB 264,000 times less than that of the sulfonamide used.

These are all the facts, still rather fragmentary, as may be seen, that we possess as to the role played by factors of the B vitamin group in the metabolism of the Trypanosomidae.

III. The Nutrition of the Trichomonads

The trichomonads are flagellates which are for the most part parasitic. As a matter of fact, very few free-living species are known: *Trichomonas keilini* discovered in 1935 by Ann Bishop in a pond in Lincolnshire, England, and *Cœlotrichomastix convergus* described by A. Hollande, in France (1939). The parasitic forms exploit very varied hosts, invertebrates (mollusks, insects, leeches), cold-blooded vertebrates (amphibians, fishes, reptiles), or warm-blooded vertebrates (birds and mammals) in which they live more often in simple commensalism. But they can cause more or less serious trouble, as does *Trichomonas columbae* in the pigeon, *T. foetus* in the cow, and *T. vaginalis* in women.

The interested reader will find in papers by Wenrich (1944, 1946, 1947) and Kirby (1947) much data on the morphology, the biology, and the culture of diverse species of this interesting family.

In 1918 Chatton obtained a bacteriologically pure culture of a *Trichomastix* from a lizard, the gecko (*Tarentola mauritanica*), and, since then, many other species which had formerly been cultured in the presence of bacteria, were isolated with greater ease thanks to the use of antibiotics. With the aid of these pure cultures, study of trichomonad nutrition has made rapid progress; it is in full development. Rather disparate information is therefore on hand, punctuated here and there by a few definite facts. As often happens, the pathogenic species have awakened the greatest interest, and one will see that the principal results have been obtained with experiments on *T. columbae, T. foetus,* and *T. vaginalis.* If one is to picture—still somewhat sketchily—the nutritive requirements of the flagellates of this group, interesting from all points of view, these findings will have to be extended. It is evident, for example, that the study, from this point of view, of Ann Bishop's *Trichomonas keilini,* or of any other free-living species, would provide material for comparison of the utmost interest.

In general, the trichomonads can only be cultured in complex media, such as those based on beef broth, or uncoagulated blood serum, egg, etc., at low oxygen tensions. Temperature optima vary with species and strains.

Thus, for *Trichomonas foetus,* Witte's strain IV, the optimal temperature is 28° C. in a medium of beef broth + horse serum, whereas for another, Witte's strain V, it is 32–37° as well as for Riedmüller's strain

78 and Glaser's strain G. These last two strains do not multiply at 28° (Cailleau, 1937). Certain individuals of the same species can withstand 52° C. for a time and can remain alive up to 14 days at 4° C. Generally the optimum appears to be 37° C. (St. Lyford, 1941). *Trichomonas columbrorum* also multiplies faster at 37° C. than at 18–22° (Chatton, 1918). The limits are 32° to 40° C. for *Trichomonas columbae,* with again the optimum likewise at 37°.

Trichomonas vaginalis can multiply at temperatures between 25° and 42° C. according to Johnson and Trussell (1944). At unfavorable

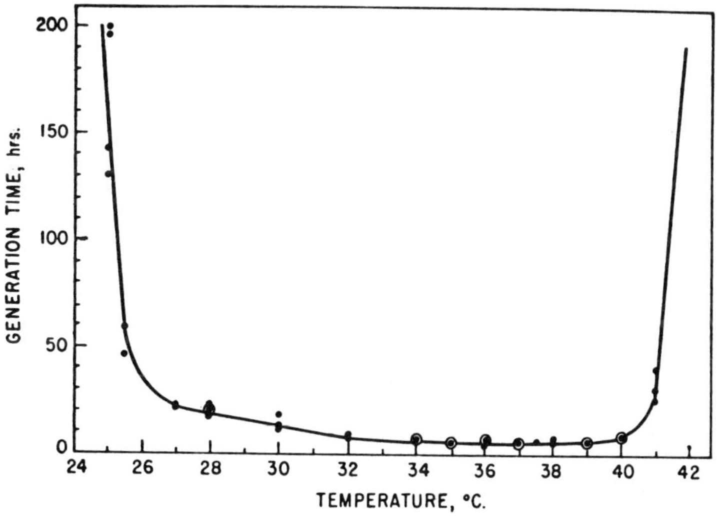

FIG. 2. Influence of temperature on *Trichomonas vaginalis* in bacteria-free culture.

temperatures such as the coagulation point the survival time is 96 to 144 hours; at 24° to 25° C. it is 120 to 154 hours. At −72° C. the flagellate is killed within one minute; at 50° C. it is killed in four minutes (see Fig. 2). Here again the optimal temperature varies with the strain: some have a thermal optimum of 28° C. (Johnson, 1942).

For each temperature the generation time, g, can be calculated from the relation $g = t/(3.3 \log b/B)$ where t is the time in hours, b is the number of flagellates per cubic millimeter at the time t, and B is the density of the population at the time of inoculation, a figure calculated from the relation established by Buchanan and Fulmer's equation $g = t \log 2/(log\ b - log\ B)$ for bacteria.

The minimal time required for one division is 5 to 7 hours, an average established for temperatures from 35° to 39° C.

No precise correlation therefore exists between the nature of the host and the optimum temperature for growth. The differences noted are not specific but valuable for each strain considered.

These subjects will be taken up successively: the physical or physicochemical factors which affect multiplication and growth, carbon and nitrogen metabolism, and finally, growth factors.

A. Physicochemical Factors

1. The pH Factor

The trichomonads vary in their sensitivity to acidity or alkalinity, depending on species and on strains. *Trichomastix colubrorum,* for example, does best in slightly acid medium, pH 6.5–6.6. *Trichomonas columbae* is very sensitive to acidification of the medium. In a buffered medium ($CaCO_3$) at pH 7, the cultures can live as long as 34 days at 32° C. In an unbuffered medium, the acids produced (pH 4.5 after the fifth or sixth day) rapidly brings about the death of the flagellates (Cailleau, 1937).

According to Morgan and Whitehair (1943), the pH of pyometra infected with *Trichomonas foetus* varies between 6.6 and 8.6. At 37° C., a culture of *Trichomonas foetus* can multiply at pH 5.5 to 8.5, with an optimum between 7 and 7.6 (Witte, 1933). The acidification produced in carbohydrate media (the final pH being about 4.8) is apparently not the cause of death of the flagellates, as they can perfectly well withstand a pH of 3.5 for about 2 hours and a pH of 5 for over 10 hours (St. Lyford, 1941). Nevertheless, multiplication no longer occurs in these acid media.

Morgan observed (1942) that, depending on the strains used, a maximum density was reached after 96 hours, when the initial pH was 6.1 or 6.3, with a final pH of 5.4 (strain A); or 5.4 to 5.8, final pH 5.2 (strain B), or between 5.7 and 5.9, final pH 5.3 (strain C), or between 6.1 and 6.3, final pH 5.4 (strain D).

A thorough study (Daniel, 1948) of pH variations in a coagulated egg medium with added liver extract and physiological saline showed, by removing aliquots from an uninoculated tube containing the above medium, that there was a spontaneous pH variation over a period of time, but it seemed that these variations were due to the egg itself, as it did not occur when egg was omitted. Such a medium, consequently, should not be used for experiments of this kind. In spite of this, one can still say that in egg media *T. foetus* can adjust itself to a wide range of pH.

pH has an effect on the morphology of *T. vaginalis*, particularly on the size of the organism as well as on the density of the culture.

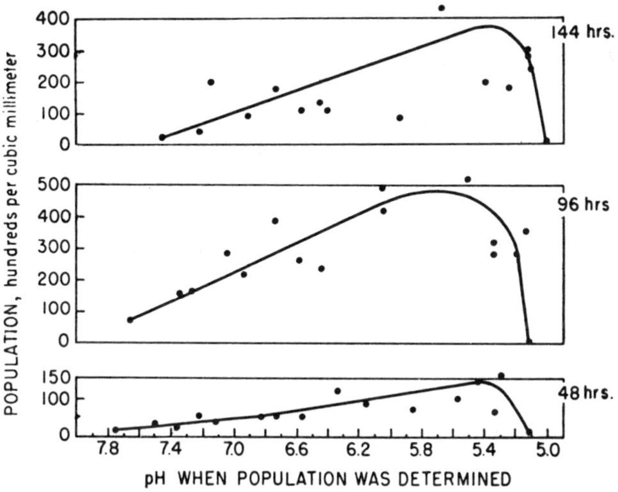

FIG. 3. Influence of pH on the multiplication of *Trichomonas vaginalis*. (After Johnson, 1942.)

In a buffered, "diphasic" medium, consisting of a liver infusion with gelatin and a solution resembling Ringer's, the optimum pH is between 5.4 and 5.8–6.0. Multiplication is hindered when the pH is higher or lower than these limits. Media should therefore be adjusted to values between pH5.5 and 6.0 (Johnson, 1942). The size of the flagellates increases in acid medium. It is smallest at the pH optimum, 5.40.

2. OXIDATION-REDUCTION POTENTIAL

As early as 1918, Chatton had observed that partial anaerobiosis, obtained by layering paraffin oil on the surface of a blood broth medium, definitely favored the multiplication of the parasite of the gecko, *Trichomastix colubrorum*, that he had isolated in pure culture. In polybacterial cultures, it is very likely that the bacteria contribute to the lowering of the oxidation-reduction potential, thus rendering the medium more suitable for the growth of the flagellates. Trichomonads, under natural conditions, live in highly contaminated environments: buccal cavity, digestive tract, vagina; and it is possible that one of the effects of bacterial metabolism could be to lower the oxidation-reduction potential of the medium to an optimal level.

From a study of *T. vaginalis,* Johnson (1942) concluded that the presence of oxygen is definitely unfavorable. The experiments were done with the following medium:

Liver infusion (Bacto-liver Difco)	80 ml.
Heart-infusion powder (Bacto)	25.6 g.
Gelatin	0.4 g.
Distilled water	720 ml.
1*N* HCl	6 ml.

To each 8 ml. of solid medium 2 ml. of whole human serum was added which had been adjusted to pH 6. The final pH in each tube was 6.1 to 6.2. The experiments were done under strict anaerobic conditions at 35° C., at varying oxygen tensions. *Trichomonas vaginalis* behaved like a facultative aerobe; the optimum condition was found to be complete anaerobiosis, with only insignificant multiplication in the presence of oxygen, no matter how little is present. A few figures will serve to demonstrate this point (see Fig. 4).

The curve in the figure shows the rate of multiplication under com-

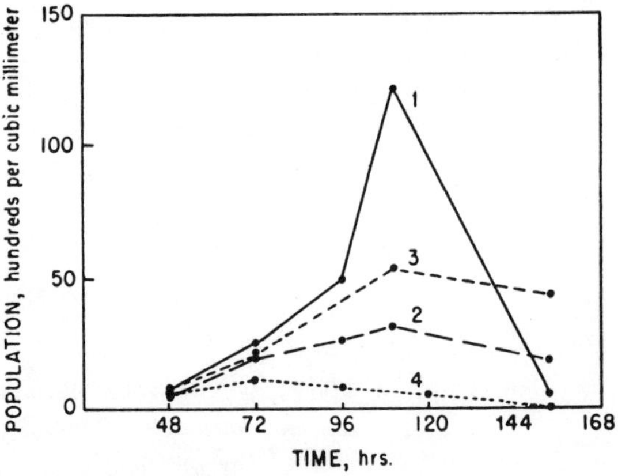

FIG. 4. Influence of oxygen tension on the multiplication of *Trichomonas vaginalis.* (After Johnson, 1942.)

Curves: 1—complete anaerobiosis
2—100 mm. oxygen tension
3—controls
4—500 mm. oxygen tension

plete anaerobiosis: after 100 to 120 hours of growth, the number of flagellates found was 12,000/mm.^{3}, while under an oxygen tension of 100 mm., the number is 3,000/mm.^{3}, and at 500 mm. it was barely 1,000/mm.^{3}. The control contained about 5,000 flagellates/mm.^{3}.

The curves in Fig. 5 make it possible to compare growth under complete anaerobiosis (1); under an oxygen tension of 100 mm. (2); in the presence of pure oxygen (3); and in an oxygen tension corresponding to atmospheric pressure (4). It is perfectly clear that anaerobic conditions

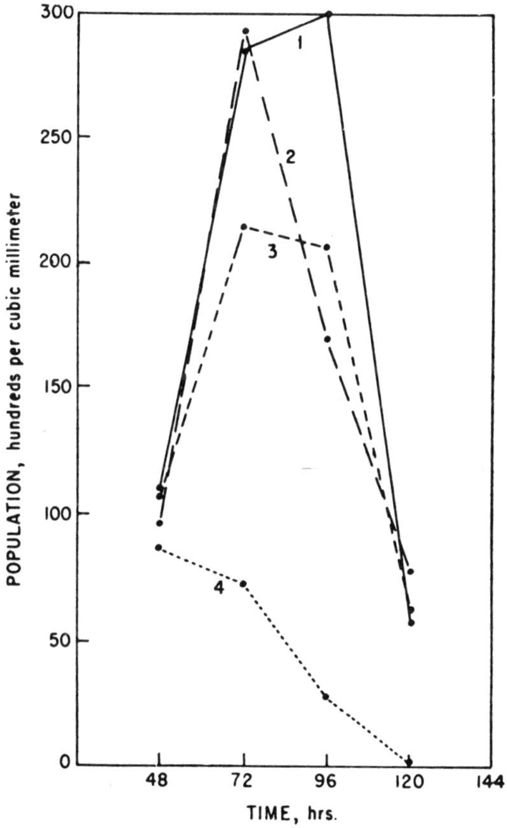

FIG. 5. Influence of oxygen tension on the multiplication of *Trichomonas vaginalis.* (After Johnson, 1942.)

Curves: 1—complete anaerobiosis
2—100 mm. oxygen tension
3—controls
4—barometric oxygen tension

are the most favorable for the multiplication of *T. vaginalis.* This finding is not surprising in a parasite whose natural habitat is definitely poor in oxygen.

B. Nitrogen and Carbon Metabolism

The nitrogen metabolism of *T. colubrorum* as well as that of *T. foetus* is very slight according to Cailleau (1937), as measured by changes in the composition of media containing protein or nonprotein nitrogen, and in amino and ammonia nitrogen. It is true that in media rich in nitrogen, only a very small but significant amount of living material is formed, and it would consequently be difficult to effect a significant decrease in the nitrogen content of the medium. There is no production of indole (*T. foetus,* Futumura, 1935).

The trichomonads do not possess proteolytic enzymes, and do not excrete, so to speak, ammonia nitrogen.

As for the carbon nutrition and the utilization of sugars, there exists a fairly consistent resemblance among the species studied (see Table VI).

Table VI. Sugar Consumption by Trichomonads

	T. colubrorum	*T. columbae*	*T. foetus*	*T. vaginalis*
Arabinose	0	0	0	
Xylose	0	0	0	
Rhamnose	0	0	0	
Glucose	+	+	+	+
Galactose	+	+	+	0
Mannose			+	
Fructose	+	±	+	0
Lactose	±	±	+	
Maltose	+	+	+	+
Sucrose	+	+	+	
Raffinose	+	0	+	
Inulin	0	±	±	
Dextrin	0	+	+	+
Soluble glycogen	0	+	+	+
Glycerol	0	0	0	
Erythritol	0	0	0	
Sorbitol	0	0	0	
Dulcitol	0	0	0	
Mannitol	0	0	0	

Trichomonas columbae uses glucose, galactose, maltose, sucrose, dextrin, and soluble starch; to a lesser extent it uses fructose, lactose, and inulin, while arabinose, xylose, rhamnose, raffinose, glycerol, erythritol, sorbitol, mannitol, and dulcitol are not utilized at all.

In ordinary media enriched with horse serum and layered with paraffin oil, *T. foetus* was shown to use glucose, galactose, lactose, fructose, maltose, sucrose, raffinose, dextrin, and soluble starch. Growth in the presence of these sugars lowers the pH of the medium from 7.0 to 5.7–6.0,

TABLE VII. Glucose Consumption by *T. foetus*
(After Andrews and von Brand, 1938)

			T. foetus		Glucose		
Ex.	Per cent glucose added	Age of culture (days)	Number (1,000/cc.)	Average number in 24 hr.	Content (mg. %)	24 hr. consumption (mg. %)	28 hr. consumption p. billion (mg.)
19	0.0	4	31.0		52		
		5	77.0	54	51	1	185
		6	216.0	147	49	2	136
22	0.0	2	23.2		52		
		3	309.0	166	50	2	120
		4	1305.5	838	33	17	203
12 A	0.1	1	58.3		132		
		2	496.6	277.5	122	10	360
		3	2200.0	1349.0	58	64	474
		4	453.0	1327.0	46	12	90
17	0.1	2	62.5		131		
		3	803.0	433	115	16	370
		4	1534.0	1169	57	58	496
		5	193.0	864	43	14	162
21	0.1	1	90.0		140		
		2	2710.0	1400	111	29	207
		3	1625.0	2168	46	65	300
19	0.2	2	28.1		229		
		3	351.5	190	222	7	368
		4	2171.6	1262	160	62	491
		5	3344.0	2758	84	76	275
		6	1719.4	2527	45	39	154
21	0.2	1	133.0		230		
		2	2173.0	1153	189	41	356
		3	906.0	1540	126	63	409
21	0.5	1	167.5		536		
		2	2430.0	1296	490	46	355
		3	1810.0	2120	429	61	288
		4	66.0	938	410	19	203
22	0.5	2	44.0		544		
		3	1009.0	527	525	19	361
		4	3490.0	2250	432	93	413
		5	971.0	2232	398	34	152
		6	430.7	701	386	12	171

while in a control tube, the final pH is 6.8–6.9 in a comparable length of time. The C_5 sugars, as well as the alcohols, glycerol, erythritol, sorbitol, mannitol, and dulcitol, are not utilized (Cailleau, 1934).

Futumura (1935) came to the same conclusions as Cailleau and noted in addition acid production in the presence of mannose. This was also observed by Riedmüller (1936) who likewise noted the particularly favorable action of glucose, the fermentation of glycogen, and the lack of fermentation of inositol.

On the whole, *T. colubrorum* utilizes the same sugars as *T. foetus* and *T. columbae* (Table VI) with the exception of inulin and dextrin.

Trichomonas vaginalis utilizes only glucose and its polymers: maltose, soluble starch, glycogen, and dextrin. Fructose and galactose are not utilized (Trussell and Johnson, 1941).

A quantitative study of the utilization of glucose by *T. foetus* has shown (Andrews and von Brand, 1938) that a concentration of 0.1 per cent glucose in a serum + broth medium suffices to meet the requirements of the flagellates. In 24 hours, one billion flagellates consume 350 mg. of sugar during active multiplication; in staling cultures, less than 200 mg. glucose/billion flagellates is consumed (see Table VII).

C. Power of Synthesis and Growth Factors

Before discussing the growth factors recognized as necessary for *Trichomonas,* it will be necessary to retrace our steps a bit in order to get a comprehensive view of the culture conditions necessary for the growth of these protozoa. For no species are the growth requirements perfectly known. The study of *T. vaginalis* is progressing rapidly, but it is still incomplete. As already said the media in which *Trichomonas* can be induced to multiply are extremely complex. For the sake of clarity in the following exposition, it is necessary to trace at first the main lines of the history of each *Trichomonas* species which has been the object of significant investigations.

D. Isolation in Pure Culture and Degree of Complexity of Ordinary Medium

Trichomastix colubrorum is a parasite of the digestive tract of a lizard, the gecko, *Tarentola mauritanica.* It was easily isolated in pure culture by passaging it through the blood, where it was usually found pure. The first culture was obtained in blood agar (N.N.N. medium), and later, in a base containing animal organs or blood. *Trichomastix colubrorum* cannot be maintained in agar-ascitic fluid medium, in ascitic fluid, or in ordinary broth, but the addition of a fragment of some organ (liver, spleen) is necessary. Finally, acidified nutrient broth enriched with fresh rabbit blood is an excellent culture medium. Rabbit serum alone added to this same broth is ineffective. But equine serum, under the same conditions, is suitable for the multiplication of the flagellate (Cailleau, 1936).

Trichomonas columbae is the agent of necrosis of organs of the pigeon. In 1932, Bos isolated for the first time a bacteriologically pure culture from the heart blood or from the hepatic sinuses of pigeons sacrificed at the height of the infection. Culture is possible in diphasic media of the type L.E.S. (Locke's solution, coagulated egg, horse serum) or in liquid media containing particles in suspension (nutrient broth-liver, for example). A medium of this type was chosen by Cailleau as a point of departure for her work on the nutrition of *T. columbae,* which culminated in the identification of cholesterol as a necessary factor for the growth of this flagellate (see p. 162).

Trichomonas foetus is the causative agent of bovine trichomoniasis. As a parasite of the genitals of the bull and the cow, it is the cause of various disorders: vaginitis, endometritis, but mostly abortive disorders. As it is known to be transmitted by coition, it is, strictly speaking, a venereal disease of cattle. Morgan, who studied this infection from a clinical, epidemiological, and therapeutic point of view, wrote a comprehensive survey in 1947. This trichomonas was observed for the first time in France by Kunstler in 1888, but this work seems to have passed unnoticed. In 1900, Mazzanti rediscovered the flagellate in Italy. As the attention of pathologists up to 1925, was exclusively held by the abortion caused by Bang's bacterium (*Brucella abortus*), the discovery of a trichomonas in the abomasum of a foetus in 1924, by Hopfengartner (*in* Drescher, 1925), was hardly noticed. The classic work of Riedmüller (1928) rehabilitated, so to speak, the flagellate and initiated a long sequence of researches when he discovered an abortion caused specifically by *Trichomonas* to be set alongside the one caused by *Brucella.*

A bacteriologically pure culture was obtained by Witte in 1933. The first one was obtained from the pus of a cow with pyometra; the flagellate was found in a state of purity in the discharge. It grew from the start in an egg medium with added blood, or in broth with added horse serum, but did best in blood broth (5–10 per cent blood), with a pH optimum of 7 to 7.6 and a temperature of 37°. Even if the flagellate is not in a pure state after the initial isolation, it can be separated from contaminating bacteria. Glaser and Coria (1935) did it by means of a migration method in tubes designed for this purpose (V tubes), Sedlmeier (1942), Plastridge (1943), and more recently Mahmoud (1944), did it by various methods and particularly through the use of penicillin. The action of penicillin on *T. foetus* was studied by Cole (1947). He reported an antibiotic action which varied with the type of penicillin used (amorphous or crystalline) and with the commercial source. *Trichomonas foetus* actually shows very little sensitivity toward crystalline penicillin, and one had to use up to 10,000 units/ml. to inhibit 50 per cent of the population. The sulfonamides could be used equally well. All the necessary directions for purification of cultures by means of antibiotics can be found in a review

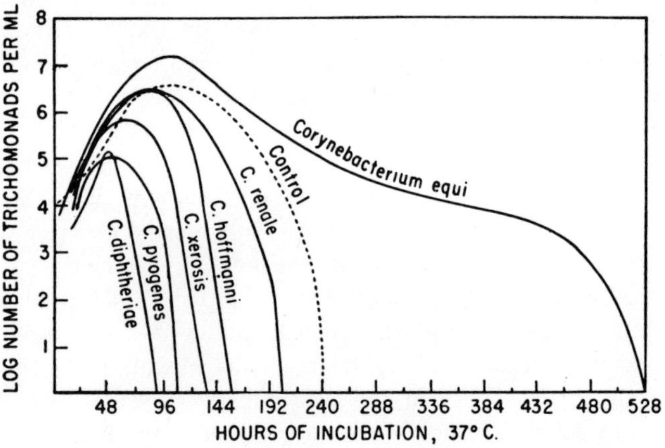

FIG. 6. Growth of *Trichomonas foetus* with various species of *Corynebacterium*. (After Johansson, Morgan, and Winkler, 1947.)

by Morgan (1946a) and by Williams and Plastridge (1946). Schneider likewise obtained a pure culture (1942) in a complex egg medium containing sodium citrate. Under natural conditions (fluids from pyometra, the foetus, seminal vesicles), *Trichomonas* is most likely to be found as a commensal of various bacteria: diplococci, staphylococci, streptococci, and, probably most frequently, *Corynebacterium pyogenes.* It was very early noticed that some of these bacteria favored while others hindered multiplication (Zeetti, 1940; Morgan, 1942; Plastridge, 1943; Williams and Plastridge, 1946). The study that Johansson, Morgan, and Winkler (1947) devoted to the effect of bacteria on *T. foetus, in vitro,* did not lead to clear-cut conclusions in regard to the factors governing their stimulating and inhibitory actions. Among the inhibitory bacteria, there are many which acidify the medium by dissimilating glucose, but a certain number which have a favorable action on multiplication possess this same property. Bacteria which produce exotoxins are all inhibitory, but certain nontoxic bacteria are more so. The most favorable bacteria are the diphtheroids, among them *Corynebacterium equi,* but many diphtheroids are inhibitory (*C. diphtheriae, C. pyogenes, C. xerosis*). *Corynebacterium hoffmanii* and *C. renale* are slightly favorable. The favorable action of the corynebacteria and in particular *C. equi* might be due to the liberation of substances in the medium as a result of metabolism. In this particular instance there was an intense liquefaction by the bacteria of the

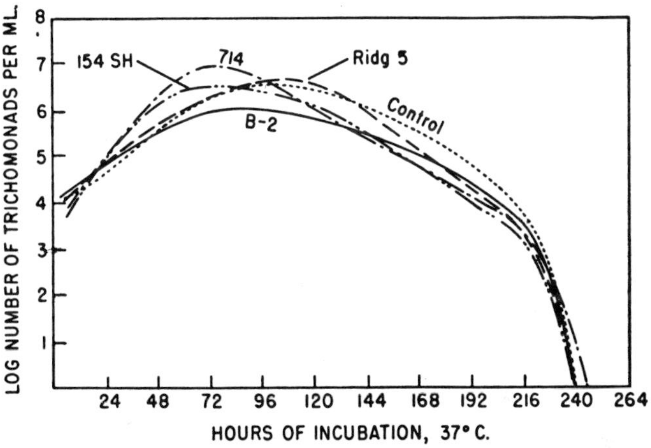

FIG. 7. Growth of *Trichomonas foetus* with various diphtheroids isolated from the genital tract of bulls. (After Johansson, Morgan, and Winkler, 1947.)

egg which was present in the basal medium. *Corynebacterium equi* prolonged the life of cultures 288 hours beyond the average limit of pure cultures (see Figs. 6 and 7). One notes in passing that diphtheroids are the bacteria found most frequently associated with *T. foetus* in the genital organs of bovines and that they play a part in the etiology of bovine trichomoniasis. *Brucella abortus* likewise has a stimulating action.

Trichomonas vaginalis is a parasite of the genito-urinary organs of women, principally of the vagina. It is also found, although less frequently, in men urethritis. Its validity as a species and its pathogenic role have been the subject of many long discussions but can no longer be placed in doubt. The most recent works in America (Shelansky; Hesseltine, Wolters, and Campbell; Wenrich; Feo; Stabler; Trussell and his colleagues, etc.) and in Europe (Käser, Jírovec, Jírovec and Peter, etc.) agree on the reality of the so-called *Trichomonas* vaginitis. Pure cultures had been obtained in 1940 by Trussell, but it was not until the use of antibiotics that pure culture could easily be achieved (Adler and Pulvertaft, 1944; Johnson, Trussell, and Jahn, 1945). *Trichomonas vaginalis* indeed is hardly sensitive to penicillin.* In a complex medium containing human serum, peptone, liver infusion, maltose, and cysteine (CPLM medium of Johnson and Trussell), penicillin is added to a final concentration of 500–1,000 units/ml. This medium is seeded with one or two loopfuls of vaginal discharge infected with flagellates and placed at 35–37° C. for 60 hours. A microscopic examination then reveals an abundant population of trichomonads. They are then inoculated into penicillin-free medium. In this manner one can obtain with ease cultures of trichomonads which are free of bacteria, as these experiments show. In their first experiments, Johnson and his collaborators succeeded in seven out of seven attempts.

These pure cultures were used for nutrition experiments and particularly for the study of blood serum and its constituents.

In 1936 Riedmüller had shown that in a broth + blood medium, serum was the fraction necessary for the multiplication of *T. foetus*, while the red cells, washed or hemolyzed, were devoid of activity. He had also shown that the active factor in serum was heat stable and capable of withstanding heating one hour at 90° C.

But this favorable action of serum is far from being constant. It varies with each strain, and for a given strain it varies with the serum. Thus Cailleau (1937) found that Riedmüller's strain 78 of *T. foetus*, multiplied very well in the presence of the sera of man, cat, and guinea pig; sera from sheep, rats, pigeons, and poultry were definitely unfavorable. Horse and rabbit sera were most often inert, but occasionally a culture grew in their presence: experiments by Cailleau succeeded one out of four times with each of these two animal sera. Glaser's *Trichomonas* strain 25 was quite different, as it could only be maintained on human serum, and then only successfully one out of four tries (sera from different individuals used).

Analysis of the causes of the lack of activity of certain sera showed that it was not due to an agglutinating action (chicken serum), as is sometimes the case for *T. columbae*. Sera heated at 56° are hardly more active than fresh sera; sera heated to 70° (horse serum) had occasionally some slight stimulating activity. According to St. Lyford (1941), the

* It is sensitive to tyrothricin (Weinman, 1943).

growth factors remained active in serum for at least five months if stored at 4° C.

No doubt the results of Daniel (1940) with *T. foetus,* which differ somewhat from those of Cailleau, can be explained by variations which occur from strain to strain, by the specific source of the sera, and even by variations which occur from animal to animal. Daniel found equine, bovine, and ovine sera to be equally good.

The search for growth factors contained in serum was not pursued any farther by Cailleau, and even their existence remains in doubt. The fact that heating to 70° renders equine serum somewhat more suitable for the culture of the flagellates makes one suspect the presence of a toxic serum factor which could be partly destroyed by heat. But this remains to be proved. It is nevertheless true that there are factors in blood serum necessary for the growth of Trichomonadinae as is shown by the cholesterol requirement of *T. columbae* and the linoleic acid requirement of *T. vaginalis* (see p. 169). Moreover, the two propositions are in no way contradictory: serum could very well bear both a toxic factor and a growth factor at the same time. However the case may be, the problem should be investigated further. The data obtained by Johansson, Morgan, and Winkler (1947) (see above, p. 158) lead one to suppose that *Trichomonas* lacks the ability to liberate the growth-promoting substances contained in serum or egg. It should not be forgotten that Cailleau (1938) showed that the addition of ascorbic acid allowed serial cultures in a medium of broth plus bovine serum (a medium which otherwise became rapidly unsuitable for the growth of *T. foetus*). In this instance it is possible that ascorbic acid does not act as a growth factor, as Cailleau thought, but rather as a reducing substance (see p. 168). Serum (human serum this time) is likewise indispensable for *T. columbae* and for *T. vaginalis,* and it will be seen later that, according to the needs of the particular species, it is a vehicle for supplying quite different substances.

Another interpretation may also be envisaged which would take into account both the variable action of serum according to source, and the favorable action of heat. Serum contains lipases, which vary in concentration according to the species of animal; human serum is low in them while horse serum is rich. These lipases, by liberating acids in the serum, could render it inhibitory. After heating and subsequent destruction, fatty acids are no longer liberated and the toxic action disappears. Lipase-poor sera would be more favorable for the growth of *Trichomonas.* These results and hypotheses serve merely to display the complexity of the problems.

E. Growth Factors

1. cholesterol

A systematic study of the indispensable factors present in media containing serum or organ fragments was carried out for *Trichomonas columbae* by Cailleau (1937).

The indefinite multiplication of *T. columbae* in the absence of bacteria is feasible in a medium made of ordinary nutrient broth with small fragments of fresh calves' liver added (one-sixth of the total volume). If the medium, in addition, is buffered with $CaCO_3$ to prevent an acidification injurious to *Trichomonas,* transplants need be made only every 14 to 15 days, at 37° C.

After extraction with water, alcohol or ether, the hepatic tissue is still suitable for inducing the growth of *Trichomonas;* the extracts are all in-

Table VIII. Growth of *Trichomonas columbae* in Nutrient Broth Supplemented with Various Substances Obtained by Extraction of Liver (After Cailleau, 1937)

Nutrient broth	0
Nutrient broth + calf liver	+
Nutrient broth + water-extracted liver	+
Nutrient broth + alcohol-extracted liver	+
Nutrient broth + ether-extracted liver	+
Nutrient broth + acetone-extracted liver	0
Nutrient broth + water extract of liver	0
Nutrient broth + alcohol extract of liver	0
Nutrient broth + ether extract of liver	0
Nutrient broth + acetone extract of liver	0
Nutrient broth + acetone-extracted liver + acetone extract of liver	+

Table IX. Growth of *T. columbae* in Nutrient Broth Supplemented with Various Substances and Cholesterol (After Cailleau, 1937)

Nutrient broth + acetone extract of liver	Subcultures 1	2	3	4	5	Total number of subcultures
+ 0 (control)	±	0				
+ egg yolk	+++	+++	+++	+++	+++	8
+ lecithin hydrolyzate	+++	+++	+++	+++	+++	13
+	+++	+++	+++	+++	+++	8
+ cholesterol (in alcohol)	+++	+++	+++	+++	+++	9
+ choline hydrochloride	+++	++	+	0		
+ ergosterol	++	+	0			

active. But after acetone extraction liver loses its growth-promoting qualities. The addition of the acetone extract reactivates the hepatic tissue (Table VIII, p. 162). Indispensable substances are present in both the tissue after extraction and the acetone extract. The acetone extract can be replaced by various substances, among them an alcoholic solution of cholesterol (Table IX); very pure cholesterol was specially prepared for the purpose.

Without question the most interesting part of Cailleau's work is the study of the specificity of the action of cholesterol. Numerous sterols were tested and certain conclusions were drawn concerning the relation between structural configuration and growth factor activity.

A certain number of modifications of the structure of cholesterol are compatible with maintenance of activity.

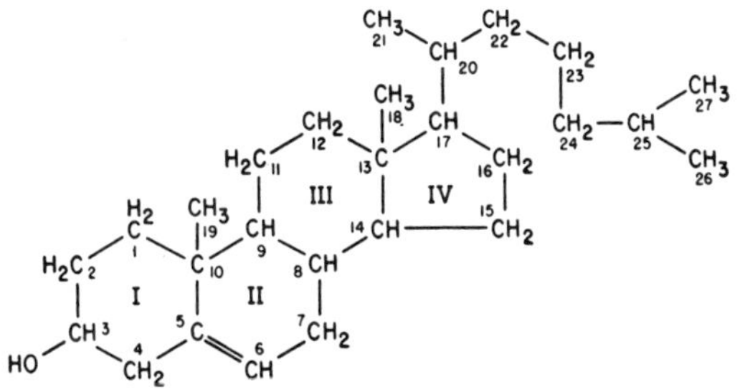

First of all, saturation of the double bond in the 5,6 position does not affect the activity, since cholesterol can be replaced by cholestanol, ergostanol, or sitostanol, as well as *cis*-cholestane-3,4-diol.

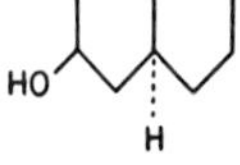

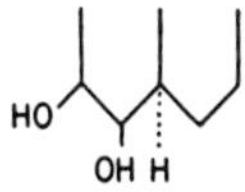

Neither does shifting the double bond from the 5,6 to the 4,5 position affect the activity.

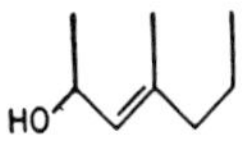

Esterification of the hydroxyl group of cholesterol has no effect; the acetate and the palmitate are active:

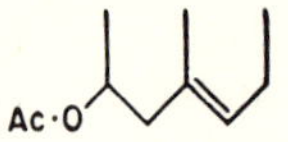

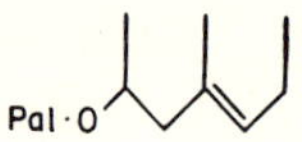

An additional hydroxyl group has no effect:

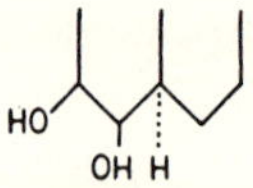

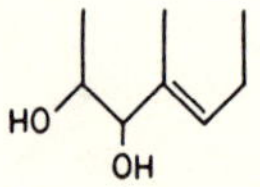

The presence of a second double bond in the cyclic portion does not inactivate the molecule: ergosterol, 7-dehydrocholesterol, and 22-dihydrocholesterol are active.

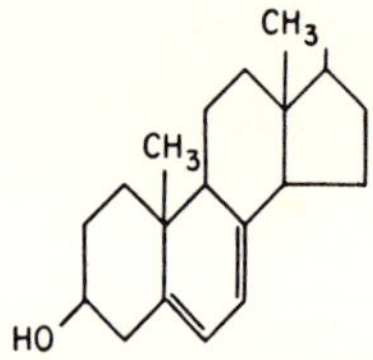

The presence of a double bond in the side chain has no effect: ergosterol and γ-dihydroergosterol are active:

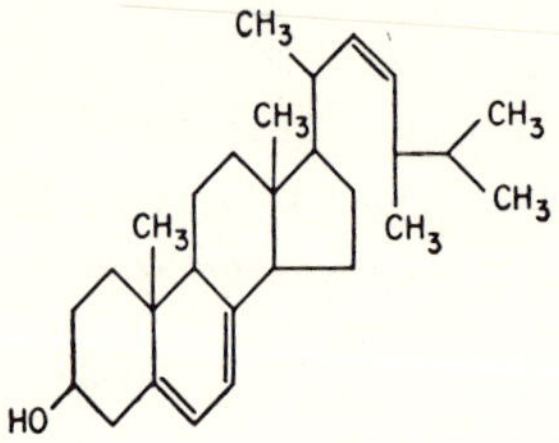

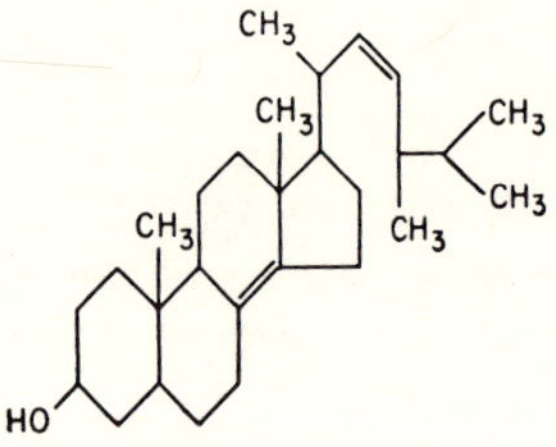

Activity is likewise maintained when the modifications of the side chain affect only the nature or position of the branches: ergosterol, 22-dihydroergosterol, ergostenol, ergostanol, α-sitosterol, α-sitostanol, cinchol.

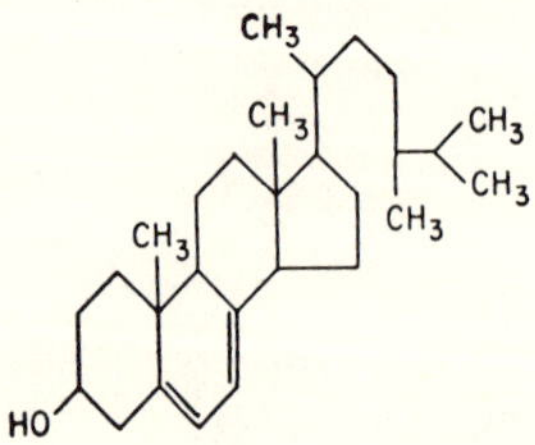

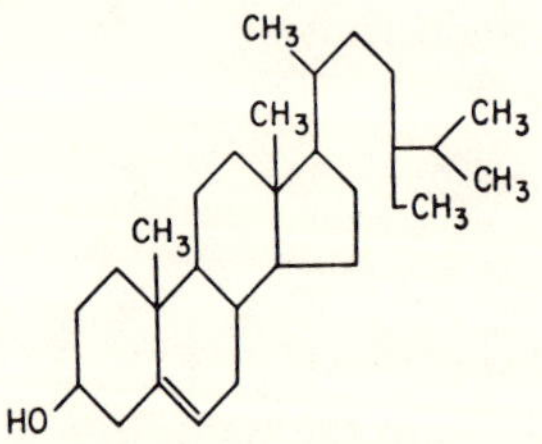

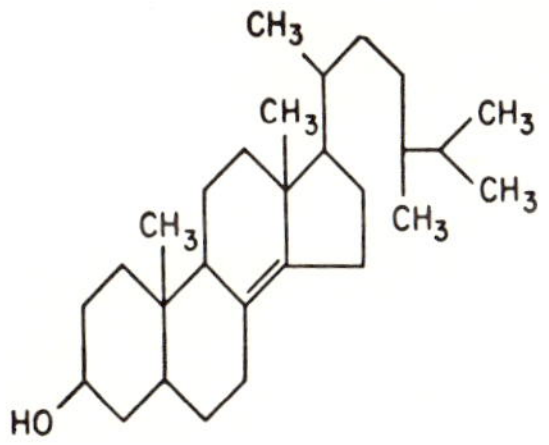

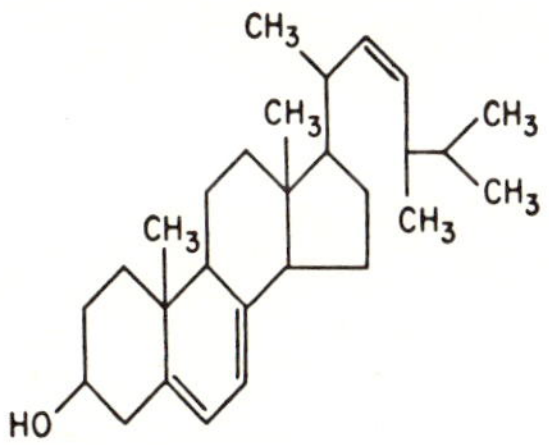

Which modifications of the molecule result in a loss of activity?

First, modifications affecting the alcohol group: its conversion to a keto group destroys activity: Δ-5,6 cholestenone, cholestane-3-one.

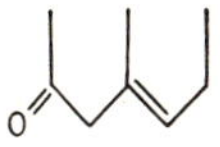

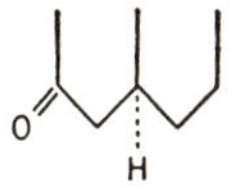

The presence of alcohol in position 6, α-3,5,6-cholestanetriol is inactive:

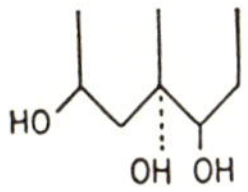

Epimerization of the alcohol group in position 3: *epi*-cholestanol is inactive, or in position 4:

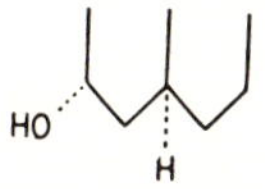

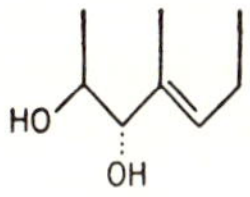

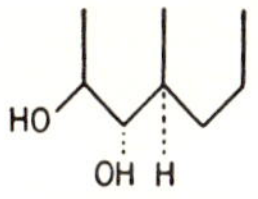

The presence of a ketone group in position 6 can also inactivate the molecule: inactivity of 3,6 diacetate of 3,5,6-cholestanetriol as well as a ketone group in position 7:

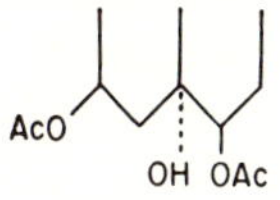

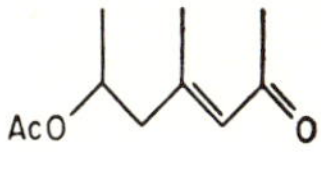

As for the hydroxyl group in position 3, its suppression is enough to inactivate the molecule:

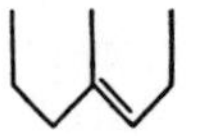

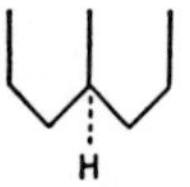

The same thing happens if a methyl radical or a Cl atom is substituted in position 3:

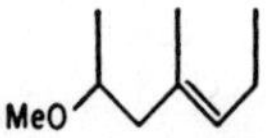

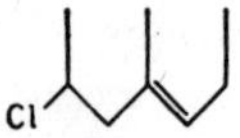

The degradation of the side chain leading to the bile acids produces inactivation, as does complete elimination of the side chain [replacement by O_2(dehydroandrosterone), or by OH(androstene-3,17-diol)].

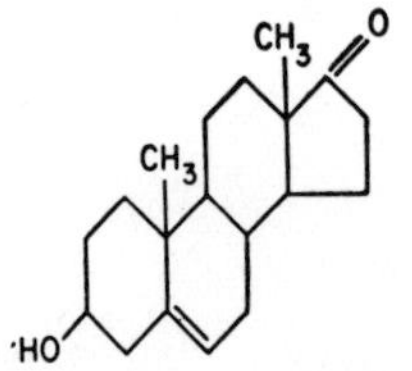

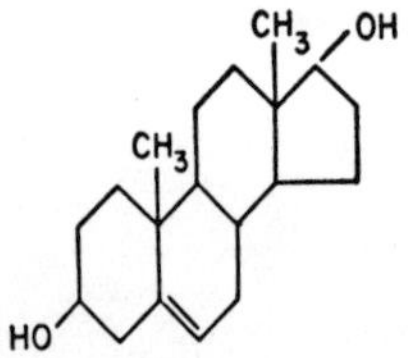

It is also evident that other substances which involve in addition the elimination of the side chain or changes in the nuclear configuration—testosterone, estrone, equilin, equilenin—are all inactive. The same is true of irradiated ergosterol.

In summary the following are noted: the importance of the OH in position 3, suppression of which inactivates the molecule, the importance of the alcohol group and its position, the importance of the side chain, while the double bonds, either in the nucleus or in the side chain seem to have no effect on the activity of cholesterol.

Cholesterol is not only essential for *T. columbae,* but for *T. foetus* and *Trichomastix colubrorum.*

Multiplication of *T. foetus* proceeds perfectly in a broth medium with a few fragments of rabbit or guinea pig liver, which are sterilized all together at 120° C. (Cailleau, 1938). Extraction of the liver with acetone and alcohol removes several growth factors, one of which is cholesterol. In this manner the cholesterol requirement could only be shown in the presence of living bacteria (Shiga's bacillus) because of the complexity of the nutritional requirements of *T. foetus.* This question ought to be taken up again as it is possible that the bacteria play a part in the utilization of cholesterol by the flagellates, as, according to Johansson, Morgan, and Winkler (1947), they play a part in the utilization of the egg medium.

Work on the necessity of cholesterol for *T. colubrorum* has also been done in an ordinary broth medium with guinea pig liver added (Cailleau, 1938). This medium is very good. If the liver is extracted with acetone-alcohol it loses its growth-promoting properties. This activity is not restored by the addition of cholesterol alone, but is if both cholesterol and egg albumin are added. Albumin alone is without action. Therefore cholesterol is needed by *T. colubrorum* as it is also needed by *T. columbae* and *T. foetus,* but other factors brought in by the egg albumin are also indispensable. Finally, cholesterol is also required by *T. batrachorum* (Cailleau, 1939b).

a. Remarks on Cholesterol as a Growth Factor. The role of cholesterol and certain other sterols as growth factors was first demonstrated with *Trichomonas columbae.* At about the same time, Hobson (1935) found it necessary for the nutrition of larvae of the fly *Lucilia sericata,* and later Rees and his collaborators have shown it to be very important for the multiplication of *E. histolytica* (see p. 237). The presence of sterols in animals and higher plants has been demonstrated; they have not been investigated in the protozoa. As for bacteria, sterols seem to be present in some (*Azotobacter,* Greaves 1935; Siffend and Anderson, 1936) and absent in others (*Mycobacterium tuberculosis, E. coli, Corynebacterium diphtheriae*). Furthermore, acetone extracts of these bacteria reputed to be lacking in sterols (*M. tuberculosis, E. coli*) cannot support the multiplication of *T. columbae* (Cailleau). It is known also that cholesterol can detoxify a medium containing fatty acids, thus making it suitable for the multiplication of certain bacteria. This is the case with *Moraxella lacunata.* The bacterium cannot grow in ordinary nutrient broth; if serum or cholesterol is added multiplication takes place (Lwoff, 1947). But, in this case, cholesterol does not act as a growth factor. It acts by neutralizing the toxicity of the fatty acids in broth. This role is discussed further in connection with the dysentery ameba (pp. 248, 249): cholesterol can intervene by neutralizing the toxic action of the unsaturated fatty acids of the medium.

There is no reason to suppose that the same holds for insects and protozoa. For *T. columbae,* cholesterol can only be replaced by compounds very closely related to cholesterol and of known composition. And this bespeaks a true vitamin-like action which responds to a definite physiological function. Observations made by Sobel and Plaut (1949) show that there are bacteria that can use cholesterol, at least under certain conditions. The smegma bacillus, *Mycobacterium smegmatis,* can attack cholesterol in suspension in the medium, preferably a complex medium such as beef heart broth. If this broth is replaced by glycine or cysteine the utilization is decreased. Curiously enough, there is also growth in the presence of chlorocholesterol and progesterone, but these substances are not attacked. Dehydroisoandrosterone, which differs from cholesterol by the absence of the side chain, has, instead, an inhibitory action. Thus here, as for *T. columbae,* the presence of the side chain seems to be related to the activity. It must be pointed out here that sex hormones (estradiol, estradiol benzoate) have no effect on the multiplication of *Trichomonas vaginalis* (Kupferberg and Johnson, 1941). The physiological role of cholesterol as a growth factor is still unknown.

2. ASCORBIC ACID

It was shown (see p. 160) that multiplication of *Trichomonas* in nutrient broth media enriched with serum was very irregular. Particularly aging of the serum seems to entail the destruction of indispensable substances. When a medium of this type has been unsuitable for the growth of *T. foetus,* its original properties can be restored by the addition of L-ascorbic acid (Cailleau, 1939). A medium consisting of peptonized broth, albumin, cholesterol, and peptone, already highly complex, is not suitable for the culture of *Trichomastix colubrorum* until ascorbic acid is added. In the medium consisting of broth + egg albumin + alcohol-acetone-extracted liver, the ascorbic acid is confined to the extracted liver (Cailleau, 1938).

In a broth medium enriched with serum from a pigeon deficient in ascorbic acid, *T. columbae* does not multiply. When ascorbic acid is added, it is possible to obtain serial cultures (Cailleau, 1939b). All the same, an acetone-extracted sample of calves' liver exposed to air at room temperature for 14 months, no longer promotes growth of the flagellates, but supports growth if ascorbic acid is added. It is therefore certain that ascorbic acid, under certain conditions, can promote the multiplication of *Trichomonas* in medium deficient in ascorbic acid. Ascorbic acid is also important for *T. vaginalis,* and Johnson (1947) noted its stimulating action, but it has not been shown to be indispensable.

The activity of ascorbic acid is not specific (Cailleau, 1939a). D-Isoascorbic, D-glucoascorbic, and D-glucoheptoascorbic, and reductinic acids, and reductone, are equally as active as L-ascorbic acid. This absence of specificity renders questionable the role of ascorbic acid and similar substances as a growth factor for trichomonads. This particular point, which merits detailed attention, is discussed elsewhere, in relation to Trypanosomidae (see p. 168). It will be seen that the problem is quite different for each of the two groups of flagellates and that the evidence shows in one case (*Trichomonas*) that the ascorbic acid is simply a reducing agent, whereas in the other case (*Trypanosomide*) it seems to act in a truly specific manner.

3. LINOLEIC ACID

When bacteriologically pure cultures of *T. vaginalis* became relatively easy to obtain with the help of penicillin (see pp. 159, 160), a more thorough study of the growth requirements of this flagellate became feasible.

When certain points on the physiology of the trichomonads, such as temperature limits, pH, E_h, source of energy (see p. 150), were established

TABLE X. Composition of Complete Trypticase Nutrient Medium for *T. vaginalis**
(The amounts listed are the calculated values per 10 ml. of final medium)
(After Sprince and Kupferberg, 1947)

Trypticase (BBL)	200 mg.	Riboflavin	8.0 μg.
Sodium acetate·$3H_2O$	48 mg.	Thiamine HCl	1.6 μg.
Cysteine·HCl	15 mg.	Pyridoxine HCl	6.4 μg.
Maltose	10 mg.	Pyridoxamine HCl	1.6 μg.
Difco agar	10 mg.	Pyridoxal HCl	1.6 μg.
$NaHCO_3$†	5 mg.	Ca pantothenate	3.2 μg.
Asparagine	2 mg.	Nicotinic acid	3.2 μg.
Ascorbic acid†	1,000 μg.	*p*-Aminobenzoic acid	1.6 μg.
Choline chloride	80 μg.	Biotin	0.8 μg.
Inositol	80 μg.	Folic acid	0.8 μg.
Ribose	40 μg.	Methylene blue (optional)	24.0 μg.
Adenine SO_4	40 μg.	Adjusted to pH 6.0, and Ringer's solution added to make:	8.0 ml.
Guanine HCl	40 μg.		
Xanthine	40 μg.	Human blood serum diluted with equal volume of Ringer's solution†	2.0 ml.
Uracil	40 μg.		
		Final volume	10.0 ml.

* For a basal medium, serum is omitted.
† Added as sterile solutions after the medium was autoclaved.

by preliminary experiments, Johnson and Trussell (1943) developed a peptone base containing liver infusion, cysteine, maltose, and the indispensable unheated blood serum. Methylene blue served as the oxidation-reduction indicator (CPLM medium). Then in 1945 Johnson and Trussell were able to replace the liver infusion, at least much of it, with a mixture of growth factors containing ascorbic acid, glutamic acid, choline, folic acid, and xanthopterin. In such a basal medium, growth was luxuriant enough so that it served as a basal medium for studying the indispensable elements in the human host, a study undertaken by Sprince and Kupferberg (1947). They modified the initial medium and finally worked out one with the composition shown in Table X.

Peptone was replaced by "trypticase," a pancreatic digest of casein devoid of carbohydrate. The liver infusion was replaced by a mixture of B vitamins, and of purines and pyrimidines. Sodium acetate, asparagine, ascorbic acid, and ribose completed the medium.

In the medium of Sprince and Kupferberg, the multiplication of *T. vaginalis* is slightly less abundant than in the original CPLM medium, but the flagellates are very active and of uniform size and appearance.

With the use of this medium, two fractions obtained from human serum were studied for their growth-promoting activity: a fraction soluble in ether, and the residue, insoluble in ether but soluble in water. A preliminary series of experiments showed that both these fractions were necessary, as each singly added to the basal medium was incapable of in-

TABLE XI. *In vitro* Cultivation of *Trichomonas vaginalis* in Fractions Prepared from Human Blood Serum
(After Sprince and Kupferberg, 1947)

Ml. of diluted serum control* or diluted serum fraction* added to 8 ml. of trypticase basal medium to give a total volume of 10 ml. per tube	Growth of *Trichomonas vaginalis* expressed in number of cells per mm.³ after 48 hr. in each serial transfer†					
	First culture	Serial transfer 1	Serial transfer 2	Serial transfer 3	Serial transfer 4	Serial transfer 5
Control (2 ml. of diluted intact serum)	1,550	1,040	740	1,895	1,305	1,510
1.0 ml. diluted aqueous emulsion of ether extract + 1.0 ml. sterile Ringer's	0					
1.0 ml. diluted ether-insoluble phase + 1.0 ml. sterile Ringer's	990	540	0			
1.0 ml. diluted aqueous emulsion of ether extract + 1.0 ml. diluted ether-insoluble phase	1,080	685	1,205	2,335	1,755	1,145

* All samples of intact serum or serum fractions were diluted with an equal volume of Ringer's solution, as indicated in a preceding paragraph.

† Each value is the average of duplicate determinations.

suring the growth of the *Trichomonas*. These results are shown in Table XI, p. 170.

As the analysis was carried further, it was seen that a water emulsion of the ether extract added before autoclaving could replace the ether extract. A dialyzate of this aqueous emulsion, dried and then restored to the original volume with distilled water and filtered, could also replace the ether-soluble fraction. This was, it is understood, in the presence of the active ether-insoluble fraction. Proceeding further, a mixture of various fat-soluble substances, namely, lecithin, cholesterol, oleic acid, linoleic acid, ergosterol, α-estradiol, α-tocopherol, α-carotene, and vitamin A, completely replaced either the ether-soluble fraction, or the heated aqueous water emulsion, or the dialyzate of this emulsion (Table XII).

TABLE XII. The Nature of the Material in the Ether-Soluble Phase Essential for the Sustained Growth of *Trichomonas vaginalis*
(After Sprince and Kupferberg, 1947)

Fraction representing ether-soluble phase added*	Growth of *Trichomonas vaginalis* expressed in number of cells per mm.³ after 48 hr. in each serial transfer†					
	First culture	Serial transfer 1	Serial transfer 2	Serial transfer 3	Serial transfer 4	Serial transfer 5
None (intact serum, control)	1,040	1,140	1,565	1,655	1,580	1,585
None (aqueous ether-insoluble phase, control)	990	0				
Autoclaved aqueous emulsion of ether extract	1,660	1,555	1,985	1,110	665	1,005
Dialyzate of aqueous emulsion of ether extract	2,075	1,340	1,895	1,695	1,600	1,640
Replacements mixture of pure compounds (described above)	710	1,135	450	950	1,560	1,120

* All fractions equivalent to serum in concentration were diluted with an equal volume of Ringer's solution. One milliliter of the diluted material was then assayed in test tubes containing 8.0 ml. of trypticase basal medium plus 1.0 ml. of diluted ether-insoluble phase.

† Each value is the average of duplicate determinations.

In the above mixture, the active substance seems to be linoleic acid. Cholesterol, in contrast, was inactive, as was estradiol. *Trichomonas vaginalis* in respect to this particular point apparently has different requirements from those of *T. columbae* studied by Cailleau (see p. 162).

In any case, the ether-soluble serum factor was thermostable, dialyzable, and could be replaced by linoleic acid.

As for the ether-insoluble fraction, it could be replaced by serum albumin, again only in the presence of the ether-soluble fraction but not when the latter was replaced by the mixture mentioned above containing linoleic acid. It is noteworthy that multiplication was much less in the presence of albumin than when total serum was present (Table XIII). It is possible that other necessary factors were present in the ether-insoluble fraction. It is interesting to note that linoleic acid is a growth factor for *T. vaginalis.*

TABLE XIII. Replacement of Ether-Insoluble Phase by Human Serum Albumin (After Sprince and Kupferberg, 1947)

Fraction representing ether-soluble phase added*	Growth of *Trichomonas vaginalis* expressed in number of cells per mm.³ after 48 hr. in each serial transfer†					
	First culture	Serial transfer 1	Serial transfer 2	Serial transfer 3	Serial transfer 4	Serial transfer 5
None (intact serum, control)	1,180	1,170	1,170	915	1,565	1,655
None (human serum albumin, control)	900	0				
Aqueous emulsion of ether extract	355	460	615	450	395	335
Replacement mixture of pure compounds (described above)	570	0				

* All fractions equivalent to serum in concentration were diluted with an equal volume of Ringer's solution. One milliliter of the diluted material was then assayed in test tubes containing 8.0 ml. of trypticase basal medium plus 1.0 ml. of diluted human serum albumin.

† Each value is the average of duplicate determinations.

It was mentioned earlier that Senekjie and Lewis (1945) showed the existence of a dialyzable factor in blood serum necessary for trypanosomes. This factor, however, is thermolabile (rapid destruction at 100° C.). A more thorough comparative study of these two factors would doubtless be very interesting.

4. PANTOTHENIC ACID

As a result of recent work by Kupferberg, Johnson, and Sprince (1948), pantothenic acid was found to play an important role in the metabolism of *Trichomonas.*

The medium of Sprince and Kupferberg (1947) for *T. vaginalis* contains two natural substances: human serum and trypticase (see Table X, p. 169). The effective concentration for human serum is 0.05 ml. in 10 ml. of base. In the presence of serum and trypticase, which no doubt contain numerous growth factors, many factors have no effect: thiamine, *p*-aminobenzoic acid, inositol, pyridoxine and its derivatives, pyridoxamine, and pyridoxal, as well as various compounds such as asparagine, Na acetate, ribose, purine bases, and uracil. If the amount of human serum used is reduced under certain conditions in the medium of Sprince and Kupferberg, it becomes necessary to add pantothenic acid. If the amount of trypticase is reduced, a need for phosphate is revealed.

Similarly, one finds (Johnson and Kupferberg, 1948) that the toxic action of certain compounds studied for therapeutic purposes, such as α-8-dihydroxy-β,β_1-dimethyl-N-[2-(phenyl-mercapto)-ethyl]-butyramide, is antagonized by calcium pantothenate and that they are truly competitive. Analogous phenomena are observable with *T. foetus* and *T. gallinae* which show somewhat different sensitivities toward therapeutic agents studied. The failure of attempts to treat—generally or locally—*Trichomonas* vaginitis can be explained, at least in part, by the existence of an *in vivo* competitive inhibition of this type, due to the blood level of pantothenic acid.

5. UNKNOWN FACTORS

While seeking to replace trypticase in the medium of Sprince and Kupferberg (1947), Sprince, Gilmore, and Lowy (1949) obtained good growth in the presence of pancreatic extracts. The factor, still unidentified, which is now being studied, called "*s*," possesses the following characteristics: it is water soluble, dialyzable, and thermostable. It does not seem to be related to the anti-anemia factors already known since it cannot be replaced by crystalline vitamin B_{12}. One of the extracts studied ("lipocaic," from Lilly) possesses much greater activity than trypticase, as it is still active at 25 mg./ml., whereas trypticase has no action at concentrations lower than 500 mg./ml. No compound so far has been found that will replace these pancreatic extracts.

REFERENCES

Adler, S. (1934). *Trans. Roy. Soc. Trop. Med. Hyg.* **28**, 201.

Adler, S., and Ashbel, R. (1934). *Arch. zool. Torino* **20**, 521.

Adler, S., and Pulvertaft, R. J. V. (1944). *Ann. Trop. Med. Parasitol.* **38**, 188.

Andrews, J., and von Brand, T. (1938). *Am. J. Hyg.* **28**, 138.

Berrebi, J. (1936). *Arch. inst. Pasteur Tunis* **25**, 89.

Bishop, Ann (1935). *Parasitology* **27**, 246.

Bishop, Ann (1936). *Parasitology* **28**, 443.

Bos, A. (1932). *Zentr. Bakt.* **130**, 221.

Buchanan, R. E., and Fulmer, E. I. (1928). Physiology and Biochemistry of Bacteria, Vol. I, Williams and Wilkins, Baltimore, Maryland.
Cailleau, R. (1934). *Bull. soc. path. exotique* **27**, 943.
Cailleau, R. (1936). *Compt. rend. soc. biol.* **121**, 108.
Cailleau, R. (1937). *Ann. inst. Pasteur* **59**, 1.
Cailleau, R. (1938). *Compt. rend. soc. biol.* **127**, 861, 1421.
Cailleau, R. (1939a). *Compt. rend. soc. biol.* **130**, 1089.
Cailleau, R. (1939b). *Compt. rend. soc. biol.* **131**, 964.
Chang, S. L. (1948). *J. Infectious Diseases* **82**, 107.
Chatton, E. (1918). *Compt. rend. soc. biol.* **33**, 405.
Clark, P. F. (1950). *Annual Rev. Microbiol.*, **4**, 343.
Cole, B. A. (1947). *J. Parasitol.* **33**, 405.
Daniel, G. E. (1940). *J. Parasitol.* **26**, 85.
Daniel, G. E. (1948). *J. Parasitol.* **34**, 496.
Dobell, C. (1939). *Parasitology* **31**, 138.
Drescher (1925). *Tierärztl. Rundschau* **31**, 936.
Feo, L. G. (1944). *Am. J. Trop. Med.* **24**, 195.
Fildes, P. (1922). *Brit. J. Exptl. Path.* **3**, 210.
Futumara, H. (1935). *J. Japan Soc. Veter. Sci.* **14**, 392.
Glaser, R., and Coria, N. A. (1935). *Am. J. Hyg.* **22**, 221.
Goat, H., and Mora, C. (1947). *Anales soc. biol. Bogota* **2**, 188.
Granick, S., and Gilder, H. (1946). *J. Gen. Physiol.* **30**, 1.
Greaves, J. (1935). *J. Bact.* **30**, 143.
Hesseltine, C. L., Wolters, S. L., Campbell, S. and A. (1942). *Infectious Diseases* **71**, 127.
Hobson, R. (1935). *Biochem. J.* **29**, 2023.
Hollande, A. (1939). *Bull. soc. zool. France* **64**, 114.
Hopfengartner (1925). *Tierärztl. Rundschau* **31**, 936.
Jirovec, O., and Peter, R. (1949). *Bull. soc. path. exotique* **42**, 148.
Johansson, K. R., Morgan, B. B., and Winkler, C. H. (1947). *J. Bact.* **53**, 271.
Johnson, J. G. (1942). *J. Parasitol.* **28**, 369.
Johnson, J. G. (1947). *J. Parasitol.* **33**, 189.
Johnson, J. G., and Kupferberg, A. B. (1948). *Proc. Soc. Exptl. Biol. Med.* **67**, 390.
Johnson, J. G., and Trussell, R. E. (1943). *Proc. Soc. Exptl. Biol. Med.* **54**, 245.
Johnson, J. G., and Trussell, M. H. (1944). *Proc. Soc. Exptl. Biol. Med.* **57**, 252.
Johnson, J. G., Trussell, M. H., and Jahn, F. (1945). *Science* **102**, 126.
Käser, O. (1943). *Schweiz. med. Wochsch.* No. **34**, 1021.
Kirby, H. B. (1947). *J. Parasitol.* **33**, 214.
Krishman, K. V. (1938). *Trop. Diseases Bull.* **37**, 744.
Kunstler, J. (1888). *Compt. rend.* **107**, 953.
Kupferberg, A. B. (1940). *Proc. Soc. Exptl. Biol. Med.* **45**, 220.
Kupferberg, A. B., and Johnson, J. G. (1941). *Proc. Soc. Exptl. Biol. Med.* **48**, 516.
Kupferberg, A. B., Johnson, J. G., and Sprince, H. (1948). *Proc. Soc. Exptl. Biol. Med.* **67**, 304.
Lotze, H. (1938). *Arch. Schiffs-Tropen-Hyg.* **42**, 287.
Lwoff, A. (1933). *Compt. rend. soc. biol.* **113**, 231.
Lwoff, A. (1934). *Zentr. Bakt. Parasitenk. Abt. I Orig.* **130**, 497.
Lwoff, A. (1947). *Ann. inst. Pasteur* **73**, 735.

Lwoff, A., and M. (1937). *Ann. inst. Pasteur* **59**, 129.
Lwoff, A., and M. (1938). *Compt. rend. soc. biol.* **127**, 1170.
Lwoff, M. (1931). *Compt. rend. soc. biol.* **107**, 1234.
Lwoff, M. (1933a). *Ann. inst. Pasteur* **51**, 55.
Lwoff, M. (1933b). *Ann. inst. Pasteur* **51**, 707.
Lwoff, M. (1935). *Compt. rend. soc. biol.* **119**, 969.
Lwoff, M. (1936). *Compt. rend. soc. biol.* **121**, 419.
Lwoff, M. (1937). *Compt. rend. soc. biol.* **126**, 771.
Lwoff, M. (1938a). *Compt. rend.* **206**, 540.
Lwoff, M. (1938b). *Compt. rend. soc. biol.* **128**, 241.
Lwoff, M. (1939). *Compt. rend. soc. biol.* **130**, 406.
Lwoff, M. (1940). Recherches sur le pouvoir de synthèse des flagellés trypanosomides, Masson, Paris.
Lwoff, M., and A. (1945). *Ann. inst. Pasteur* **71**, 206.
Lwoff, M., and Chorine, V. (1943). *Ann. inst. Pasteur* **69**, 158.
Lwoff, M., and Nicolle, P. (1945). *Compt. rend. soc. biol.* **139**, 879.
Lwoff, M., and Nicolle, P. (1947a). *Compt. rend.* **225**, 147.
Lwoff, M., and Nicolle, P. (1947b). *Bull. soc. path. exotique* **40**, 467.
McKee, R. W., and Geiman, Q. M. (1946). *Proc. Soc. Exptl. Biol. Med.* **63**, 313.
Mahmoud, A. H. (1944). *Ann. Trop. Med. Parasitol.* **38**, 219.
Mazzanti, E. (1900). *Giorn. soc. accad. vet. Ital.* **49**, 629.
Mohr, W. (1941). *Deut. Tropenmed. Z.* **45**, 404.
Morgan, B. B. (1942). *Proc. Soc. Exptl. Biol. Med.* **51**, 380.
Morgan, B. B. (1946a). Bovine trichomoniasis. Burgess Publishing Co., Minneapolis, Minnesota. 165 pp.
Morgan, B. B. (1946b). *Anat. Record* **94**, 95.
Morgan, B. B. (1947). *J. Parasitol.* **33**, 201.
Morgan, B. B., and Campbell, H. M. (1946). *Am. J. Vet. Research* **7**, 45.
Morgan, B. B., and Whitehair, C. K. (1943). *North Am. Veterinarian* **24**, 729.
Nicolle, C. (1908). *Compt. rend.* **146**, 498, 842.
Novy, F. G., and MacNeal, W. J. (1903). *J. Am. Med. Assoc.* **45**.
Nyden, Shirley J. (1948). *Proc. Soc. Exptl. Biol. Med.* **69**, 206.
Olsen, O. (1921). *Zentr. Bakt. I.* **85**, 12.
Pfeiffer, R. (1892). *Z. Hyg.* **13**, 357.
Pfeiffer, R. (1893). *Z. Hyg. Bakt.* **13**, 357.
Plastridge, W. N. (1943). *J. Bact.* **45**, 196.
Ray, J. C. (1932). *Indian J. Med. Research* **20**, 355.
Reiner, L., and Paton, J. B. (1932). *Proc. Soc. Exptl. Biol. Med.* **30**, 345.
Riedmüller, L. (1928). *Zentr. Bakt. Parasitenk. Abt. I Orig.* **108**, 103.
Riedmüller, L. (1936). *Zentr. Bakt.* **137**, 428.
Rose, D., and Peterson, R. (1949). *Can. J. Research* **27**, 428.
Roskin, G., and Nastiukowa, O. (1941). *Compt. rend. acad. sci. U.R.S.S.,* **32**, 566.
St. Lyford, H. (1941). *Am. J. Hyg.* **33**, 69.
Saïto, Y. (1937). *J. Oriental Med.* **26**, 9.
Salle, A. J., and Schmidt, C. L. A. (1928). *J. Infectious Diseases* **43**, 378.
Schneider, M. D. (1942). *J. Parasitol.* **28**, 428.
Sedlmeier, H. (1942). *Berlin. u. Münch. tierärztl. Wochschr.* 240.
Senekjie, H. A. (1943). *Am. J. Trop. Med.* **23**, 523.

Senekjie, H. A., and Lewis, R. A. (1945). *Am. J. Trop. Med.* **25**, 345.
Shelansky, H. A. (1940). *J. Parasitol.* **26**, 21.
Siffend, R. H., and Anderson, R. S. (1936). *Z. physiol. Chem.* **239**, 270.
Sobel, H., and Plaut, A. (1949). *J. Bact.* **57**, 377–382.
Sprince, R. H., Gilmore, E. L., and Lowy, R. S. (1949). *Arch. Biochem.* **22**, 483.
Sprince, R. H., and Kupferberg, A. B. (1947). *J. Bact.* **53**, 435, 441.
Stabler, R. M. (1941). *J. Parasitol.* **27**, 32.
Stabler, R. M., and Feo, L. G. (1942). *J. Parasitol.* **28**, 14.
Stabler, R. M., Feo, L. G., and Rakoff, A. E. (1940). *J. Parasitol.* **26**, 22.
Stabler, R. M., Feo, L. G., and Rakoff, A. E. (1941). *Am. J. Hyg.* **34**, 114.
Thjötta, T. (1921). *J. Exptl. Med.* **33**, 763.
Thjötta, T., and Avery, O. T. (1921a). *J. Exptl. Med.* **34**, 97, 455.
Thjötta, T., and Avery, O. T. (1921b). *Proc. Soc. Exptl. Biol. Med.* **18**, 197.
Trussell, R. E. (1940). *J. Iowa State Med. Soc.* **30**, 66.
Trussell, R. E., and Johnson, J. G. (1941). *Proc. Soc. Exptl. Biol. Med.* **47**, 176.
Weber, M. M., Cowperthwaite, J., and Hutner, S. H. Unpublished.
Weinman, D. (1943). *Proc. Soc. Exptl. Biol. Med.* **54**, 38.
Welch, A. D., and Wilson, M. F. (1949). *Arch. Biochem.* **22**, 486.
Wenrich, D. H. (1944). *Am. J. Trop. Med.* **24**, 39.
Wenrich, D. H. (1946). *J. Parasitol.* **32**, 40.
Wenrich, D. H. (1947). *J. Parasitol.* **33**, 25, 62, 177.
Williams, L. F., and Plastridge, W. N. (1946). *J. Bact.* **51**, 127.
Witte, J. (1933). *Arch. wiss. u. prakt. Tierheilk.* **66**, 333; *Zentr. Bakt. Orig.* **128**, 189.
Zeetti, R. (1940). *In* Morgan (1942). *Proc. Soc. Exptl. Biol. Med.* **51**, 380.
Zotta, G. (1923). *Compt. rend. soc. biol.* **88**, 913, 1350.

Metabolism of Trypanosomidae and Bodonidae

THEODOR VON BRAND

Laboratory of Tropical Diseases,
National Institutes of Health,
Bethesda, Maryland

CONTENTS

I. Introduction

This chapter falls into two very unequal parts. The first deals with the chemical physiology of Trypanosomidae and is lengthy; the second deals with the same aspects of the Bodonidae and is extremely brief. This difference, of course, is due to the fact that the Trypanosomidae are of immense importance as causative agencies of many important diseases of man and livestock. They consequently have attracted many investigators. The Bodonidae, on the other hand, are free living protozoa or parasites of lower animals without economic importance. They have been studied so far almost exclusively from the taxonomic and cytological viewpoints; data on their chemical physiology are hardly available as yet.

It is of some interest to trace briefly the course of development of our knowledge of the physiology of Trypanosomidae, in particular the trypanosomes. The first line of research was developed during the first two decades of this century from the standpoint of chemotherapeutic considerations. Inevitably many interesting physiological observations were made on the mode of drug action, drug resistance, and similar problems. The chief physiological advances of these early years are primarily associated with the names of Ehrlich, Yorke, and Voegtlin. As can be expected from its practical importance, this line of work has been continued, though with varying degrees of vigor, up to the present time.

During the 1920's and 30's the main interest centered around the question of how the trypanosomes injure their hosts. Following the initial observation by Schern that the carbohydrate metabolism of infected animals shows signs of being abnormal, a flood of papers dealing with various phases of the metabolic interrelationships between parasites and hosts appeared for a period of about 10 years. Interest in this particular aspect subsided somewhat in later years.

Overlapping to some extent with this development, and partly inspired by it, the study of the metabolism of the trypanosomes themselves began in the late 1920's, the first reliable observations being linked to the names of Yorke, von Fenyvessy, and their collaborators. The interest centered first around quantitative aspects of the sugar metabolism of the parasites. It has shifted in later years to the intermediate processes of carbohydrate utilization and the various enzyme systems involved in the metabolism of the flagellates. It is, at present, the field attracting the majority of investigators.

A quite recent development is the attempt at synthesis of the various specialized fields of knowledge, for example, the search for an explanation of drug activity in terms of interference with metabolic enzymes of the parasites. Undoubtedly this approach will receive increasing attention in the future.

II. Trypanosomidae

A. Metabolism of Trypanosomidae

1. Respiration

a. Intensity of Oxygen Consumption

Indications that the bloodstream forms of trypanosomes consume oxygen at a rapid rate were first obtained by Nauss and Yorke (1911) who observed that blood containing *Trypanosoma rhodesiense in vitro* rapidly

TABLE I. Rate of Oxygen Consumption of Various Trypanosomidae

Species	Form*	Temp. °C.	Average oxygen consumption mm.³/100 mill./1 hr.	Reference
Strigomonas oncopelti†	C	28	41	Lwoff, 1934
Strigomonas fasciculata†	C	28	37	Lwoff, 1934
Leptomonas ctenocephali†	C	28	27	Lwoff, 1934
Leishmania tropica	C	28; 32	45; 39	von Brand and Johnson, 1947; Chang, 1948
Leishmania brasiliensis	C	28; 32	42; 32	von Brand and Johnson, 1947; Chang, 1948
Leishmania donovani	C	28; 32	18; 27	von Brand and Johnson, 1947; Chang, 1948
Trypanosoma lewisi	B	37	50; 62	Reiner, Smythe, and Pedlow, 1936; von Brand, Tobie, and Mehlman, 1950
Trypanosoma lewisi	B (young)	37	50	Moulder, 1948b
Trypanosoma lewisi	B (old)	37	69	Moulder, 1948b
Trypanosoma cruzi	C	28; 32	25; 43	von Brand and Johnson, 1947; Chang, 1948
Trypanosoma cruzi	B	37	124	von Brand, Tobie, and Mehlman, 1950
Trypanosoma conorhini	C	28	26	von Brand and Johnson, 1947
Trypanosoma pipistrelli	C	30		von Brand, Tobie, and Mehlman, 1950
Trypanosoma congolense	B	37	153	von Brand, Tobie, and Mehlman, 1950
Trypanosoma evansi	B	37	166	von Brand, Tobie, and Mehlman, 1950
Trypanosoma equinum	B	37	166	von Brand, Tobie, and Mehlman, 1950
Trypanosoma equiperdum	B	37	185	von Brand, Tobie, and Mehlman, 1950
Trypanosoma hippicum	B	38	200	Harvey, 1949
Trypanosoma rhodesiense	B	37	194; 194	Christophers and Fulton, 1938; von Brand, Tobie, and Mehlman, 1950
Trypanosoma gambiense	B	37	170	von Brand, Tobie, and Mehlman, 1950
Trypanosoma gambiense	C	28	14	von Brand and Johnson, 1947

* C = culture forms; B = bloodstream form.

† Calculated on the basis of 150 million organisms per 1 mg. dry weight.

assumed the color of reduced hemoglobin. The oxygen consumption of various culture and bloodstream Trypanosomidae has since been studied repeatedly from a quantitative standpoint and representative data have been assembled in Table I.

In so far as the bloodstream trypanosomes are concerned, it is of interest to note that the African pathogenic species consume considerably more oxygen than *T. cruzi* and especially more than *T. lewisi.* Calculated on the basis of wet weight which, in an approximate way, is possible in view of the weight determinations published by Reiner, Leonard, and Chao (1932a) and Hawking (1944), figures of around 60 cc. oxygen per hour are obtained for 1 g. African trypanosome. Christophers and Fulton (1938) give a figure of 285 cc. oxygen per hour per 1 g. dry weight for *T. rhodesiense.* To appreciate the enormous intensity of the oxygen consumption of these forms, it must be remembered that a resting young rat uses only about 1.5 cc. oxygen/ g. wet weight/ hour (von Brand and Krogh, 1935), or about 4.5 cc./ gm. dry weight/ hour. If calculated on the basis of relative surface, however, trypanosomes consume about 400 times less oxygen per unit surface than a rat. These calculations show that it is impossible to find a common denominator for the rate of metabolism when animals as far apart in the animal kingdom as protozoa and mammalians are concerned. This has been emphasized by von Brand (1935) and has recently been reaffirmed by Zeuthen (1947).

The rate of oxygen consumption of the developmental stages of Trypanosomidae is considerably lower than that of most bloodstream forms. Besides differences in experimental temperatures, differences in type of metabolism, size, and perhaps degree of motility may be responsible. A detailed analysis of the respiratory metabolism of a culture trypanosome *versus* that of the bloodstream form of the same species is still lacking, but ought to represent an attractive research problem. The few available data having a bearing on this point will be reviewed in a later section.

b. Environmental Factors Influencing the Rate of Oxygen Consumption

The medium in which respiratory studies are carried out has a considerable influence on the intensity of oxygen consumption. Lwoff (1934) has shown that *Strigomonas fasciculata* suspended in broth has a constant but relatively low rate of oxygen consumption which rises rapidly upon the addition of some blood. He has shown conclusively that this increase is due to the synthesis of a deficient component of the catalytic respiratory system from the added hemin. This problem is only touched upon here since it will be treated more fully in another chapter.

A considerable decrease in oxygen consumption has been observed (Marshall, 1948a; Moulder, 1948a) when the bloodstream forms of *T. evansi* and *T. lewisi* were studied in blood-free media as compared to the rates obtained in the presence of blood. However, the above explanation probably does not apply to these cases. In the case of *T. evansi,* most of the "lost" respiration was recovered upon the addition of 5 per cent mouse plasma to the medium while the addition of lysed mouse erythrocytes had no such effect.

Of single constituents of the medium, glucose is by far the most important one in determining the rate of oxygen consumption of the bloodstream form of the African trypanosomes and of *T. lewisi.* As soon as the sugar of the medium becomes depleted, the respiratory rate begins to decline and rapidly sinks to negligible values (Christopers and Fulton, 1938; Marshall, 1948a; Moulder, 1948a; Harvey, 1949). *Trypanosoma cruzi,* the bloodstream form of which has not yet been studied in this respect, may well react differently since it does not consume appreciable amounts of sugar.

Variations in inorganic constituents, on the other hand, have little influence on the oxygen consumption of *T. lewisi,* as long as their overall molecular concentration is kept within reasonable physiological limits (Moulder, 1948a). The same is essentially true for pH changes, unphysiologically low and high values leading to a declining rate (Christophers and Fulton, 1938; Moulder, 1948a).

The oxygen tension also has but little influence as has been shown in the case of the culture form of *T. cruzi* (von Brand, Johnson, and Rees, 1946). This is not unexpected since in such small organisms the surface-volume ratio is very favorable to the diffusion of oxygen into the bodies even at low tensions.

Temperature, on the other hand, is an extremely important factor. Lwoff (1934) found that the oxygen consumption of *S. fasciculata* rose regularly between 13° and 34° C., the application of Arrhenius' equation giving a straight line relationship. Above 34° C. the respiration began to decline. Von Brand, Johnson, and Rees (1946), working with the culture form of *T. cruzi,* found a similar decline only above 40° C. Arrhenius' equation gave in the range 13° to 40° C. a bisecting line, the break occurring at 28° C. These differences in response to temperature changes may be related to the fact that *S. fasciculata* is a parasite of the cold-blooded *Culex,* whereas *T. cruzi* normally alternates between a cold-blooded and a warm-blooded host.

c. *Internal Factors Influencing the Rate of Oxygen Consumption*

As in all other animals, the metabolic level of Trypanosomidae may be correlated with the rate of motility, but no exact data are available on this point. It must be realized that all rate determinations have been carried out with flagellates moving about with greater or less vigor; determination of their basal metabolism is not feasible. The observation by Lwoff (1934) that gregarinoid forms of *Leptomonas ctenocephali* consume less oxygen than the monadine forms may be related to a greater locomotor activity of the latter.

Von Brand, Johnson, and Rees (1946) reported a declining metabolic rate in aging cultures of *T. cruzi* without being able to correlate the decline with any specific factor. However, a recent observation by Chang (1948) may have a bearing on this point. He reports that two types of crithidias develop in cultures of this parasite, one predominating in young, the other in older cultures. It would seem possible that both have different metabolic rates.

Moulder (1947, 1948b) has made the interesting observation that the oxygen consumption of young, reproducing *T. lewisi* is significantly lower than that of older undividing specimens, while the reverse is true for the rate of sugar consumption. The malonate inhibition of oxygen consumption increased with the age of the parasites. Obviously then, a change in metabolism has taken place. He could correlate this change with the appearance of ablastin in the host's blood, which apparently interferes with the oxidative glucose metabolism of the flagellates. Although, therefore, the ultimate cause for the change in metabolism appears to be an external one, the direct cause is probably an internal one, correlated with the suspension of reproduction. A shift between synthetic and energy-producing processes may be the main reason for the observed differences.

d. *The Respiratory Quotient*

The available data on respiratory quotients of Trypanosomidae have been assembled in Table II. The data show that the RQ of culture forms is relatively high and that its level is primarily determined by the availability of significant amounts of sugar in the medium. The relationship is the same as found in many other organisms; if sugar is available the RQ is high, if sugar is missing, it is lower.

It should be emphasized, however, that an RQ of 1 in these cases does not necessarily indicate a pure carbohydrate metabolism of a completely oxidative type as one would be inclined to assume in other cases. As will be shown in a following section, all hitherto studied Trypanosomidae are aerobic fermenters; that is, they do not oxidize the carbohydrate completely. It is the balance between various metabolic processes that determines the RQ level, and speculations concerning the relative participation of various carbon dioxide sources without simultaneous biochemical studies are hazardous.

TABLE II. Respiratory Quotient of Trypanosomidae

Species	Form*	Sugar available	Average RQ	Reference
Strigomonas oncopelti	C	Yes	1.0	Lwoff, 1934
Strigomonas fasciculata	C	Yes	1.0	Lwoff, 1934
Leptomonas ctenocephali	C	Yes	0.88	Lwoff, 1934
Leishmania tropica	C	Yes	0.95	Soule, 1925
Leishmania tropica	C	No	0.88	Soule, 1925
Trypanosoma lewisi	C	Yes	0.94	Soule, 1925
Trypanosoma lewisi	C	No	0.82	Soule, 1925
Trypanosoma lewisi	B	Yes	0.98	Reiner, Smythe, and Pedlow, 1936
Trypanosoma lewisi	B (young)	Yes	0.74	Moulder, 1948b
Trypanosoma lewisi	B (old)	Yes	0.91	Moulder, 1948b
Trypanosoma cruzi	C	Yes	1.0	von Brand, Johnson, and Rees, 1946
Trypanosoma cruzi	C	No	0.7	von Brand, Johnson, and Rees, 1946
Trypanosoma equiperdum	B	Yes	0.06	Reiner, Smythe, and Pedlow, 1936
Trypanosoma hippicum	B	Yes	0.0	Harvey, 1949
Trypanosoma rhodesiense	B	Yes	0.16	Christophers and Fulton, 1938

* C = culture forms; B = bloodstream form.

A very interesting situation prevails in the case of the bloodstream trypanosomes. The RQ of *T. lewisi,* although depending in some measure on the age of the specimens (Moulder, 1948b), is always relatively high, while that of *T. rhodesiense, T. equiperdum,* and *T. hippicum* is extremely low. In view of the similarity in metabolism between members of a given trypanosome group, it is probable that these findings are characteristic for the various groups involved. It seems obvious that these differences arise from differences in glucose utilization. In *T. equiperdum* glucose degradation stops at the pyruvate stage and, theoretically, no respiratory carbon dioxide at all is evolved. In *T. lewisi,* on the other hand, the sugar degradation proceeds farther along the glycolytic chain, and carbon dioxide is produced. *T. rhodesiense* probably resembles *T. equiperdum* qualitatively, but the quantitative aspects of the various processes involved have been less studied. Details concerning these phases will be discussed later.

It would be of great interest to study the RQ of *T. congolense* since in some metabolic relationships it is intermediate between the *lewisi* and the *evansi-brucei* groups. Also determinations of the RQ of culture forms of flagellates of the *brucei* group would be profitable. Valuable leads as to the question of metabolic similarities or differences between various stages in the life cycle could be obtained in this way.

e. Respiratory Enzyme Systems

The respiration of culture forms of leptomonads, strigomonads, leishmanias, and trypanosomes of the *lewisi* group is very strongly inhibited by low concentrations of cyanide, and the same is true in the case of the bloodstream form of the *lewisi* group (Lwoff, 1934; von Brand, Johnson, and Rees, 1946; von Brand and Johnson, 1947; Moulder, 1948a). This proves beyond question that their respiratory enzyme system contains one or more components depending for their activity on heavy metals. Lwoff (1934) demonstrated spectroscopically in suspensions of *S. fasciculata* the occurrence of two absorption bands, one lying around 530 mμ, the other fairly broad one around 555 mμ, both bands disappearing upon ventilation with oxygen. He therefore concluded that cytochrome was present in this organism.

Whether cytochrome oxidase is a component of the system has not been proved with the same degree of certainty. The respiration of *Strigomonas* is strongly inhibited by carbon monoxide as well as by cyanide (Lwoff, 1934); that of the culture form of *T. cruzi* by azide and hydrogen sulfide, but not by pyrophosphate (von Brand, Johnson, and Rees, 1946); and that of the bloodstream form of *T. lewisi* by azide (Moulder, 1948a). These observations, while compatible with the presence of cytochrome oxidase, are not quite sufficient for its final identification. An investigation of whether the carbon monoxide inhibition is reversible by illumination is extremely desirable in this connection.

At the present time, therefore, it can be stated definitely only that the respiration of the above forms depends for oxygen transfer on enzyme systems catalyzed by heavy metals, possibly iron, and that the cytochrome system may be important.

A totally different situation prevails in respect to the bloodstream trypanosomes of the *evansi* and *brucei* groups. Their respiration is not inhibited at all by cyanide, but actually rather stimulated (von Fenyvessy and Reiner, 1924, 1928; Christophers and Fulton, 1938; von Brand and Johnson, 1947; von Brand and Tobie, 1948; Marshall, 1948a,b). Two possibilities then exist: Either these organisms do not possess enzyme systems depending on heavy metals, or they have an alternating pathway which substitutes fully in case the first one is blocked by cyanide. Neither of these alternatives can be excluded with certainty at present, but it may be significant that Harvey (1949) found no cytochrome oxidase in *T. hippicum*.

The cyanide stimulation consists essentially in a stimulation of the sugar consumption resulting in a higher oxygen consumption (von Brand and Tobie, 1948a; Marshall, 1948a). Von Brand and Tobie (1948) consider it as possible that the keto-binding properties may lie at the root of the stimulatory effect of cyanide. The assumption that it combines with pyruvic acid, the chief metabolic end product of these forms, may be sufficient to explain the observed effects.

The question then arises as to what mechanism accounts for the oxygen consumption in the absence of functioning heavy metal systems. The only possible mechanism, as Marshall (1948a) pointed out, is by the transfer of hydrogen from a given compound through a series of reducing processes until it reaches the stage where it can be oxidized by atmospheric oxygen. In other words, dehydrogenase systems must be operating. Their presence in African pathogenic trypanosomes has repeatedly been proved either by the Thunberg technique (Reiner and Leonard, 1932; Singer, 1936; Christophers and Fulton, 1938), or by the inhibitory action of compounds like iodo- or bromo-acetic acid and others (Marshall, 1948a; von Brand and Tobie, 1948b; Harvey, 1949). It must be emphasized, however, that no definite system has been identified as yet. Indeed, Marshall (1948b) and Harvey (1949) pointed out that those dehydrogenase systems commonly associated with respiration, such as succinic dehydrogenase or the coenzyme I oxidation-reduction system or malic dehydrogenase, are apparently absent in these trypanosomes.

Dehydrogenase systems operate also in forms that have a cyanide-sensitive respiration as their response to inhibitors indicates (von Brand, Johnson, and Rees, 1946; von Brand and Tobie, 1948b; Moulder, 1948a). However, recent studies have shown that the bloodstream forms of trypanosomes belonging to the *evansi* and *brucei* groups are considerably more sensitive to dehydrogenase inhibitors than the bloodstream form of the *lewisi* group (von Brand, Tobie, and Mehlman, 1950). This observation probably indicates that the former are more dependent on

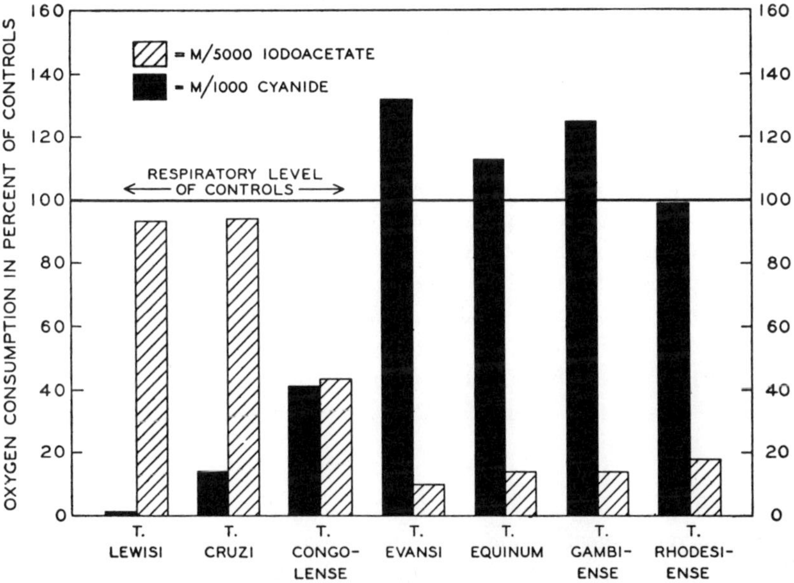

FIG. 1. Oxygen consumption of the bloodstream form of various trypanosomes under the influence of cyanide and iodoacetate.

dehydrogenase systems than the latter and might serve as an additional argument in favor of assuming a lack of heavy metal containing oxidizing systems in the *brucei* and *evansi* groups. It is worth noting in this connection that the catalase activity of the bloodstream form of African trypanosomes is very small (Strangeways, 1937b; Harvey, 1949), or may sometimes be even completely missing (Krijgsman, 1936b), although this iron-containing enzyme is widely distributed in most animal cells. In *T. lewisi*, bird and frog trypanosomes, a stronger catalase activity seems to have been observed (Tasaka, 1935).

Trypanosoma congolense is intermediate between the *lewisi* and the *evansi-brucei* groups in its sensitivity to cyanide and, somewhat less distinctly, to dehydrogenase inhibitors (Fig. 1). This is of considerable theoretical interest since Hoare (1948) in his recent reclassification of mammalian trypanosomes has placed *T. congolense* in just this intermediate position on the basis of morphological and developmental grounds. A summary of our present knowledge concerning the metabolism of mammalian trypanosomes (von Brand, 1951) has shown that the known physiological facts agree very well with Hoare's classification.

Another interesting point is that the respiration of the proventricular form of *T. gambiense* is sensitive to cyanide (von Brand and Johnson, 1947), in contrast to that of the bloodstream form. This observation clearly indicates that the enzyme systems of the developmental stages of this parasite are different from those of the adult stage. If future studies should conclusively reveal the actual absence in the latter of heavy metal systems, this finding would represent a powerful argument in favor of Lwoff's (1940) concept of the loss of enzyme-synthetic powers in Trypanosomidae or the somewhat similar views developed by Krijgsman (1936b) concerning the loss of enzymes as an adaptation to parasitic life. It must be noted in this connection that the culture forms of trypanosomes correspond to the stages they assume in the insect vectors and can thus be considered as more primitive than the bloodstream form, since it is generally assumed that the latter are descended from the former.

f. Summary

The oxygen consumption of the bloodstream form of trypanosomes is characterized by its extreme intensity. The RQ of *T. lewisi* and of the developmental stages of the hitherto-studied forms is high, that of the bloodstream form of the African pathogenic species extremely low.

The respiration of the trypanosomes of the *lewisi* group is catalyzed by a heavy metal system, perhaps the cytochrome system. There is no indication that such systems play a role in the respiration of the African forms, although the fact that cyanide does not inhibit their respiration does not completely rule out the coexistence of a cyanide sensitive and insensitive mechanism. Differences in the respiratory systems between the bloodstream form and the developmental stages of the same species also occur, and it appears possible that further studies along such lines would reveal interesting phylogenetic relationships.

2. CARBOHYDRATE METABOLISM

a. Utilization of Various Carbohydrates

The observation by Schern (1912) that serum or liver extract of normal animals but not of animals heavily infected with trypanosomes restored the motility of trypanosomes which had become quiescent *in vitro* is known as Schern's reviving phenomenon. It formed the starting point for the analysis of the metabolic processes in this group of flagellates.

It was recognized that Schern's phenomenon could be elicited by the addition of glucose, fructose, mannose, or glycerol to the medium whereas galactose and maltose were much less active in this respect. Many other substances tested failed to restore the motility of the trypanosomes (Schern, 1925, 1928; Kudicke and Evers, 1924; von Fenyvessy, 1926; Dubois, 1926, 1930; von Jancsó and von Jancsó, 1935b; Ivanov and Umanskaya, 1945). It was also soon recognized that both *T. lewisi* and pathogenic trypanosomes survived *in vitro* better in the presence than in the absence of sugar (Biot, Biot, and Richard, 1911; Fleig, 1911;

TABLE III. Glucose Consumption of the Bloodstream Form of Some Mammalian Trypanosomes

Species	Average glucose consumption mg./100 million/1 hour	Reference
Trypanosoma lewisi	0.1; 0.1	Regendanz, 1930; von Brand, 1933
Trypanosoma lewisi (young)	0.4	Moulder, 1948b
Trypanosoma lewisi (old)	0.2	Moulder, 1948b
Trypanosoma cruzi	Not demonstrable	von Brand, 1933; von Brand, Tobie, Kissling, and Adams, 1949
Trypanosoma congolense	0.8; 0.7	von Brand, 1933; von Brand and Tobie, 1948a
Trypanosoma evansi	0.9; 0.6	Geiger, Kligler, and Comaroff, 1930; von Brand and Tobie, 1948a
Trypanosoma equinum	0.8	von Brand and Tobie, 1948a
Trypanosoma equiperdum	2.0	Chen and Geiling, 1945
Trypanosoma brucei	0.3; 0.8	von Issekutz, 1933a; von Brand, 1933
Trypanosoma gambiense	0.6; 1.0	Yorke, Adams, and Murgatroyd, 1929; von Brand and Tobie, 1948a
Trypanosoma rhodesiense	1.0; 1.5	von Brand and Tobie, 1948a

Poindexter, 1935), and Hagemeister (1914) was the first to show that glucose favored the development of trypanosomes in culture. These qualitative observations suggested that the trypanosomes consume sugar and subsequent quantitative determinations proved this assumption to be correct.

The data summarized in Table III show that the intensity in sugar metabolism is higher in the *congolense, evansi,* and *brucei* groups than in the *lewisi* group. An approximate calculation on weight basis, again based on the weight determinations by Reiner, Leonard, and Chao (1932a) and Hawking (1944), shows that the African pathogenic trypanosomes consume in 1 hour sugar corresponding to 15 to 30 per cent of their wet weight, or 50 to 100 per cent of their dry weight.

Obviously, in view of this enormous need for carbohydrate, the accumulation of a glycogen reserve within the trypanosome body would be of little value since it could last only a very short time in the absence of an exogenous carbohydrate source. Schern and Bozzolo (1930) did find small amounts of glycogen in African trypanosomes, Krijgsman (1936b) did not, and Moulder (1948b) found but little indication of carbohydrate storage in both young and adult *T. lewisi.*

It is at the present time not quite clear as to what extent the same situation prevails in the case of the developmental stages of Trypanosomidae. It has been shown that they do consume sugar from the medium. This has been demonstrated for *Leishmania* by Salle and Schmidt (1928), for *T. cruzi* by von Brand, Tobie, Kissling, and Adams (1949), and for *T. gambiense* and *rhodesiense* by Tobie, von Brand, and Mehlman (1950). It is of interest to note at this point that the culture forms of *T. cruzi* consume more sugar than the bloodstream form while the opposite is apparently true in the case of the proventricular forms of *T. gambiense* and *T. rhodesiense.* Whether these latter forms utilize an endogenous carbohydrate source for the production of energy has not yet been studied. A polysaccharide fraction with immunological properties has been isolated by Senekjie (1941) and Muniz and de Freitas (1944) from culture forms of the following flagellates: *Leishmania tropica, L. donovani, L. brasiliensis, Leptomonas oncopelti, L. culicidarum, Trypanosoma cruzi,* and *Endotrypanum schaudinni.*

It should be recognized that a definite contrast in carbohydrate requirement exists between the bloodstream form of the African pathogenic trypanosomes and their developmental stages. The former derive most, if not all, their energy from the degradation of carbohydrate and are therefore absolutely dependent on the presence of an adequate supply of sugar in the medium. The concentration of the sugar is of minor importance; they can extract adequate amounts of glucose even from fairly dilute solutions (von Brand, 1933), a point important in their biology since the sugar content of the liquor cerebrospinalis is much lower than that of blood. The developmental stages of the African species, on the other hand, seem not to be dependent in the same way on sugar. Reichenow (1937) has pointed out that thriving cultures develop even if no sugar at all is added to the medium. It must be realized, however, that his medium, as well as all other media used in the cultivation of these species, are not absolutely devoid of glucose; traces are always introduced together with the blood which is an indispensable ingredient. It is probably justifiable to assume that in nature, within the tsetse flies, the parasites will at least from time to time find themselves in surroundings containing but little if any sugar, since the sugar contained in the blood meal ought soon to pass from the alimentary canal of the fly to its tissues.

The carbohydrate most readily available to most Trypanosomidae in nature is doubtless glucose, an exception being possibly some plant-parasitizing species. It is consequently not surprising that glucose is the sugar most commonly employed in metabolic and cultural experiments. The qualitative studies by Colas-Belcour and Lwoff (1925), Noguchi (1926), Kligler (1926), Ivanov and Jakovlev (1943), and Chang (1948), carried out with a variety of Trypanosomidae, have shown that the hexoses fructose and mannose, but not galactose, are usually good sources of energy. In many cases the disaccharides maltose and saccharose are also utilized, while lactose is not. Pentoses and polysaccharides are consumed only in isolated cases. In so far as the latter are concerned, the reason may be a lack of polysaccharide-splitting digestive enzymes. The literature concerning this point is somewhat contradictory, however. Thus far only the bloodstream form of *T. evansi* has been studied in some detail in this respect. Krijgsman (1936a,b) found no amylase, while Mannozzi-Torini (1940) reports the organism capable of hydrolyzing glycogen, starch, and inulin. Glycerol is a good substrate (Harvey, 1949).

The literature concerning the quantitative aspects of the consumption of different carbohydrates is scanty. The available data are shown in Table IV. They seem to indicate rather pronounced differences between various species.

TABLE IV. Rate of Utilization of Various Carbohydrates by the Bloodstream Form of Some Trypanosomes

Species	Carbohydrate utilization in per cent of glucose utilization						Reference
	Glucose	Fructose	Mannose	Galactose	Maltose	Lactose	
T. lewisi	100	50	132	0	218 (?)	0	Mercado, 1947
T. equiperdum	100	75	57	0	87	21	Plunkett, 1946
T. brucei	100	21	86	9	50	0	von Brand, 1933

b. End Products of Carbohydrate Metabolism

A characteristic feature of the carbohydrate metabolism of all hitherto studied Trypanosomidae is that they do not completely oxidize the sugar to carbon dioxide and water. As the data summarized in the section on the respiratory quotient show, carbon dioxide is produced by all culture forms investigated and by *T. lewisi,* while the bloodstream form of *T. rhodesiense* produces little and that of *T. equiperdum* and *T. hippicum* evolves practically no carbon dioxide. In these latter forms in a very pronounced way, in the former less markedly, only partially oxidized end products are formed and excreted. The Trypanosomidae, therefore, belong among the aerobic fermenters in von Brand's (1946) classification of metabolic types occurring in the animal kingdom.

The end products other than carbon dioxide formed during the sugar degradation processes have so far been identified only in isolated cases. These are shown in Table V and indicate that in various species the sugar degradation does not proceed equally far along the glycolytic chain. More details concerning this point will be presented in the following section.

Many papers deal with the problem at hand in a more general way, proving the production of organic acids either by showing that the pH of the medium drops during incubation, or by demonstrating that carbon dioxide has been evolved from bicarbonates of the medium. Corresponding data for the genera *Leptomonas* and *Strigomonas* will be found in the papers by Colas-Belcour and Lwoff (1925) and Lwoff (1934); for the genus *Leishmania* in those by Colas-Belcour and Lwoff (1925), Noguchi (1926), Salle and Schmidt (1928), Salle (1931), and Adler and Ashbel (1934, 1940); for the genus *Trypanosoma* in those by von Fenyvessy (1926), von Fenyvessy and Reiner (1928), Kligler, Geiger, and Comaroff (1930), von Brand, Regendanz, and Weise (1932), von Brand,

TABLE V. End Products of Aerobic Fermentations in Trypanosomidae

Species	Form*	Pyruvic acid	Lactic acid	Formic acid	Acetic acid	Oxalic acid	Succinic acid	Ethyl alcohol	Glycerol	Reference
Leishmania brasiliensis	C	+	+	+			+			Chang, 1948
Leishmania donovani	C	+	+	+			+			Chang, 1948
Leishmania tropica	C	+	+	+			+			Chang, 1948
Trypanosoma cruzi	C	+	+	+			+			Chang, 1948
Trypanosoma lewisi	B			+	+		+	+		Reiner, Smythe, and Pedlow, 1936
Trypanosoma evansi	B	+								Marshall, 1948a
Trypanosoma hippicum	B	+								Harvey, 1949
Trypanosoma equiperdum	B	+								Reiner, Smythe, and Pedlow, 1936
Trypanosoma brucei	B	+	+			+				Glowazky, 1937
Trypanosoma rhodesiense	B	+	+	+	+		+	+	+	Fulton and Stevens, 1945

* C = culture forms; B = bloodstream form.

Johnson, and Rees (1946), and von Brand, Tobie, Kissling, and Adams (1949).

c. *Intermediate Carbohydrate Metabolism*

(1) *African Pathogenic Trypanosomes.* Reiner and Smythe (1934) and Reiner, Smythe, and Pedlow (1936) found that under anaerobic conditions *T. equiperdum* decomposed one molecule of glucose into one molecule of glycerol and one molecule of pyruvic acid. The same reaction occurred also under aerobic conditions, but it was followed by the oxidation of glycerol to pyruvic acid and water. In the end then, aerobically two molecules of pyruvic acid and two molecules of water were formed from one molecule of glucose. Essentially similar results were obtained by Marshall (1948a) with *T. evansi* who found the production of 1.75 molecule pyruvic acid from one molecule of glucose, and by Harvey (1949) with *T. hippicum.*

While Reiner, Smythe, and Pedlow (1936) were doubtful concerning the occurrence of phosphorylation processes, Marshall's (1948a) studies proved their occurrence conclusively. He observed that in incubates of the parasites glucose-6-phosphate was rapidly utilized while glucose-1-phosphate, triosephosphate, and adenosinetriphosphate (ATP) accumulated. From these observations he draws the conclusion that the main pathway of glucose utilization follows the usual Embden-Meyerhof-Parnas sequence of reactions.

The accumulation of glucose-1-phosphate is somewhat puzzling since this compound is usually an intermediate of glycogen but not of glucose degradation. In Marshall's (1948a) experiments, however, the flagellates did not utilize glycogen. It is not likely that the 1-phosphate was produced from the 6-phosphate via the reversible phosphoglucomutase reaction because it accumulated also when the formation of the 6-phosphate was inhibited by arsenicals.

Harvey (1949) fractionated *T. hippicum* and found all the usual phosphorylated intermediates.

Lactic acid, this common end product of aerobic and anaerobic fermentations in many animals, did not accumulate in incubates of African pathogenic trypanosomes (von Brand, Regendanz, and Weise, 1932; Reiner, Smythe, and Pedlow, 1936; Marshall, 1948a); rather, some lactic acid was consumed if the normal glucose metabolism was interrupted by iodoacetic acid. According to Marshall (1948a) this observation suggests that endogenous lactic acid is converted to pyruvic acid.

In so far as enzymes are concerned, it can be assumed that these parasites have a full set of enzymes governing the glucose breakdown to the pyruvic acid stage. Direct evidence concerning this point has been presented by Chen (1948b) who worked with lysed preparations of *T. equiperdum.* He studied the hexokinase activity, the 3-phosphoglyceraldehyde dehydrogenase system, and adenosinetriphosphatase, the occurrence of all of which he could definitely demonstrate. He pointed out that the presence of an active adenosinetriphosphatase may at least in part explain the failure of Christophers and Fulton (1938) to observe glycolysis with lysed *T. rhodesiense.* Whether this observation also explains the negative results of Moulder (1948a) with lysed *T. lewisi* remains to be studied.

Harvey (1949) was able to demonstrate the presence of the following enzymes in *T. hippicum:* Hexokinase, adenosinetriphosphatase, aldolase, triosephosphate dehydrogenase with coupled oxidation-reduction, glycerol dehydrogenase, and β-glycerophosphate dehydrogenase, and alkaline phosphatase.

It is probable that the above findings will be found characteristic also for other members of the *evansi* group, and possibly also largely for those of the *brucei* group. In this latter, however, the glucose degradation may go somewhat farther than in the former. This seems indicated by their slightly higher RQ and the more varied end products (Table V). No data whatever are so far available for the *congolense* group.

(2) *Trypanosoma lewisi.* In *T. lewisi* glucose breakdown follows a different pattern from that in *T. evansi* or *T. equiperdum.* It was first studied in detail by Reiner, Smythe, and Pedlow (1936). Under anaerobic conditions 1 molecule of glucose was broken down into one molecule of succinic acid and one molecule of probably glycol. This latter compound was then changed into acetic acid and ethyl alcohol, acetaldehyde being an intermediate. The same reactions took place under aerobic conditions, but they were followed by the oxidation of the acetaldehyde to formic acid, carbon dioxide, and water. Most of the formic acid in turn was also oxidized to carbon dioxide and water. The above authors found no indications of phosphorylation processes, but this phase should definitely be reinvestigated.

It does appear that the glucose utilization by *T. lewisi* stops short of the Krebs tricarboxylic acid cycle. This has been emphasized by Moulder (1948a) who found that neither pyruvate, succinate, fumarate, or α-ketoglutarate were oxidized to a significant degree. The appearance of succinic acid, an intermediate of the Krebs cycle, as one end product of sugar degradation, does not in this case suggest the presence of an incomplete Krebs cycle, since it is formed during the anaerobic phase of sugar consumption. Its formation via the Krebs cycle, on the other hand, would involve oxidative steps.

A question requiring further study concerns the possible role of carbon dioxide in the sugar degradation of *T. lewisi.* Searle and Reiner (1940, 1941) state that carbon dioxide stimulated both the anaerobic and aerobic glucose consumption and that carbon dioxide was actually fixed during anaerobiosis. They do not claim to have elucidated fully the mechanisms involved; they do consider it as possible that the carbon dioxide assimilation is related to oxidation-reduction processes and that it involves the formation of glucose carbonate and glycerol carbonate. This concept has been criticized by Thomas (1942) who showed by means of tracer studies that in the ciliate *Tetrahymena* the carbon dioxide fixed is incorporated exclusively into the carboxyl group of succinic acid. He visualizes the process as proceeding from pyruvic acid to oxaloacetic acid (this step requiring the carbon dioxide fixation), malic acid, fumaric acid, and then succinic acid. If, however, the flagellates would contain oxaloacetic carboxylase, as liver and bacteria do (Krampitz, Wood, and Werkman, 1943; Krampitz and Werkman, 1941), oxaloacetic acid could be transformed directly into succinic acid. Moulder (1948a) does not deny a possible participation of carbon dioxide in the glucose metabolism of *T. lewisi,* but his careful studies did not confirm Searle and Reiner's (1941) claim that hardly any glucose is utilized in the absence of carbon dioxide or bicarbonate.

d. Summary

The carbohydrate metabolism of all hitherto studied Trypanosomidae is characterized by incomplete oxidations. In various groups of trypanosomes the sugar degradation stops at different points along the glycolytic chain, proceeding least far in the *evansi* group and farthest in the *lewisi* group. Some glycolytic enzymes have been identified, and the occurrence of phosphorylation has been established. The nature of the dehydrogenase systems operating in the trypanosomes is as yet unknown. Their characterization unquestionably represents a point of major interest for further research.

3. FAT METABOLISM

Landsteiner and Raubitschek (1907) were the first to show, by means of qualitative methods, that *T. equiperdum* contains alcohol-soluble lipoids. Kligler and Olitzki (1936) reported the surprisingly high figure of 60 per cent alcohol- and ether-soluble material in dried *T. evansi.* One of the lipoid fractions occurring in *T. equiperdum* has been identified as phospholipid by Moraczewski and Kelsey (1948). They found 4.41 μg. of phospholipid phosphorus in 100 million parasites and report experiments proving that it is synthesized *in vivo* at a rather rapid rate from inorganic phosphate.

No indication has been found so far that any Trypanosomidae are capable of deriving energy from fats, that is, from triglycerides of higher fatty acids, or from lipoids in general. Neither Califano and Gritti (1930) nor Krijgsman (1936a,b) found a lipase in *T. brucei* or *T. evansi* respectively.

In so far as lower fatty acids are concerned, Mannozzi-Torini (1940) states that the oxygen consumption of *T. evansi* was increased if the substrate contained the sodium salts or methyl esters of formic, acetic, propionic, or butyric acid. The reported increases were on an average small, as was the number of experiments carried out with each compound. Consequently, a reinvestigation of this problem appears desirable. *Trypanosoma lewisi* was reported by Moulder (1948a) as incapable of oxidizing formate or acetate.

4. PROTEIN METABOLISM

The developmental stages of Trypanosomidae are definitely capable of deriving energy from protein, the only identified process being deamination. Salle and Schmidt (1928) found an increase in ammonia nitrogen in cultures of *Leishmania tropica* if the medium contained no sugar, while the increase was insignificant in the presence of sugar. Similarly, in cultures of *T. cruzi* containing no available sugar initially, or in which the medium had been depleted of sugar by the biological activities of the developing flagellates, a rise in pH occurred (von Brand, Tobie, Kissling, and Adams, 1949), and this was due to an accumulation of ammonia (unpublished experiments). In similar cultures of *T. gambiense* and *T. rhodesiense,* on the other hand, at most traces of ammonia were formed (Tobie, von Brand, and Mehlman, 1950). This may indicate protein utilization by another process than deamination. All culture media for Trypanosomidae are complex. It is at the present time impossible even to hazard a guess as to the protein fraction, or fractions, that may serve as source for energy.

There is still less evidence about the utilization of proteins by the bloodstream form of trypanosomes for energetic purposes. According to von Brand's (1933) experiments with *T. brucei,* the nonprotein nitrogen of incubates increased by 0.51 mg. per cent, corresponding to an approximate protein turnover of 3 mg. per cent, while during the same time 60 mg. per cent glucose was consumed. He concluded therefore that the protein metabolism of this form was not well developed. It is a general experience that the bloodstream forms of African trypanosomes die rapidly in the absence of sugar, and it is consequently evident that they cannot derive sufficient energy from the breakdown of proteins to sustain their life processes. It is regrettable that so far no data whatever are available concerning the protein metabolism of the bloodstream form of *T. cruzi*. Its extremely low rate of sugar consumption may indicate a higher rate of protein utilization.

Mannozzi-Torini (1940) tested the influence of 14 amino acids on the oxygen consumption of *T. evansi*. He found the greatest increase, 111 per cent, with cysteine, but he does not mention any control experiments which would have been desirable with this compound which is so readily oxidized spontaneously. Other amino acids increased the oxygen consumption variably between 11 and 74 per cent, the more effective ones being histidine, asparagine, and valine. His efforts to demonstrate ammonia as an end product of the surmised utilization of these compounds remained inconclusive. He leaves the question open as to whether they gave rise to the increased oxygen consumption directly, that is, by being oxidized, or indirectly by being transformed into protoplasm, the energy for the synthetic processes being derived from the oxidation of additional sugar.

Moulder (1948a), working with the bloodstream form of *T. lewisi* in a glucose-free medium, found that DL-alanine was not oxidized at all, L-aspartate and L-asparagine were oxidized very slowly, while L-glutamate and L-glutamine were utilized at a rapid rate. In the presence of these last two substrates the oxygen consumption was 97 and 121 per cent respectively of the rate with glucose alone. No significant oxidation occurred, on the other hand, in the presence of α-ketoglutarate or equal amounts of α-ketoglutarate and ammonium chloride, which makes it unlikely that the former compound represents an intermediate of glutamate and glutamine utilization.

Experiments concerning the occurrence of protein-splitting enzymes have so far been carried out only with the bloodstream form of *T. evansi*. Krijgsman (1936a,b) found a cathepsin-like proteinase capable of splitting casein and peptone but which had hardly any effect on the serum proteins. Its pH optimum was 4.8. A somewhat higher optimum, pH 5.9, for apparently the same enzyme was found by Mannozzi-Torini (1938), who also described some activity of the enzyme on gelatine and blood proteins. Krijgsman (1936a,b) also presented evidence for the occurrence of a carboxypolypeptidase with a pH optimum of 4.49, an aminopolypeptidase with an optimum of pH 8.4, and a dipeptidase most active at pH 7.8. He did not find pepsin- or trypsin-like enzymes, nor urease or deaminases. He emphasized, however, that his technique does not conclusively prove the lack of hydrolytic or oxidative deamination.

In view of the apparently low rate of protein utilization for energetic purposes, it may be possible that the main advantage of the above set of enzymes lies in the preparation of protein for synthesis of new protoplasm. Obviously, in an acute infection of a small laboratory animal with a pathogenic trypanosome, or even with *T. lewisi,* considerable protein synthesis must take place. There is practically no evidence available on this point, however. The nature of the digestive enzymes found in trypanosomes makes it likely that the lower proteins are mainly utilized for this purpose. This has already been postulated by Mueller and Simons (1920) on the basis of their observation that starvation of the host retards the multiplication of the parasites. It would appear that the development of a medium in which the bloodstream form of trypanosomes can be grown successfully is a prerequisite for further progress along this line. At the present it can only be stated that possibly connections between protein synthesis and sugar metabolism exist. Marshall (1948a) has pointed out that in incubates of *T. evansi* only 1.75 molecule of pyruvic acid, instead of the theoretical 2, are derived from the breakdown of 1 molecule of glucose. The unaccounted part of the acid may have been drawn into the cycle of protein synthesis, and he presents some evidence that the pyruvate may undergo preliminary decarboxylation.

Moraczewski and Kelsey (1948) found 5.1 micrograms of nucleic acid phosphorus and 0.79 microgram of phosphoprotein phosphorus in 100 million freshly isolated *T. equiperdum.* They observed, by means of radioactive phosphorus, that the turnover of these fractions was relatively low in experiments carried out *in vitro,* but high *in vivo.* In these latter experiments the specific activities were significantly higher in the flagellates than in the host tissue or the blood plasma. This indicates that inorganic phosphate served for the synthesis of these compounds.

B. Pathological Physiology of Infections with Trypanosomidae

The effect of infections with leishmanias or trypanosomes on the metabolism of the host has proved to be a fertile field of study, occupying the attention of many workers and leading to interesting speculations concerning the mechanism or mechanisms by which the parasites damage their hosts. In the following sections questions dealing with the gaseous exchanges, the metabolism of carbohydrates, fats, proteins, and inorganic substances will be considered separately as will be the puzzling question of toxin production.

1. PATHOLOGICAL RESPIRATORY PHYSIOLOGY

a. Influence of Trypanosomes on the Gaseous Exchanges of Their Hosts

The data on the influence of an infection with African pathogenic trypanosomes on the gaseous exchanges of the host are rather contradictory. Von Fenyvessy (1926) found the respiratory rate of infected rats increased, the increase roughly paralleling the numbers of parasites in the blood. The respiratory quotient of the animals remained unchanged, and all irregularities disappeared upon elimination of the parasites by means of Bayer 205.

Similarly, Scheff and Rabati (1938) observed an increased oxygen consumption in infected mice. Premature death of the mice occurred if the metabolic rate of the infected animals was further increased by the injection of dinitrophenol. This latter increase was due solely to action of dinitrophenol on the host tissues. The compound has no influence on the trypanosomes (Scheff and Rabati, 1938; Tainter and Cutting, 1933).

Kligler, Geiger, and Comaroff (1929), on the other hand, found no increase in the oxygen consumption of infected rats. Indeed, they described the occurrence of a lowered respiratory rate in the terminal stages of the infection.

It was also observed that in rats infected with *T. equiperdum*, beginning 72 to 96 hours after infection, the oxygen content of the blood declined from the normal level of 16 to 18 volume per cent until in the terminal stages (128 hours after infection) such low values as 2.6 to 6.5 volume per cent were reached (Andrews, Johnson, and Dormal, 1930).

b. Asphyxiation Theories of Trypanosome Injuries

All the above authors assume that asphyxiation is responsible for the death of the infected animals, but their views concerning the mechanism bringing about this asphyxiation are widely divergent.

Scheff and Rabati (1938) seem to place the responsibility directly on the oxygen consumption of the flagellates. From a quantitative standpoint, however, such an assumption appears untenable. They found an increase in oxygen consumption of about 30 per cent in infected animals over the resting value of noninfected ones. Such an increase appears clearly insufficient to bring about asphyxiation if one considers the fact that activity can increase the oxygen consumption much more without interfering with the well being of animals.

Kligler, Geiger, and Comaroff (1929) also considered the metabolism of the parasites directly responsible for the surmised asphyxiation, but by another mechanism. They assumed that the trypanosomes produced great amounts of lactic acid during their sugar metabolism which, owing to a specific influence on the hemoglobin, would interfere with the oxidative processes of the host's tissues and thus result in asphyxiation. In view of the fact brought out in a previous section that lactic acid does not appear in significant amounts as an end product of the metabolism of pathogenic trypanosomes, this assumption now has only historical interest.

It is true, however, that the lactic acid content of the host's blood is increased, at least during the terminal stages of an infection. This has been demonstrated conclusively both by direct and indirect methods (lactic acid determinations, lowering of the pH or alkaline reserve of the blood) by numerous authors (Kligler and Geiger, 1928; Scheff, 1928; Kligler, Geiger, and Comaroff, 1929; Dominici, 1930; Linton, 1930a; Andrews, Johnson, and Dormal, 1930; von Brand, Regendanz, and Weise, 1932; and Krijgsman, 1933). Its formation is obviously due to the tissues of the host and is indicative of a metabolic disturbance. But even with this modified view of the genesis of the lactic acid, it is most unlikely that it would bring about asphyxiation. Krijgsman (1936b) has emphasized that the accumulation of lactic acid is insufficient to cause injury to the host.

This then leaves the theory of Andrews, Johnson, and Dormal (1930). They observed agglutination of parasites in the blood vessels of heart and lungs. This, in their view, prevents the proper aeration of the blood. Anoxemia follows, resulting in an uncompensated fatal acidosis. Although this theory is perhaps plausible in cases of sudden death of small laboratory animals, it certainly is insufficient to explain the injurious effect of trypanosome infections in animals like rabbits where the parasites develop only in small numbers. In these latter cases no significant agglutination of larger masses of parasites has ever been observed as far as the present writer is aware.

2. PATHOLOGICAL PHYSIOLOGY OF CARBOHYDRATE METABOLISM

a. Blood Sugar

The influence of *Leishmania* infections on the blood sugar of the host has not been studied extensively. In experimental animals the blood sugar concentration seems not to be changed materially (Stein and Wertheimer, 1942). In humans, Banerjee and Saha (1923) reported a considerable decline in adult kala-azar patients, while Auricchio (1924), on the contrary, observed a slight hyperglycemia and a prolonged alimentary blood sugar curve in infected children.

Much more attention has been given to the blood sugar during the course of trypanosome infections. Animals infected with *T. lewisi* (Linton, 1929) and with *T. cruzi* (von Brand, Tobie, Kissling, and Adams, 1949) usually show no abnormalities in this respect. A marked hypoglycemia, however, has occasionally been observed in infections with *T. lewisi,* either in the rare cases where it is pathogenic (Regendanz and Tropp, 1927) or in hypophysectomized hosts (Molomut, 1947).

The situation is quite different in the case of the African pathogenic trypanosomes. A very pronounced lowering of the blood sugar, amounting in the terminal stages to almost complete depletion, has been observed both in large and small experimental animals by numerous investigators. It has been found during the last days, or more often, only during the last hours before death (Schern, 1925, 1928; Regendanz and Tropp, 1927; Cordier, 1927; Dubois and Bouckaert, 1927; Bruynoghe, Dubois, and Bouckaert, 1927; Knowles and DasGupta, 1927/28; Zotta and Radacovici, 1929a,b; Scheff, 1928, 1932; Regendanz, 1929b; Linton, 1930; Locatelli, 1930; von Brand and Regendanz, 1931; Tubangui and Yutuc, 1931; von Brand, Regendanz, and Weise, 1932; Krijgsman, 1933; von Jancsó and von Jancsó, 1935a; French, 1938e; Hudson, 1944; and Hoppe and Chapman, 1947).

In fewer cases of either acute or chronic infections, an abnormally low blood sugar was found during longer periods, occasionally even during the greater part of the infection (von Fenyvessy, 1926; Dubois and Bouckaert, 1927; Cordier, 1927; Scheff, 1932; Poindexter, 1935; and Browning, 1938).

On the other hand some cases have been described in which the course of the disease was characterized by a hyperglycemia (Schern, 1928; Andrews and Sanders, 1928; Angolotti and Carda, 1929; Scheff, 1932; and Bell and Jones, 1946), and others in which the blood sugar level stayed essentially normal (Cordier, 1927; Savino, 1927; and Tubangui and Yutuc, 1931). The human infections studied by Walravens (1931) and Wormall (1932) also belong to this last group. These authors, however, give data only for patients in the early stages of the disease.

No great significance can be accorded to the terminal decline in blood sugar since it is not specific for trypanosome infections. It occurs also, for example, in bacterial and *Bartonella* infections (Zotta and Radacovici, 1929a; Linton, 1929; Regendanz, 1929a; Hoffenreich, 1932; and von Brand, Regendanz, and Weise, 1932). The fluctuations in blood sugar level preceding the terminal hypoglycemia, on the other hand, do indicate a distinct disturbance in blood sugar regulation.

This point is further emphasized by the fact that the tolerance for sugar and the alimentary blood sugar curves are frequently abnormal. Schern and Citron (1913) observed excretion of fructose in the urine of infected animals upon administration of amounts readily and completely assimilated by normal ones. Scheff (1932) obtained typical blood sugar curves upon feeding of sorbite to infected guinea pigs while similar treatment did not affect the blood sugar level of noninfected animals. Abnormalities in the shape of the alimentary blood sugar curve after oral introduction of glucose or saccharose have been described by Scheff (1932) and, although only in a minority of their cases, by von Brand and Regendanz (1931). Especially convincing are the data of Bell and Jones (1946) who worked with Zebu cattle chronically infected with *T. congolense.* They found a distinctly protracted blood sugar curve following administration of 0.2 g. glucose per kilogram body weight. In human trypanosomiasis patients, on the other hand, glucose tolerance tests did not indicate the presence of gross disturbances (Wormall, 1932). It must be remembered, however, that these patients were in the early stages of the disease.

b. Carbohydrate Reserves

Some of the observations summarized above, pointing to a disturbance of the host's carbohydrate metabolism, led directly to a study of the behavior of glycogen in infected animals, since the glycogen stores of the body represent the normal reservoir for the blood sugar.

Schern and Verokay (1925) and Schern and Bozzolo (1930) showed, by means of morphological methods, that little or no glycogen could be found in various organs of animals having died from trypanosomiasis. Other workers, using quantitative chemical methods, arrived at essentially similar conclusions. The polysaccharide content of liver and muscles, the chief storage places for glycogen in the mammalian body, was in almost every case found to be much lower in infected animals than in controls, and occasionally not even traces could be found (Regendanz and Tropp, 1927; Bruynoghe, Dubois, and Bouckaert, 1927; Linton, 1930a; von Brand and Regendanz, 1931; Scheff, 1932; and Krijgsman, 1933). An appreciable glycogen content of the liver was, however, found in one dying monkey (Regendanz, 1929b) and Bruynoghe, Dubois, and Bouckaert (1927) even found an increased amount of glycogen in muscles of infected mice and rabbits. In *T. cruzi* infections, on the contrary, the glycogen relationships of the host appeared normal (von Brand, Tobie, Kissling, and Adams, 1949).

On the whole, the evidence summarized above points to an impairment of the glycogen functions of the liver in infections with pathogenic African trypanosomes. A conclusive proof has been recorded by von Brand and Regendanz (1931). After feeding large amounts of saccharose to rabbits, they found an average of 10.90 and 0.57 per cent glycogen respectively in the liver and muscles of noninfected animals, while infected rabbits were capable of building their carbohydrate stores up to only 4.80 and 0.37 per cent respectively. In some of the infected animals no glycogen synthesis at all seemed to have taken place.

c. *Sugar Consumption Theory of Trypanosome Injury*

Following Schern's (1925, 1928) lead, several investigators (von Fenyvessy, 1926; Scheff, 1928, 1932; Knowles and DasGupta, 1927/28; Schern and Artagaveytia-Allende, 1936; and Hoppe and Chapman, 1947) assume that the sugar consumption of the trypanosomes is directly responsible for the above-described pathological changes in the physiology of the host and also for the latter's death. Briefly summarized, their view is as follows. The parasites rapidly consume the sugar present in the blood of the host, and it must be replenished from the glycogen stores of the liver. This results in a continuous strain on the liver whose functions finally break down. The consequence is a glycopryvic intoxication of the body finally causing death. Schern (1930) and Andrews, Johnson, and Dormal (1930) reconcile the occasional presence of glycogen in dying animals with this view by the assumption that in these cases the glycogen can no longer be mobilized.

There is no experimental evidence available concerning this specific point, but sufficient evidence is available to warrant the statement that not all sources of blood sugar are exhausted when hypoglycemia develops.

It has been found by Regendanz and Tropp (1927), Regendanz (1929b), and Krijgsman (1933) that the blood sugar level of hypoglycemic animals rises after injection of adrenalin, an observation which, it is true, was not confirmed by Scheff (1932). But the latter author and Bruynoghe, Dubois, and Bouckaert (1927) observed a rapid rise of the blood sugar level to normal or even above normal upon elimination of the parasites by means of drugs, this rise occurring even in starving animals (Scheff, 1932).

From a theoretical standpoint, it is clear that Schern's hypothesis requires that the feeding of sugar to infected hosts should prevent the development of the symptoms ascribed to the sugar consumption of the parasites. This, however, is by no means the case. Infected animals die even if fed with sugar (Cordier, 1927; Bruynoghe, Dubois, and Bouckaert, 1927; Andrews, Johnson, and Dormal, 1930). It is true that in a number of cases death occurred somewhat later than in animals having received no sugar (Dubois, 1928; Kligler, Geiger, and Comaroff, 1930; Angolotti and Carda, 1929; and Hoppe and Chapman, 1947) but at least one investigator found his sugar-fed animals dying even earlier than his controls (Poindexter, 1933).

A point frequently overlooked is that exactly identical symptoms of disturbed carbohydrate metabolism with ensuing death occur in experimental animals like rabbits where the parasites never occur in large numbers. It is inconceivable from a quantitative standpoint that here the simple withdrawal of sugar by the parasites should put a considerable strain on the liver. But even in a heavily infected rat only about 30 per cent of the calories required by a resting rat are withdrawn by the parasites during the last days of an infection, and in man this figure would not surpass 2 to 3 per cent (von Brand, 1938). It must furthermore be kept in mind that *T. cruzi* is often very pathogenic although it does not consume any appreciable amount of sugar. On the basis of these considerations then, Schern's hypothesis does not appear to explain the pathogenicity of trypanosomes in an acceptable manner.

3. PATHOLOGICAL PHYSIOLOGY OF NITROGEN METABOLISM

According to the observations of Staehelin (1904), Fellmer (1909), Scheff (1928), and French (1938a), experimental animals infected with pathogenic trypanosomes excreted an increased amount of nitrogen in the urine while the nitrogen excreted in the feces remained about normal or was even occasionally reduced. The nitrogen balance was negative in practically all cases studied, especially during the later stages of the disease. Identical symptoms appear, as is well known, in many febrile diseases; they are not characteristic only for trypanosome infections.

Considerable changes in the nitrogen constituents of the blood have been observed both in trypanosomiasis and leishmaniasis. The most characteristic sign was a diminished albumin-globulin ratio. It was due primarily to an increase in globulin, often especially euglobulin. The albumin level was often, though less conspicuously, lowered. This has been observed in human and animal infections with pathogenic trypanosomes (Mayer, 1905; Wiechmann and Horsters, 1927; Sicé, Boisseau, Provost, and Deniel, 1931; Scheff, 1932; Trensz and Jardon, 1933; Wilde and French, 1945, French, 1938b; and Ikejiani, 1946a), as well as in corresponding *Leishmania* infections (Lloyd and Paul, 1928a,b; Wu, 1922, and Stein and Wertheimer, 1942).

The total protein level of the blood in trypanosome infections generally showed only inconstant variations (Sicé, Boisseau, Provost, and Deniel, 1931; Krijgsman, 1933; Launoy and Lagodsky, 1937; French, 1938b). The nonprotein nitrogen was usually about normal throughout the greater part of an infection, but increased during the last stages of fatal cases (Scheff, 1928; Linton, 1930a; Randall, 1934; Launoy and Lagodsky, 1937; French, 1938b). Blood urea behaved in an essentially similar manner, increasing only in moribund hosts (Jones, 1933; Launoy and Lagodsky, 1936, 1937; French, 1938c). In dogs with kala-azar only normal values were observed (Stein and Wertheimer, 1942).

It is a well-known fact that in the late stages of human sleeping sickness an increase in the protein content of the liquor cerebrospinalis occurs. The medical casuistics concerning this point cannot be reviewed here. It may be mentioned that Zschucke (1932) found the increase due largely to an increase in the globulin fraction.

4. PATHOLOGICAL PHYSIOLOGY OF LIPOID METABOLISM

Linton (1930a,b) found a normal cholesterol and lecithin level in the blood of rats infected with *T. lewisi*. In *T. equiperdum* infections only the cholesterol content of the blood was normal while the lecithin level was increased.

Scheff (1932), on the other hand, observed no striking irregularities in either the cholesterol or phospholipid content of the blood in guinea pigs infected with *T. equiperdum*, but he described a considerable increase in neutral fat. In normal animals the neutral fat level in the blood varied from 64 to 101 mg. per cent whereas in infected animals values ranging from 174 to 421 mg. per cent were found. Exactly the same picture prevailed in the liver. According to Scheff and Horner (1932), the sterol and phosphatid content was identical in livers taken from normal and infected animals, while the livers from the latter showed a neutral fat content of 6.91 per cent as contrasted to 0.78 per cent in normal livers. Finally, a similar progressive fat accumulation in the liver of infected animals was observed by Scheff and Csillag (1936).

5. PATHOLOGICAL PHYSIOLOGY OF GLUTATHIONE AND ASCORBIC ACID METABOLISM

Scheff and Csillag (1936) observed that the amounts of reduced glutathione present in blood plasma, liver, and adrenals of guinea pigs infected with *T. equiperdum* were greatly reduced while the oxidized glutathione was present in larger amounts than normal. The ascorbic acid content of these same organs was greatly reduced. Nyden (1948) came to essentially similar conclusions concerning the ascorbic acid content of rats infected with *T. hippicum.* She found a significant decrease in the reduced ascorbic acid content of liver, spleen, and adrenals. The dehydroascorbic acid content of the muscles was slightly increased, while both the reduced ascorbic acid and dehydroascorbic acid of the blood plasma were increased.

These observations indicate abnormalities in the oxidation reduction processes in the host. Their exact significance, however, has not yet been clarified.

6. PATHOLOGICAL PHYSIOLOGY OF METABOLISM OF INORGANIC SUBSTANCES

a. Excretion of Inorganic Substances

Fellmer (1909) observed an increased urinary excretion of chlorine and phosphorus in rabbits infected with *T. brucei,* French (1938a), an increased excretion of calcium, potassium, and phosphate in cattle and sheep infected with *T. congolense* and *T. brucei.* The excretion of sodium was more pronounced in the *congolense* than the *brucei* infections. Whether an increased chlorine excretion, or on the contrary, a retention took place depended upon whether the animals were kept on a low or high salt-containing ration respectively. No decided change in magnesium balance occurred.

In view of this increased excretion of many important inorganic components, it is not surprising that Kligler, Geiger, and Comaroff (1929) noted a decreased survival time of *T. evansi*-infected rats upon removal of any important cation from the diet. Along comparable lines, Henderson (1931) and Le Roux (1933) observed geophagia in trypanosome-infected cattle, a condition which could be relieved by supplementing the ration with salt and bone meal.

b. Inorganic Constituents of the Blood

French (1938d) studied some inorganic constituents of the whole blood during trypanosome infections of cattle, sheep, and donkeys. The blood calcium was unchanged in some cases, while it decreased in others. The potassium content fell, while sodium and chlorine increased. These changes are interpreted as being due to the anemic condition which reduces the proportion of blood cells to blood plasma.

Linton (1930a) found no significant change in the chlorine content of the blood of rats infected with *T. equiperdum,* nor did Hudson (1944) in *T. vivax* infections of cattle.

Zwemer and Culbertson (1939) found a rather marked increase in serum potassium in rats infected with *T. equiperdum.* This observation was confirmed by Ikejiani (1946b,c) for infections with *T. brucei* and *T. equiperdum,* while no similar increase occurred in *T. lewisi* infections. The increased serum potassium seems to be derived from the red blood cells, or perhaps also from other cells.

Changes in inorganic constituents occur also in the cerebrospinal fluid of human sleeping sickness patients. Sicé (1930) described a lowering of its chloride concentration.

The inorganic blood constituents during leishmaniasis have received but little attention. Stein and Wertheimer (1942) found normal chlorine and calcium values in kala-azar dogs. Cacioppo (1947) found a decreased serum iron content in the majority of his cases of infantile visceral leishmaniasis. Some cases, however, showed normal values, and two even abnormally high values.

c. Potassium Theory of Trypanosome Injury

Zwemer and Culbertson (1939) thought that the increase in serum potassium mentioned above may play an important role in bringing about the death of trypanosome-infected animals because of the well-known toxicity of potassium. However, Ikejiani (1946b,c) pointed out that the potassium increase is only a terminal phenomenon becoming distinct only a short time before death. Furthermore, Scheff and Thatcher (1949) developed potassium tolerance in rats according to the procedure of Thatcher and Radike (1947) and found that these rats died as rapidly of a trypanosome infection as nontolerant rats. The potassium theory of trypanosome injury therefore appears untenable.

7. TOXIN THEORY OF TRYPANOSOME INJURY

Various theories concerning the mechanism by which pathogenic trypanosomes damage their hosts have been mentioned in the foregoing sections, and serious objections have been leveled against each one. A last theory remains to be considered, the view that one or more hitherto unrecognized metabolic end products of the trypanosomes, usually called toxins, are responsible for the injurious action of the flagellates. It has been proposed by a number of investigators (Reichenow, 1921; Regendanz and Tropp, 1927; Zotta and Radacovici, 1929a; Locatelli, 1930; von Brand and Regendanz, 1931; Krijgsman, 1933, 1936b; von Brand, 1938; French, 1938e).

The evidence for the occurrence of endotoxins in African pathogenic trypanosomes is quite contradictory, however. Some investigators (Laveran, 1913; Laveran and Roudsky, 1913; Schilling and Rondoni, 1913; Schilling, Schreck, Neumann, and Kunert, 1938) have described toxic phenomena upon injection of dried or lysed trypanosomes into experimental animals. Others, however (Braun and Teichmann, 1912; Kligler, Geiger, and Comaroff, 1929; Andrews, Johnson, and Dormal, 1930), reported completely negative results.

The question whether exotoxins may be involved has not yet been approached seriously. In this connection it should be pointed out that the development of culture media in which the bloodstream form of trypanosomes would multiply is a very desirable project. With it one could test the merit of Krijgsman's (1936b) assumption that the surmised toxin belongs to the amines.

Equally contradictory is the evidence concerning the occurrence of endotoxins in *T. cruzi*. Roskin and Romanova (1938) and Klyueva and Roskin (1946) claimed that extracts of this flagellate have a cancerolytic effect. These findings, however, have not been confirmed by other workers (Engel, 1944; Hauschka, Saxe, and Blair, 1947; Cohen, Borsook, and Dubnoff, 1947; Spain, Molomut, and Warshaw, 1948; Belkin, Tobie, Kahler, and Shear, 1949).

8. SUMMARY

Animals infected with Trypanosomidae show considerable metabolic disturbances. The most dramatic changes occur, in the case of the African pathogenic trypanosomes, in the carbohydrate metabolism. This phase has been studied most thoroughly. This concentration on one phase led some workers to overlook the disturbances in protein and fat metabolism, as well as those related to the turnover of inorganic substances. They do indicate clearly, however, that the changed carbohydrate metabolism is only one of many symptoms brought about by the infections.

The factor, or factors, responsible for these changes are none too clear. Of the various theories discussed, the sugar consumption theory, the various asphyxiation theories, and the potassium theory do not explain the experimental facts satisfactorily. The toxin theory does, but it in turn lacks definite experimental proof. It is quite evident that the entire field of trypanosome injury waits for a new approach.

C. Physiological Basis of Chemotherapy

In the following sections a discussion will be presented of the mode of action of drugs which have been analyzed to a greater or lesser extent by biochemical methods. Drugs like phenanthridinium compounds or antrycide, which may be of considerable value in practical chemotherapy but for which no such data are available, will not be mentioned further. It will also not be possible to go deeply into the question of correlations between chemical structure and parasiticidal activity. For reviews of this latter field the reader is referred to Wright (1946) and Work and Work (1948).

1. METALLIC COMPOUNDS

a. Arsenicals

Friedheim (1949) estimates that in the last 40 years about 12,000 trypanocidal arsenicals have been synthesized of which several hundred may produce definite cures. Chemically, the arsenicals tested fall into 3 main groups: trivalent organic arsenicals, pentavalent organic arsenicals, and arsenobenzenes. The relative merit of various compounds in practical chemotherapy cannot be reviewed here. For the purposes of this discussion it will be sufficient to remember that arsenicals are inactive against *T. cruzi,* are of little value in infections with *T. congolense,* but are quite effective against trypanosomes of the *evansi* and *brucei* groups. Even here, however, most of the older arsenicals are of definite value only in the early stages of the disease before the parasites have entered the central nervous system, the fairly toxic tryparsamide being the only exception. A new drug reportedly active in all stages of human sleeping sickness and relatively nontoxic is Mel B (an alkyl mercapto derivative of melamynilphenylarsenoxide, Friedheim, 1949). We can review in more detail here only the physiological mechanism of arsenic action.

(1) *Arsenic in Trypanosomes.* Arsenicals, in order to exert trypanocidal effects, have to penetrate the bodies of the flagellates. The occurrence of arsenic in trypanosomes isolated from the blood of an experimental animal injected previously with an arsenical drug seems to have been shown first by Levaditi and von Knaffl-Lenz (1909). Later experiments along the same lines have been performed by Singer and Fischl (1934) and Singer, Kotrba, and Fischl (1934). Arsenic determinations in trypanosomes exposed to arsenicals *in vitro* have been carried out by Reiner, Leonard, and Chao (1932a). They found differences in the amount of arsenic bound correlated with the drug used. Thus sodium arsanilate was hardly bound by *T. equiperdum,* while of all the drugs tested 3-amino-4-hydroxyphenylarseniousoxide was most easily fixed. The pH of the medium also has a considerable influence, much more of an acid-substituted phenylarsenoxide being bound by *T. equiperdum* in acid than alkaline environment (Eagle, 1945b).

On the other hand, differences in arsenic-binding powers exist also between different species of trypanosomes. Pedlow and Reiner (1935) observed that *T. equiperdum* binds much more neoarsphenamine than *T. lewisi.*

Yorke and Murgatroyd (1930) devised a biological method to demonstrate the extent of arsenic uptake by trypanosomes, and their procedure seems to be more sensitive than the methods of analytical chemistry employed by the above-mentioned authors. It consists essentially in the exposure of a given number of parasites for a definite period of time to a solution of a drug in a nutritive medium. The decrease of toxicity of this same solution, after removal of the first set of parasites by centrifugation, to a new batch of flagellates serves as an index of drug uptake by the first group.

It has been found that the fixation of arsenic is a very rapid process requiring but a few minutes. It is, in its first stages, a reversible process and becomes irreversible only after some time when secondary chemical reactions within the cell have taken place (Hawking, 1938; Eagle and Magnuson, 1944).

(2) *Trypanocidal Powers of Arsenicals in vitro and in vivo.* It has been generally recognized since Ehrlich's (1909) first observations that the trivalent organic and the inorganic arsenicals are highly trypanocidal both *in vitro* and *in vivo.* The pentavalent organic arsenicals, on the

other hand, are relatively nontoxic *in vitro* while they exert a powerful trypanocidal action *in vivo.* The data for the arsenobenzenes are somewhat variable; their *in vitro* toxicity seems to be intermediate between those of the two above groups. As Yorke and Murgatroyd (1930) have pointed out, much of the older *in vitro* work (Jacoby and Schuetze, 1908; Schilling, 1909; Halberstaedter, 1912; Adler, 1921; Simic, 1923; Papamarku, 1927; and others) is unconvincing because it was carried out either under unphysiologically low temperatures, over too short periods of time, or both. Yorke and Murgatroyd (1930), using a reliable technique, present quantitative data on the relative trypanocidal power of various drugs *in vitro.* They found concentrations of 1:1,600 of pentavalent arsenicals were required to kill all the flagellates within 24 hours, while the corresponding figures for trivalent arsenicals varied between 1:102,400,000 and 1:204,800,000, and those for arsenobenzenes between 1:3,200,000 and 1:102,400,000.

Ehrlich (1909) was the first to point out that the discrepancy between *in vitro* and *in vivo* experiments can best be explained by assuming a conversion of the pentavalent arsenicals into trivalent compounds within the body of the host. Only these latter would be chemotherapeutically active.

This view has been put on a secure basis mainly by the careful studies of Voegtlin and his collaborators (Voegtlin, Dyer, and Miller, 1924; Voegtlin, Dyer, and Leonard, 1923; Voegtlin, 1925) and is generally accepted in so far as pentavalent arsenicals are concerned, while some doubts persist whether the same assumption is necessary to explain the trypanocidal activity of thioarsinites and arsphenamines, as a recent review by Wright (1946) shows.

Voegtlin and collaborators point out that arsenic exerts a direct trypanocidal action only if present in the form of an arsenoxide, $R \cdot As{=}O$, wherein R can be represented by an aliphatic or aromatic radical. Pentavalent arsenicals or arsenobenzenes are converted to this form through reduction or partial oxidation, as shown in the following formulae:

$$\underset{\text{I}}{R \cdot As(OH)_2{=}O} \rightarrow \underset{\text{II}}{R \cdot As{=}O} \leftarrow \underset{\text{III}}{R \cdot As{=}As \cdot R}$$

There is general agreement that the reaction I to II does not proceed to any noticeable degree within the bodies of the trypanosomes themselves but is accomplished by the tissues of the host. This indirect action then accounts for the latent period of activity observed with this type of drug *in vivo*. While a sufficiently high dose of a trivalent arsenical clears the trypanosomes from the blood of an infected animal in a very short time, several hours may elapse before the same effect is obtained with pentavalent drugs. The same latent period is observed upon injection of arsenobenzenes, but it appears that in this case the oxidation to the active compound can be accomplished either on the surface or within the body of the trypanosomes themselves, as Strangeways (1937a) has found for neoarsphenamine.

(3) *Ehrlich's Theory of Arsenic Action.* The views concerning the point of attack of arsenic within the parasite cell have undergone considerable changes over the years. Ehrlich's (1909) well-known theory, in its original form, postulating the existence of specific side chains or arsenoreceptors and other receptors responsible for the fixation of hydroxyl and amino groups of the drugs, now has only historical interest. As Voegtlin (1925) pointed out, it had to be abandoned because "protoplasm" is not a single large molecule but a complicated system composed of many diverse chemical substances "undergoing constant change but being adjusted to a certain (dynamic) equilibrium with regard to the various components."

(4) *Voegtlin's Theory of Arsenic Action.* Voegtlin, Dyer, and Leonard (1923) stressed the point that arsenoxide reacts readily with sulfydryl compounds, a reaction of this type being shown in the following formula:

$$\mathrm{As}\begin{cases}\mathrm{OH}\\ \mathrm{OH}\\ \mathrm{OH}\end{cases} + \begin{array}{l}\mathrm{HS\cdot CH_2\cdot COOH}\\ \mathrm{HS\cdot CH_2\cdot COOH}\\ \mathrm{HS\cdot CH_2\cdot COOH}\end{array} \rightarrow \mathrm{As}\begin{cases}\mathrm{S\cdot CH_2\cdot COOH}\\ \mathrm{S\cdot CH_2\cdot COOH}\\ \mathrm{S\cdot CH_2\cdot COOH}\end{cases} + 3H_2O$$

They demonstrated by means of the nitroprusside reaction the occurrence of a sulfydryl compound within the body of *T. equiperdum.* They consider it as likely that glutathione or a similar sulfur-containing substance reacts with arsenoxide in the parasite body. This binding of the sulfur compounds would result in a disruption of the cellular oxidations and reductions which Voegtlin (1925) considers as largely controlled by the system

$$\begin{array}{l}\mathrm{R\cdot SH} \rightharpoonup \mathrm{S\cdot R}.\\ \mathrm{R\cdot SH} \leftharpoondown \mathrm{\overset{|}{S}\cdot R}\end{array}$$

Voegtlin's views are partly based on the observation that sulfydryl compounds like glutathione and cysteine protect trypanosomes and vertebrate cells against the toxic action of arsenicals both *in vitro* and to some extent *in vivo* (Voegtlin, Dyer, and Leonard, 1923; Voegtlin, Rosenthal, and Johnson, 1931). The concentration of sulfydryl compounds required to afford best protection *in vitro* is fairly high, about 10 times the drug concentration. This is due to the fact (Strangeways, 1937a) that thioarsinites, compounds of the type $R\cdot As\langle^{\text{glutathione}}_{\text{glutathione}}$, are readily hydrolyzed into free sulfydryl compounds and arsenoxide as follows:

$$R\cdot As\begin{matrix} \diagup SR' \\ \diagdown SR' \end{matrix} + 2H_2O \rightleftarrows R\cdot As(OH)_2 + 2R'SH$$

An excess of free sulfydryl compounds favors the left-hand side of the above equation; that is, it stabilizes the more or less nontoxic condensation product of arsenoxide and glutathione. BAL (2-3-dimercaptopropanol) also exerts a powerful protective action against the trypanocidal action of arsenicals (Eagle, Magnuson, and Fleischman, 1946). A condensation product of oxophenarsin and BAL (Friedheim and Vogel, 1947) was definitely trypanocidal. Whether in this case the molecule acts as a whole, or whether, in accordance with Strangeway's (1937a) views, it is dissociated within the body of the host, has not yet been established.

(5) *Enzyme Theory of Arsenic Action.* Voegtlin's assumption of the central position of glutathione in the picture of arsenic poisoning has been superseded in recent years by the view that an interference with intracellular enzymes plays a more important role and is the really decisive point. It has been recognized that many metabolic enzymes contain free SH groups upon which they are dependent for proper activity (Barron, 1943; Barron and Singer, 1945a,b). Sulfydryl enzymes are involved in the degradation of both sugar and protein. Examples of glycolytic enzymes very sensitive to SH inhibitors are hexokinase and pyruvic oxidase, of proteolytic enzymes, *d*-amino acid oxidase and transaminase.

The evidence proving an interference of arsenicals with specific intracellular glycolytic enzymes of trypanosomes is just beginning to accumulate. Leonard (1946) pointed out that the known sensitivity of pyruvic oxidase may incriminate it in the case of the flagellates. Marshall (1948a), however, observed that in incubates of *T. evansi*, exposed to phenylarsenoxide, very little glucose disappeared while ATP accumulated. ATP derived from dephosphorylation processes did not therefore serve to phosphorylate more glucose. This fact together with the observation that intermediates like glucose-6-phosphate were rapidly consumed, suggests that in *T. evansi* glycolysis was stopped at its first step, the transformation of glucose to glucose-6-phosphate. This then would point to hexokinase as the most vulnerable point of the glycolytic chain.

Chen (1948b), using systems of lysed *T. equiperdum,* demonstrated the occurrence of hexokinase, phosphoglyceraldehyde dehydrogenase, and adenosine triphosphatase, all SH enzymes, and he found that arsenicals (mapharsen and tryparsamide) had a powerful influence on all three. In intact trypanosomes the interference with hexokinase alone appears to be sufficient to explain the trypanocidal effect of arsenicals, since under their influence apparently no significant amounts of intermediate sugar degradation products are formed on which other arsenic sensitive SH enzymes could be active.

(6) *Evaluation of Voegtlin's Theory and the Enzyme Theory.* The question now arises whether Voegtlin's "glutathione theory" or the enzyme theory explains better the differential sensitivity to arsenic as observed between parasites and host tissues and also the different sensitivity of various trypanosome species.

Voegtlin, Dyer, and Leonard (1923) assume that the trypanosomes contain a smaller absolute amount of SH compounds than the cells of the host. The oxidation-reduction systems of the parasites would, therefore, be more easily disrupted than that of the host cells. The above authors leave the possibility open that other factors, e.g., a differential permeability, may also be important. In a later paper Voegtlin (1925) suggests that the lesser susceptibility of the host tissues may be due to a greater oxidation potential facilitating the conversion of the toxic arsenoxide, or its sulfydryl derivatives, into nontoxic and more easily excreted pentavalent compounds. Although Voegtlin does not discuss this question specifically, it is obvious that on similar assumptions the different responses to arsenicals by various trypanosome species can also be explained theoretically.

Work and Work (1948) have recently reviewed the chemotherapeutic mechanisms of SH inhibitors from the enzyme chemist's standpoint. They emphasize especially the following points. Dixon (1937) has shown that low concentrations of iodoacetate inhibit yeast fermentation by interfering specifically with alcohol dehydrogenase while other SH enzymes involved in alcoholic fermentation are inhibited only by much higher concentrations of this compound. Hellerman, Chinard, and Deitz (1943) demonstrated a differential reactivity of various sulfydryl groups of urease to *p*-chloromercuribenzoic acid. Different SH enzymes show a considerable degree of variability in their reactions to arsenicals (Barron, Miller, Bartlett, Meyer, and Singer, 1947). Furthermore, some SH enzymes are to some extent protected from the influence of inhibitors by naturally occurring cell components, for example, phosphoglyceraldehyde dehydrogenase by its coenzyme, diphosphopyridine nucleotide (Rapkine, 1938), or *d*-amino acid oxidase by an excess of its coenzyme, flavine adenine dinucleotide (Hellerman, Lindsay, and Bovarnick, 1946).

It is clear that such points introduce a much greater flexibility of the enzyme theory into the question of differential toxicity as compared to Voegtlin's rather rigid concept of the prime importance of glutathione or some similar compound. It is entirely reasonable to assume that in various types of cells the relative importance of various enzyme systems differs, that the amounts of protecting substances (coenzymes, glutathione) differ, that in some cases enzymatic bypasses may exist, or that some types of cells can switch more readily than others to energy production from a source not requiring the activity of a particular system that has been blocked. Of course, differential permeability may also enter the picture.

In any event, such considerations, while obviously raising many new questions, can readily explain the variability in response to arsenicals (and other drugs) both among various species of parasites and between parasites and host tissues. In this connection, one can point to the fact that the trypanosomes of the *brucei* and *evansi* groups, which react most readily to arsenicals, depend entirely upon the degradation of glucose, whereas the same is probably not true for vertebrate cells and quite apparently not for *T. cruzi,* which does not respond to arsenicals. Among the many unsolved questions, an important one is why *T. congolense,* which in its quantitative sugar requirements is similar to the parasites of the arsenic-susceptible groups, does not react well to arsenicals.

b. Antimonials

Trivalent antimonials and pentavalent antimonials have trypanocidal properties, but their chief value in practical chemotherapy lies in their activity against leishmanias. Their mode of action has been analyzed much less than that of the arsenicals from a physiological standpoint, but a number of interesting parallels exist between both.

According to Voegtlin and Smith (1920) it appears possible that the pentavalent antimonials are changed to trivalent compounds in the body of the host and that only the latter are chemotherapeutically active.

Chen, Geiling, and MacHatton (1945) demonstrated that the trypanocidal activity of trivalent antimonials, but not of pentavalent ones, is antagonized by cysteine. Chen (1948b) found that tartaremetic and stibamine inhibit the same sulfydryl enzymes of lysed *T. equiperdum* as do arsenicals.

The possibility then exists that enzyme inhibition lies at the root of antimony activity. On the whole, however, the impression is gained from work on other parasites as well, that antimonials are somewhat less specific than arsenicals and nonspecific actions may have to be considered. In this connection it may be mentioned that Chopra (1936) assumes a combination of antimony with protoplasm for which it supposedly has a special affinity.

2. NONMETALLIC COMPOUNDS

a. Diamidines

As mentioned in a previous section, Schern (1925) had observed first that trypanosomes deprived of sugar soon lose their motility. This observation led directly to the question whether substances capable of lowering the blood sugar of the host would deprive the parasites of so much sugar that a therapeutic action would ensue.

Schern and Artegaveytia-Allende (1935, 1936) and, entirely independently, von Jancsó and von Jancsó (1935b) found that one blood sugar-lowering substance, synthalin, did have a powerful therapeutic activity. Both groups of investigators tended to accept the above interpretation concerning the mechanism of action, although they felt that a direct action on the parasites was not completely ruled out. Lourie and Yorke (1937), however, demonstrated that the compound kills trypanosomes *in vitro* in a sugar-containing medium and in dilutions which *in vivo* would not lower the blood sugar of a mammalian host. An indirect action is thus obviously ruled out. Browning (1938) supported this view by his finding that a slight hypoglycemia occurs in nontreated mice chronically infected with *T. congolense.* The degree of blood sugar change was about the same as that produced by the injection of therapeutic doses of synthalin and guanidine into normal mice; of these compounds, however, only the former is trypanocidal.

The potential usefulness of synthalin led to the testing of numerous related compounds. It was found that both straight chain compounds, such as undecane diamidine, and aromatic diamidines, such as stilbamidine or propamidine, showed good activity against trypanosomes of the *brucei* group (King, Lourie, and Yorke, 1938; Lourie and Yorke, 1939; Yorke, 1940; Fulton and Yorke, 1942), while *T. lewisi* and *T. cruzi* were hardly or not at all affected (Adler and Tchernomoretz, 1941; Lourie and Yorke, 1939). *Trypanosoma congolense* did not respond well to the majority of diamidines (King, Lourie, and Yorke, 1938; Daubney and Hudson, 1941; Yorke, 1944); only 4,4-diamidino-α, β-dimethyl stilbene proved relatively effective (Fulton and Yorke, 1943; Carmichael and Bell, 1943; Wien, 1946). We have, therefore, a fairly good parallel to the reactions of various trypanosomes to sulfydryl inhibitors. However, the diamidines are also effective against leishmanias (Adler and Tchernomoretz, 1941; Adler, Tchernomoretz, and Ber, 1945, 1948; Collier and Lourie, 1946). This has a certain bearing on the explanation of the mode of action as will be shown presently.

The field has been discussed recently by Schoenbach and Greenspan (1948), who review the action of aromatic diamidines on cells ranging from bacteria to tumor cells. They point to Bichowsky's (1944) observation that in bacteria and protozoa these drugs tend to concentrate in bodies rich in nucleic acid. Since they are basic, they may conceivably combine with nucleic acids or nucleoproteins and in this way interfere with the utilization of an essential metabolite or may block an enzymatic passway. However, on the basis of Kopac's (1945, 1947a) studies on cancer cells, another less specific mode of action has to be considered seriously, namely, the denaturation of nucleoproteins. He found that many diamidines, stilbamidine being one of the potent compounds, have such an effect on various nucleoproteins in physiological concentrations. The mechanism, in these cases, may be either an irreversible dissociation of nucleic acid from protein or the blockage of critical intramolecular linkages (Kopac, 1947a,b; Snapper, Mirsky, Ris, Schneid, and Rosenthal, 1947).

Whether these latter effects take place in trypanosomes and leishmanias remains to be studied. They are in most cases rapidly dividing organisms in which a considerable rate of nucleoprotein synthesis and turnover must occur, and it is furthermore known that they frequently store free nucleic acid in granules known under the name of volutin. On the other hand, Schoenbach and Greenspan (1948) point out with justification that the diverse action of diamidines on different Trypanosomidae, together with the chemotherapeutic effectiveness of phenoxydiamidines despite their relative ineffectiveness as protein denaturants, seems to indicate a specific action.

We are thus led back to the question of enzyme inhibition. Marshall (1948a) alone has carried out experiments on trypanosomes shedding light on this point. He studied the action of a straight chain diamidine (undecane diamidine) and an aromatic one (stilbamidine) on *T. evansi.* The former compound reduced the oxygen consumption by about 5 per cent. The glucose consumption appeared somewhat stimulated, while less pyruvate accumulated than in controls. The intermediate glucose metabolism appeared essentially normal. The exact point of attack of this compound could not be determined; it is believed to lie somewhere in the dehydrogenase system.

Stilbamidine acted quite differently. It did not interfere with the oxygen or glucose consumption; it did increase the accumulation of pyruvate and changed the levels of phosphorylated intermediates. Marshall (1948a) considers it likely that stilbamidine inhibits the decarboxylase by which pyruvic acid is drawn into the cycle of protein synthesis. It may thus cause suppression of growth. He does not consider the possibility that the pyruvate effect may be an indirect one, resulting from an interference with nucleic acid metabolism which might lead to the same overall effect. In any event, generalizations as to the mode of action of diamidines on protozoa are not yet feasible. Obviously a field of considerable interest and amenable to experimental approach is open.

b. Bayer 205

Few physiological studies with Bayer 205 shedding light on its mode of action are available. Von Fenyvessy and Reiner (1928) found no influence on either respiration or glycolysis during short-time experiments *in vitro* with *T. equiperdum.* Von Issekutz (1933a,b) reached a similar conclusion derived from *in vitro* experiments of longer duration. He reports, however, that the respiration and sugar consumption of trypanosomes exposed to the drug for several hours *in vitro* were considerably reduced.

Experiments of the latter type were also carried out by Glowazky (1937). He found that the oxygen consumption and anaerobic glycolysis of *T. brucei* was normal 30 minutes after the drug had been injected into the experimental animals. Six hours after the injection, however, both metabolic phases were reduced, the former more than the latter. In anaerobic incubates of the parasites, glycerol accumulated in contrast to incubates with untreated flagellates. His conclusion is that the anaerobic glycerol-splitting enzymes have been damaged. Recently, Town, Wills, and Wormall (1949) reported that the drug inhibits strongly the fermentation of glucose by yeast juice.

It is evident that further physiological studies are urgently required.

c. *Nitrogen Mustards*

The only representative of this group studied so far from a physiological standpoint is methyl-*bis* (β-chloroethyl) amine. According to Chen (1948a) it lowers the glucose consumption of *T. equiperdum* only in relatively high concentrations, a 10^{-2} molar solution producing an inhibition of only 20 per cent. Lower concentrations did not inhibit the glucose consumption but produced an inhibition of reproduction, an effect which could be antagonized by cysteine. Chen (1948) considers it possible that an enzyme containing an SH group may be involved in this inhibition of cellular division.

d. *Halogenated Fatty Acids*

The effect of iodoacetic acid on trypanosomes *in vivo* was studied by Smythe and Reiner (1933), von Jancsó and von Jancsó (1936), this group also employing bromoacetic acid, and Hoppe and Chapman (1947). A curious coincidence is that none of the later workers knew of the similar studies of their predecessors. Although the halogenated acetic acids proved too toxic to the host for practical use, they were highly effective in reducing or even eliminating African pathogenic trypanosomes from the bloodstream of experimental animals. It is of interest to note that the therapeutic activity of these compounds was tried because of purely physiological considerations, the realization that they might effectively interfere with the glycolytic processes of the trypanosomes in the same way as they are known to do in many other organisms. The effect of these acids on the metabolism of trypanosomes *in vitro* has been described in a previous section.

e. *Dyes*

The biochemical analysis of the effect of dyes on trypanosomes has not proceeded far as yet, despite the fact that the chemotherapeutic activity of numerous dyes has been assessed since Ehrlich (1907) first recognized the trypanocidal properties of some of them. Fulton and Christophers (1938) observed that acriflavine and methylene blue in fairly high dilutions inhibited effectively the oxygen consumption of *T. rhodesiense.*

Glowazky (1937) had previously reported that acriflavine decreased the oxygen consumption of *T. brucei* by 50 to 75 per cent but did not interfere with the anaerobic glucose consumption. Just as in the case of Bayer 205, mentioned above, he found an accumulation of glycerol in anaerobic incubates of acriflavine-treated trypanosomes.

3. SUMMARY

By far the greatest amount of knowledge concerning the mode of action of drugs on trypanosomes has been gained with arsenicals. The best theory to explain their action is the assumption of an interference with specific intracellular enzymes. On this assumption the differential toxicity of the drugs on parasites and host tissues, as well as on various species of parasites, can be understood in broad outlines at least. It is very likely that a fundamentally similar situation exists also in respect to other groups of drugs. In the great majority of cases, however, our present knowledge is insufficient to fix the point of attack of a given drug with certainty.

D. PHYSIOLOGICAL BASIS OF DRUG RESISTANCE

1. METHODS OF PRODUCING RESISTANT STRAINS

It was first observed in Ehrlich's laboratory (Ehrlich, 1907) that injection of parafuchsin caused *T. brucei* to disappear temporarily from the blood of mice. After a certain number of passages through mice treated with identical amounts of the dye, the amount had to be increased steadily in order to achieve the same result. The parasites had developed a resistance toward the drug. This method, the injection of subcurative doses of a drug, is still a much used procedure of developing resistant strains; it is, however, not the only possible way.

For this purpose Yorke, Murgatroyd, and Hawking (1931) introduced two methods of *in vitro* exposure. In their first procedure they exposed *T. rhodesiense* to a concentration of 1:12,800,000 of reduced tryparsamide, injected the parasites into animals, and repeated this alternation between *in vitro* exposure to the same concentration of the drug and infection 36 times. At the end of this series the parasites were about 20 times as resistant as the original strain. The second procedure rests on the same principle, but proved more effective. The parasites were initially exposed *in vitro* to a tryparsamide concentration of 1:800,000 and, as resistance developed, the concentration of the drug was increased. After the ninth exposure the flagellates withstood a concentration of

1:6,250. This represented the limit obtainable with this drug; the strain had increased its resistance at least 500 times.

Finally, drug-resistant strains can arise spontaneously. An example is the strain of *T. equiperdum* observed by Eagle and Magnuson (1944) which without apparent cause suddenly developed a considerable resistance to amino- and amide-substituted phenylarsenoxides.

The speed with which drug fastness can be developed experimentally in trypanosomes varies with different types of drugs. It is rapidly induced with most organic arsenicals and antimonials, and acriflavine, but only slowly with Bayer 205 or diamidines. A special situation seems to prevail in respect to tartar emetic. It is not possible to produce a tartar-emetic-resistant strain directly from a normal strain of trypanosomes, but it is easy to do so when starting from an arsenic-, antimony-, or acriflavine-resistant one (Lourie and Yorke, 1938).

2. SPECIFICITY OF DRUG RESISTANCE

The resistance produced in a trypanosome strain by a given drug in most cases is not confined to this one particular compound, but is also evident against chemically related drugs. In some cases even no evident close chemical relationship exists. Thus arsenic-resistant strains are also resistant to acriflavine and vice versa. Some data illustrating these relationships are shown in Table VI. Similar observations had already

TABLE VI. Specificity of Drug Resistance in Trypanosomes (After Data from Lourie and Yorke, 1938)

Made resistant to:	Sensitivity to (+ = resistant; − = nonresistant)						
	Aromatic arsenicals	Aromatic antimonials	Acriflavine	Tartar emetic	Bayer 205	Amidines	Guanidines
Aromatic arsenicals	+	+	+	−	−	−	−
Aromatic antimonials	+	+	+	−	−	−	−
Acriflavine	+	+	+	−	−	−	−
Bayer 205	−	−	−	−	+	−	−
Amidines	−	−	−		−	+	+
Guanidines	−	−	−		−	+	+

been made by Ehrlich (1907, 1909), who distinguished between resistance to arsenicals, basic triphenylmethane dyes, and azo dyes.

It must be clearly understood, however, that a so-called arsenic-resistant strain is not resistant against all arsenicals. For example, atoxyl-resistant strains are readily influenced by arsenophenylglycine (Roehl,

1909) show a variable increase in resistance to other organic arsenicals, and none whatever to sodium arsenite (Yorke and Murgatroyd, 1930); strains resistant to amino- and amide-substituted arsenicals are susceptible to γ-(*p*-arsenosophenyl)butyric acid (Eagle, 1945a). Indeed, Yorke and Murgatroyd (1930) emphasize that no arsenic resistance in the strict sense exists, but only a resistance to the substituted phenyl radical.

These and similar relationships may have their origin in the way the arsenicals penetrate the body of the trypanosomes. King and Strangeways (1942) and King (1943) postulate three types:

1. Compounds containing solubilizing carboxyl groups forming neutral salts and entering the parasites in the same way as very water-soluble substances, e.g., glucose. Substances of this type, like 4-carboxyphenylarsenoxide, are of equally low toxicity for normal and resistant trypanosomes because the greatest amount of the drugs always remains in the ambient fluid.

2. Substances like phenylarsenoxide which are highly toxic to both normal and resistant flagellates. They are devoid of hydrophylic groups and probably enter the trypanosomes at lipoid-water interfaces. In view of Kligler and Olitzki's (1936) observation of a high lipoid content of trypanosomes the ready uptake of this type is understandable.

3. Substances against which a high degree of resistance is shown by arsenic-fast trypanosomes, e.g., benzamide-*p*-arsenoxide. They should be taken up in the same way as oxazine or acridine dyes; chemically polar substances may play a role in this process, but no definite observation is available.

Studies on the newer arsenicals, Melarsen (Friedheim, 1940) and γ-(*p*-arsenosophenyl)butyric acid (Eagle, 1945a), support the view that, although the ultimate lethal mechanism is probably in all cases identical, the entry into the trypanosome cell may involve different mechanisms. It has been shown that both compounds are inactivated by glutathione just like many other arsenicals. The former, however, is selectively antagonized by Surfen C, a compound containing the same melamime nucleus as melarsen, and Eagle's compound is selectively inhibited by *p*-aminobenzoic acid (Williamson and Lourie, 1946, 1948).

3. STABILITY OF DRUG RESISTANCE

The stability of drug resistance is variable. It was again Ehrlich (1907) who contributed the first observations in this direction. He found that an atoxyl-resistant strain maintained this property for at least one year whereas a fuchsin-fast strain regained its susceptibility after

$4\frac{1}{2}$ months. The most thorough observations in this direction are those of Murgatroyd and Yorke (1937) and Fulton and Yorke (1941). They found that an atoxyl-resistant strain maintained the same degree of resistance over a period of $12\frac{1}{2}$ years during which time it had been passed by blood inoculation through 1,500 mice passages. They reached the conclusion that the resistance to aromatic compounds of arsenic and antimony is maintained indefinitely. The character seems not to be changed even during cyclical development. They showed this by studying a tryparsamide-resistant strain of *T. brucei* which, over a period of 4 years, was passed through 59 guinea pigs by blood inoculation, the series of syringe passages having been interrupted by 4 passages through *Glossina morsitans.*

Diamidine- and synthalin-fastness is a much less stable character. The strains studied by Fulton and Yorke (1941) had lost all traces of resistance within $3\frac{1}{2}$ years.

In the case of Bayer 205 considerable differences between strains of trypanosomes apparently exist (Fulton and Yorke, 1941) and there is some contradiction as to the influence of cyclical development. Murgatroyd and Yorke (1937) observed that the resistance of *T. brucei* to Bayer 205 survived at least one passage through *Glossina morsitans,* while Van Hoof, Henrard, and Peel (1938) report that *T. gambiense* lost its resistance completely after one passage through *G. palpalis.* It would seem possible that the use of different species of parasites and tsetse flies may have influenced the outcome of the experiments.

4. BIOCHEMICAL MECHANISM OF DRUG RESISTANCE

Considerable discussion centering around the question of arsenic resistance has arisen over the years concerning the mechanism of drug resistance in trypanosomes. Ehrlich (1909) explained it on the basis of his chemoreceptor theory by assuming that the avidity of the specific receptors for the drugs would diminish.

Voegtlin, Dyer, and Miller (1924) considered Ehrlich's hypothesis far too simple to explain the complicated phenomena of drug resistance. They discussed two mechanisms: (1) reduction in permeability of the parasites for the drug and (2) fortification of the natural physicochemical defense mechanisms of the parasite protoplasm against the toxic action of the drug. They could not find any experimental evidence supporting the first alternative, but considered the second one as likely.

In conformity with Voegtlin's explanation of arsenic toxicity, these authors assume that an interference with cell respiration causing ultimate death results if enough sulfydryl compounds of the protoplasm are bound by arsenic. If then a trypanosome contains an excess of sulfydryl compounds it should survive. The development of resistance may consist in shifting the equilibrium

$$\begin{matrix} R \cdot SH \rightharpoonup S \cdot R \\ R \cdot SH \leftharpoondown S \cdot R \end{matrix} + 2H$$

from right to left resulting in a greater supply of R·SH for combination with arsenic; or else the absolute amounts of both reduced and oxidized sulfydryl compounds may be increased. The fact that ortochinoid dyes like trypaflavin produce arsenic resistance is explained by the fact that these compounds can, in a reversible process, be reduced to leuco compounds, thus conceivably influencing the SH system of the protoplasm in a way similar to arsenic.

Recent quantitative determinations of the sulfydryl content of normal and arsenic-resistant *T. equiperdum* and *T. hippicum* by Harvey (1948) have, however, not shown a significant difference. Although Reiner, Leonard, and Chao (1932b) and Pedlow and Reiner (1935) describe an arsenic fixation of the same order of magnitude by normal and arsenic-fast trypanosomes, it must be emphasized that the more sensitive technique employed by Yorke, Murgatroyd, and Hawking (1931) and Hawking (1937) showed rather conclusively that arsenic-resistant trypanosomes actually bind much less organic arsenicals than normal ones. This observation and the fact mentioned previously, that so-called arsenic resistant strains are highly susceptible to sodium arsenite, led this last group of investigators to a rejection of Voegtlin's hypothesis. They are inclined to seek the explanation in an altered permeability of the parasites. Support for this view can also be found in von Jancsó's (1931, 1932) finding that arsenic-fast trypanosomes bind much less trypaflavin or styrylquinoline than normal ones. Schueler (1947) has recently pointed out that drug resistance may have connections with the electrical charge of ionizable groups of the drugs. He suggests that the development of drug resistance may involve changes in the isoelectric points of some protoplasmic constitutents. He found indications of such changes in differential stainability of normal and arsenic-resistant trypanosomes, primarily with purely basic dyes (methylene blue and toluidine blue O).

As mentioned in a previous section, it is reasonable to assume that the toxic action of arsenic is due to an interference with SH enzymes. The possibility that arsenic resistance may rest at least partially on the development of enzymatic bypasses has not yet received much attention. It is true that the observations of Yorke, Murgatroyd, and Hawking (1931) mentioned above and especially Hawking's (1938) finding that the actual amount of intracellular arsenic required to kill a resistant trypanosome is even smaller than required for a normal flagellate, appears not favorable to such an assumption. Harvey (1949) found no difference in metabolism between normal and arsenic resistant trypanosomes. Nevertheless, a definite ruling out of the above possibility by systematic enzymatic studies appears desirable.

5. BIOLOGICAL MECHANISM OF DRUG RESISTANCE

Whatever the final explanation of the biochemical mechanism of drug resistance may be, a last point requiring discussion is how such strains arise biologically. Voegtlin, Dyer, and Miller (1924) pointed out that upon injection of arsenoxide into an infected animal some parasites die almost instantaneously while others require a longer time; that is, a differential susceptibility exists. Moderate drug resistance could then obviously arise by the survival and multiplication of the least susceptible parasites, that is, by selection. They emphasize, however, that the origin of strains withstanding such higher concentrations of the drug as would be required to kill all specimens of the nonresistant parent strain cannot be explained on this basis. An adaptation must have occurred.

Following similar lines of reasoning, based mainly on their *in vitro* experiments, Yorke, Murgatroyd, and Hawking (1931) are inclined to explain the development of highly resistant strains on the basis of a genetic change, the occurrence of a mutation. This explanation also seems the most probable one in the case of the appearance of strains spontaneously resistant to drugs (Eagle and Magnuson, 1944).

A new viewpoint has recently been injected into the discussion concerning the biological mechanism of drug resistance by Sonneborn (1949). He points out that his concept of plasmatic inheritance may have a bearing on this matter. An experimental approach along these lines with trypanosomes has not yet been attempted, and an analysis should prove quite difficult in organisms like trypanosomes which do not show any fusion phenomena.

6. SUMMARY

Trypanosomes become resistant against many drugs, the speed with which the resistance develops and the tenacity with which it is retained varying with different compounds. The biochemical mechanism of drug resistance seems, according to our present knowledge, to consist in the development of a mechanism diminishing the drug uptake. The question whether enzymatic bypasses also develop has not yet been solved. Biologically, a genetic change of some kind, perhaps a mutation, must be assumed in those cases where higher degrees of resistance have developed.

III. Bodonidae

The following genera belong to the Bodonidae: *Bodo, Rhynchomonas, Proteromonas, Cryptobia, Trypanophis,* and *Trypanoplasma.*

The free-living members of the family are frequently found in surroundings containing but little or no free oxygen. It can be surmised that they have well-developed anaerobic mechanisms, but their nature is not yet known. As examples of such forms *Bodo glissans* and *Rhynchomonas nasuta* may be mentioned. They are found in the anaerobic layers of sewage tanks and the anaerobic mud of the sapropelic habitat respectively (Lackey, 1932; Lauterborn, 1916).

The only chemical data available are those of Lawrie (1935) who reports that *Bodo caudatus,* fed on bacteria, produced ammonia and consumed relatively large amounts of oxygen. Urea or uric acid was not produced, and no urease activity could be observed.

Some parasitic members of the family may gain energy from carbohydrates. This may be deduced from the observation by Keysselitz (1906) and Schindera (1922) of the occurrence of glycogen granules in *Trypanoplasma borreli* and *T. helicis.*

REFERENCES

Adler, S. (1921). *Ann. Trop. Med. Parasitol.* **15**, 427.
Adler, S., and Ashbel, R. (1934). *Arch. zool. ital.* **20**, 521.
Adler, S., and Ashbel, R. (1940). *Ann. Trop. Med. Parasitol.* **34**, 207.
Adler, S., and Tchernomoretz, I. (1941). *Ann. Trop. Med. Parasitol.* **35**, 9.
Adler, S., Tchernomoretz, I., and Ber, M. (1945). *Ann. Trop. Med. Parasitol.* **39**, 14.
Adler, S., Tchernomoretz, I., and Ber, M. (1948). *Ann. Trop. Med. Parasitol.* **42**, 1.
Andrews, J., Johnson, C. M., and Dormal, V. J. (1930). *Am. J. Hyg.* **12**, 381.
Andrews, J., and Sanders, E. P. (1928). *Am. J. Hyg.* **8**, 947.
Angolotti, E., and Carda, P. (1929). *Med. Paises calides* **2**, 431.
Auricchio, L. (1924). *Pediatria* **32**, 704.

Banerjee, D. N., and Saha, J. C. (1923). *Calcutta Med. J.* 17, 109. (Seen only in abstract.)
Barron, E. S. G. (1943). *Advances in Enzymol.* 3, 149.
Barron, E. S. G., Miller, Z. B., Bartlett, G. R., Meyer, J., and Singer, T. P. (1947). *Biochem. J.* 41, 69.
Barron, E. S. G., and Singer, T. P. (1945a). *J. Biol. Chem.* 157, 221.
Barron, E. S. G., and Singer, T. P. (1945b). *J. Biol. Chem.* 157, 241.
Belkin, M., Tobie, E. J., Kahler, J., and Shear, M. J. (1949). *Cancer Research.*
Bell, F. R., and Jones, E. R. (1946). *Ann. Trop. Med. Parasitol.* 40, 199.
Bichowsky, L. (1944). *Proc. Soc. Exptl. Biol. Med.* 57, 163.
Biot, C., Biot, R., and Richard, G. (1911). *Compt. rend. soc. biol.* 71, 368.
Braun, H., and Teichmann, E. (1912). Versuche zur Immunisierung gegen Trypanosomen. Jena. (Not seen.)
Browning, P. (1938). *J. Path. Bact.* 46, 323.
Bruynoghe, R., Dubois, A., and Bouckaert, J. P. (1927). *Bull acad. roy. méd. Belg.* Ser. 5, 7, 142.
Cacioppo, F. (1947). *Boll. soc. ital. biol. sper.* 23, 150.
Califano, L., and Gritti, P. (1930). *Riv. patol. sper.* 5, 9.
Carmichael, J., and Bell, P. F. (1943). *Ann. Trop. Med. Parasitol.* 37, 145.
Chang, S. L. (1948). *J. Infectious Diseases* 82, 109.
Chen, G. (1948a). *J. Infectious Diseases* 82, 133.
Chen, G. (1948b). *J. Infectious Diseases* 82, 226.
Chen, G., and Geiling, E. M. K. (1945). *J. Infectious Diseases* 77, 139.
Chen, G., Geiling, E. M. K., and MacHatton, R. M. (1945). *J. Infectious Diseases* 76, 152.
Chopra, R. N. (1936). A handbook of tropical therapeutics. Calcutta. Art Press.
Christophers, S. R., and Fulton, J. D. (1938). *Ann. Trop. Med. Parasitol.* 32, 43.
Cohen, A. L., Borsook, H., and Dubnoff, J. W. (1947). *Proc. Soc. Exptl. Biol. Med.* 66, 440.
Colas-Belcour, J., and Lwoff, A. (1925). *Compt. rend. soc. biol.* 93, 1421.
Collier, H. O. J., and Lourie, E. M. (1946). *Ann. Trop. Med. Parasitol.* 40, 88.
Cordier, G. (1927). *Compt. rend. soc. biol.* 96, 971.
Daubney, R., and Hudson, J. R. (1941). *Ann. Trop. Med. Parasitol.* 35, 175.
Dixon, M. (1937). *Nature* 140, 806.
Dominici, A. (1930). *Boll. ist. sieroterap. milan.* 9, 438.
Dubois, A. (1926). *Compt. rend. soc. biol.* 95, 1130.
Dubois, A. (1928). *Compt. rend. soc. biol.* 99, 656.
Dubois, A. (1930). *Ann. soc. belge méd. trop.* 10, 445.
Dubois, A., and Bouckaert, J. P. (1927). *Compt. rend. soc. biol.* 96, 431.
Eagle, H. (1945a). *Science* 101, 69.
Eagle, H. (1945b). *J. Pharmacol. Exptl. Therap.* 85, 265.
Eagle, H., and Magnuson, H. J. (1944). *J. Pharmacol. Exptl. Therap.* 82, 137.
Eagle, H., Magnuson, H. J., and Fleischman, R. (1946). *J. Clin. Invest.* 25, 451.
Ehrlich, P. (1907). *Berliner klin. Wochschr.* 44, 233.
Ehrlich, P. (1909). *Ber. deut. chem. Ges.* 42, 17.
Engel, R. (1944). *Klin. Wochschr.* 23, 127.
Fellmer, T. (1909). *Z. Immunitätsforsch.* 3, 474.
von Fenyvessy, B. (1926). *Biochem. Z.* 173, 289.

von Fenyvessy, B., and Reiner, L. (1924). *Z. Hyg. Infektionskrankh.* **102**, 109.
von Fenyvessy, B., and Reiner, L. (1928). *Biochem. Z.* **202**, 75.
Fleig, C. (1911). *Compt. rend. soc. biol.* **71**, 527.
French, M. H. (1938a). *J. Comp. Path. Therap.* **51**, 23.
French, M. H. (1938b). *J. Comp. Path. Therap.* **51**, 36.
French, M. H. (1938c). *J. Comp. Path. Therap.* **51**, 42.
French, M. H. (1938d). *J. Comp. Path. Therap.* **51**, 119.
French, M. H. (1938e). *J. Comp. Path. Therap.* **51**, 269.
Friedheim, E. A. H. (1940). *Ann. inst. Pasteur* **65**, 108.
Friedheim, E. A. H. (1949). *Am. J. Trop. Med.* **29**, 173.
Friedheim, E. A. H., and Vogel, H. J. (1947). *Proc. Soc. Exptl. Biol. Med.* **64**, 418.
Fulton, J. D., and Christophers, S. R. (1938). *Ann. Trop. Med. Parasitol.* **32**, 77.
Fulton, J. D., and Stevens, T. S. (1945). *Biochem. J.* **39**, 317.
Fulton, J. D., and Yorke, W. (1941). *Ann. Trop. Med. Parasitol.* **35**, 221.
Fulton, J. D., and Yorke, W. (1942). *Ann. Trop. Med. Parasitol.* **36**, 131.
Fulton, J. D., and Yorke, W. (1943). *Ann. Trop. Med. Parasitol.* **37**, 152.
Geiger, A., Kligler, J., and Comaroff, R. (1930). *Ann. Trop. Med. Parasitol.* **24**, 319.
Glowazky, F. (1937). *Z. Hyg. Infektionskrankh.* **119**, 741.
Hagemeister, W. (1914). *Z. Hyg. Infektionskrankh.* **77**, 227.
Halberstaedter, L. (1912). *Arch. Schiffs. u. Tropen-Hyg.* **16**, 641.
Harvey, S. C. (1948). *Proc. Soc. Exptl. Biol. Med.* **67**, 269.
Harvey, S. C. (1949). *J. Biol. Chem.* **179**, 435.
Hauschka, T. S., Saxe, L. H., Jr., and Blair, M. (1947). *J. Natl. Cancer Inst.* **7**, 189.
Hawking, F. (1937). *J. Pharmacol. Exptl. Therap.* **59**, 123.
Hawking, F. (1938). *Ann. Trop. Med. Parasitol.* **32**, 313.
Hawking, F. (1944). *J. Pharmacol. Exptl. Therap.* **82**, 31.
Hellerman, L., Chinard, F. P., and Deitz, V. R. (1943). *J. Biol. Chem.* **147**, 443.
Hellerman, L., Lindsay, A., and Bovarnick, M. R. (1946). *J. Biol. Chem.* **163**, 553.
Henderson, W. W. (1931). *Vet. J.* **87**, 518.
Hoare, C. A. (1948). *Proc. Fourth Intern. Congr. Trop. Med. Malaria* **2**, 1110.
Hoffenreich, F. (1932). *Arch. Schiffs. u. Tropen-Hyg.* **36**, 141.
van Hoof, L., Henrard, C., and Peel, E. (1938). *Trans. Roy. Soc. Trop. Med. Hyg.* **32**, 197.
Hoppe, J. O., and Chapman, C. W. (1947). *J. Parasitol.* **33**, 509.
Hudson, J. R. (1944). *J. Comp. Path. Therap.* **54**, 108.
Ikejiani, O. (1946a). *J. Parasitol.* **32**, 369.
Ikejiani, O. (1946b). *J. Parasitol.* **32**, 374.
Ikejiani, O. (1946c). *J. Parasitol.* **32**, 379.
von Issekutz, B. (1933a). *Arch. exptl. Path. Pharmakol.* **137**, 479.
von Issekutz, B. (1933b). *Arch. exptl. Path. Pharmakol.* **173**, 499.
Ivanov, I. I., and Jakovlev, W. G. (1943). *Biokhimiya* **8**, 229.
Ivanov, I. I., and Umanskaya, M. V. (1945). *Compt. rend. acad. sci. U.R.S.S.* **48**, 337.
Jacoby, M., and Schuetze, A. (1908). *Biochem. Z.* **12**, 193.
von Jancsó, N. (1931). *Zentr. Bakt. Parasitenk. Abt. I.* Orig. **123**, 129.
von Jancsó, N. (1932). *Zentr. Bakt. Parasitenk. Abt. I. Orig.* **124**, 167.
von Jancsó, N., and von Jancsó, H. (1935a). *Z. Immunitätsforsch.* **84**, 471.
von Jancsó, N., and von Jancsó, H. (1935b). *Z. Immunitätsforsch.* **86**, 1.

von Jancsó, N., and von Jancsó, H. (1936). *Biochem. Z.* **286**, 392.
Jones, E. R. (1933). *Vet. Record* **13**, 1062.
Keysselitz, G. (1906). *Arch. Protistenk.* **7**, 1.
King, H. (1943). *Trans. Faraday Soc.* **39**, 383.
King, H., Lourie, E. M., and Yorke, W. (1938). *Ann. Trop. Med. Parasitol.* **32**, 177.
King, H., and Strangeways, W. I. (1942). *Ann. Trop. Med. Parasitol.* **36**, 47.
Kligler, I. J. (1926). *Trans. Roy. Soc. Trop. Med. Hyg.* **19**, 330.
Kligler, I. J., and Geiger, A. (1928). *Proc. Soc. Exptl. Biol. Med.* **26**, 229.
Kligler, I. J., Geiger, A., and Comaroff, R. (1929). *Ann. Trop. Med. Parasitol.* **23**, 325.
Kligler, I. J., Geiger, A., and Comaroff, R. (1930). *Ann. Trop. Med. Parasitol.* **24**, 329.
Kligler, I. J., and Olitzki, L. (1936). *Ann. Trop. Med. Parasitol.* **30**, 287.
Klyueva, N. G., and Roskin, G. (1946). *Am. Rev. Soviet Med.* **4**, 127.
Knowles, R., and DasGupta, B. M. (1927/28). *Indian J. Med. Research* **15**, 997.
Kopac, M. J. (1945). *Trans. N. Y. Acad. Sci.* **8**, 5.
Kopac, M. J. (1947a). *Cancer Research* **7**, 44.
Kopac, M. J. (1947b). In: Approaches to tumor therapy, p. 27. American Association for the Advancement of Science, Washington, D. C.
Krampitz, L. O., and Werkman, C. H. (1941). *Biochem. J.* **35**, 595.
Krampitz, L. O., Wood, H. G., and Werkman, C. H. (1943). *J. Biol. Chem.* **147**, 243.
Krijgsman, B. J. (1933). *Z. Parasitenk.* **6**, 1, 438.
Krijgsman, B. J. (1936a). *Natuurw. Tijdschr. (Belg.)* **18**, 237.
Krijgsman, B. J. (1936b). *Z. vergleich. Physiol.* **23**, 663.
Kudicke, R., and Evers, E. (1924). *Z. Hyg. Infektionskrankh.* **101**, 317.
Lackey, J. B. (1932). *Biol. Bull.* **63**, 287.
Landsteiner, K., and Raubitschek, H. (1907). *Zentr. Bakt. Parasitenk. Abt. I.* Orig. **45**, 660.
Launoy, L. L., and Lagodsky, H. (1936). *Compt. rend. soc. biol.* **122**, 1055.
Launoy, L. L., and Lagodsky, H. (1937). *Bull. soc. pathol. exotique* **30**, 57.
Lauterborn, R. (1916). *Verhandl. Naturhist. Med. Ver. Heidelberg N.F.* **13**, 395.
Laveran, C. L. A. (1913). *Bull. soc. path. exotique* **6**, 693.
Laveran, C. L. A., and Roudsky, D. (1913). *Bull. soc. path. exotique* **6**, 176.
Lawrie, N. R. (1935). *Biochem. J.* **29**, 588.
Leonard, C. S. (1946). *Bull. Nat. Formulary Comm.* **14**, 139.
Le Roux, P. L. (1933). Ann. Rept. Dept. Animal Health, North Rhodesia, p. 43. (Not seen.)
Levaditi, C., and von Knaffl-Lenz (1909). *Bull. soc. pathol. exotique* **2**, 405.
Linton, R. W. (1929). *Ann. Trop. Med. Parasitol.* **23**, 307.
Linton, R. W. (1930a). *J. Exptl. Med.* **52**, 103.
Linton, R. W. (1930b). *J. Exptl. Med.* **52**, 695.
Lloyd, R. B., and Paul, S. N. (1928a). *Indian J. Med. Research* **16**, 203.
Lloyd, R. B., and Paul, S. N. (1928b). *Indian J. Med. Research* **16**, 529.
Locatelli, P. (1930). *Compt. rend. soc. biol.* **105**, 449.
Lourie, E. M., and Yorke, W. (1937). *Ann. Trop. Med. Parasitol.* **31**, 435.
Lourie, E. M., and Yorke, W. (1938). *Ann. Trop. Med. Parasitol.* **32**, 201.
Lourie, E. M., and Yorke, W. (1939). *Ann. Trop. Med. Parasitol.* **33**, 289.
Lwoff, A. (1934). *Zentr. Bakt. Parasitenk. Abt. I.* Orig. **130**, 498.

Lwoff, M. (1940). Recherches sur le pouvoir de synthèse des flagellés trypanosomides. Monogr. inst. Pasteur. Masson et Cie, Paris.
Mannozzi-Torini, M. (1938). *Boll. ist. sieroterap. milan.* **17**, 830.
Mannozzi-Torini, M. (1940). *Arch. sci. biol. (Italy)* **26**, 565.
Marshall, P. B. (1948a). *Brit. J. Pharmacol.* **3**, 1.
Marshall, P. B. (1948b). *Brit. Sci. News* **1**, 16.
Mayer, M. (1905). *Z. Pathol. Therap.* **1**, 539.
Mercado, T. I. (1947). Unpubl. Master's thesis. Catholic University of America, Washington, D. C.
Molomut, N. (1947). *J. Immunol.* **56**, 139.
Moraczewski, S. A., and Kelsey, F. E. (1948). *J. Infectious Diseases* **82**, 45.
Moulder, J. W. (1947). *Science* **106**, 168.
Moulder, J. W. (1948a). *J. Infectious Diseases* **83**, 33.
Moulder, J. W. (1948b). *J. Infectious Diseases* **83**, 42.
Mueller, J., and Simons, H. (1920). *Z. Biol.* **70**, 231.
Muniz, J., and de Freitas, G. (1944). *Rev. brasil. biol.* **4**, 421.
Murgatroyd, F., and Yorke, W. (1937). *Ann. Trop. Med. Parasitol.* **31**, 165.
Nauss, R. W., and Yorke W. (1911). *Ann. Trop. Med. Parasitol.* **5**, 199.
Noguchi, H. (1926). *J. Exptl. Med.* **44**, 327.
Nyden, S. J. (1948). *Proc. Soc. Exptl. Biol. Med.* **69**, 206.
Papamarku, P. (1927). *Z. Hyg. Infektionskrankh.* **107**, 407.
Pedlow, J. T., and Reiner, L. (1935). *J. Pharmacol. Exptl. Therap.* **55**, 179.
Plunkett, A. (1946). Unpublished Master's thesis. Catholic University of America, Washington, D. C.
Poindexter, H. A. (1933). *Am. J. Trop. Med.* **13**, 555.
Poindexter, H. A. (1935). *J. Parasitol.* **21**, 292.
Randall, R. (1934). *Philippine J. Sci.* **53**, 97.
Rapkine, L. (1938). *Biochem. J.* **32**, 1729.
Regendanz, P. (1929a). *Ann. Trop. Med. Parasitol.* **23**, 523.
Regendanz, P. (1929b). *Arch. Schiffs-u. Tropen-Hyg.* **33**, 242.
Regendanz, P. (1930). *Zentr. Bakt. Parasitenk. Abt. I.* Orig. **118**, 175.
Regendanz, P., and Tropp, C. (1927). *Arch. Schiffs-u. Tropen-Hyg.* **31**, 376.
Reichenow, E. (1921). *Z. Hyg. Infektionskrankh.* **94**, 266.
Reichenow, E. (1937). *Compt. rend. congr. intern. zool. Lisbonne,* 1935. **3**, 1955.
Reiner, L., and Leonard, C. S. (1932). *Arch. intern. pharmacodynamie* **43**, 49.
Reiner, L., Leonard, C. S., and Chao, S. S. (1932a). *Arch. intern. pharmacodynamie* **43**, 186.
Reiner, L., Leonard, C. S., and Chao, S. S. (1932b). *Arch. intern. pharmacodynamie* **43**, 199.
Reiner, L., and Smythe, C. V. (1934). *Proc. Soc. Exptl. Biol. Med.* **31**, 1086.
Reiner, L., Smythe, C. V., and Pedlow, J. T. (1936). *J. Biol. Chem.* **113**, 75.
Roehl, W. (1909). *Z. Immunitätsforsch.* **1**, 643.
Roskin, G. I., and Romanova, K. G. (1938). *Bull. biol. méd. exptl. U. R. S. S.* **6**, 118.
Salle, A. J. (1931). *J. Infectious Diseases* **49**, 481.
Salle, A. J., and Schmidt, C. L. A. (1928). *J. Infectious Diseases* **43**, 378.
Savino, E. (1927). *Compt. rend. soc. biol.* **96**, 220.
Scheff, G. (1928). *Biochem. Z.* **200**, 309.

Scheff, G. (1932). *Biochem. Z.* **248**, 168.
Scheff, G., and Csillag, Z. (1936). *Arch. exptl. Path. Pharmakol.* **183**, 467.
Scheff, G., and Horner, E. (1932). *Biochem. Z.* **248**, 181.
Scheff, G., and Rabati, F. (1938). *Biochem. Z.* **298**, 101.
Scheff, G., and Thatcher, J. S. (1949). *J. Parasitol.* **35**, 35.
Schern, K. (1912). *Arb. Reichsgesundh.* **38**, 338.
Schern, K. (1925). *Zentr. Bakt. Parasitenk. Abt. I.* Orig. **96**, 356, 440.
Schern, K. (1928). *Biochem. Z.* **193**, 264.
Schern, K. (1930). *Zentr. Bakt. Parasitenk. Abt. I.* Orig. **119**, 297.
Schern, K., and Artagaveytia-Allende, R. (1935). *Arch. soc. biol. Montevideo* **6**, 244.
Schern, K., and Artagaveytia-Allende, R. (1936). *Z. Immunitätsforsch.* **89**, 21.
Schern, K., and Bozzolo, E. (1930). Miessner Festschrift, p. 175. Hannover.
Schern, K., and Citron, H. (1913). *Deut. med. Wochschr.* **39**, 1356.
Schern, K., and Verokay (1925). Cited in Schern and Bozzolo (1930).
Schilling, C. (1909). *Arch. Schiffs-u. Tropen-Hyg.* **13**, 1.
Schilling, C., and Rondoni, P. (1913). *Z. Immunitätsforsch.* **18**, 651.
Schilling, C., Schreck, H., Neumann, H., and Kunert, H. (1938). *Z. Immunitätsforsch.* **87**, 47.
Schindera, M. (1922). *Arch. Protistenk.* **45**, 200.
Schoenbach, E. B., and Greenspan, E. M. (1948). *Medicine* **27**, 327.
Schueler, F. W. (1947). *J. Infectious Diseases* **81**, 139.
Searle, D. S., and Reiner, L. (1940). *Proc. Soc. Exptl. Biol. Med.* **43**, 80.
Searle, D. S., and Reiner, L. (1941). *J. Biol. Chem.* **141** (563).
Senekjie, H. A. (1941). *J. Hyg.* **34**, 63.
Sicé, A. (1930). *Bull. soc. path. exotique* **26**, 640.
Sicé, A., Boisseau, R., Provost, J., and Deniel, L. (1931). *Bull. soc. path. exotique* **24**, 181.
Simic, T. V. (1923). *Z. Hyg. Infektionskrankh.* **99**, 417.
Singer, E. (1936). *Z. Hyg. Infektionskrankh.* **117**, 752.
Singer, E., and Fischl, V. (1934). *Z. Hyg. Infektionskrankh.* **116**, 36.
Singer, E., Kotrba, J., and Fischl, V. (1934). *Z. Hyg. Infektionskrankh.* **116**, 133.
Smythe, C. V., and Reiner, L. (1933). *Proc. Soc. Exptl. Biol. Med.* **31**, 289.
Snapper, I., Mirsky, A. E., Ris, H., Schneid, B., and Rosenthal, M. (1947). *Blood* **2**, 311.
Sonneborn, T. M. (1949). *Am. Scientist* **37**, 33.
Soule, M. H. (1925). *J. Infectious Diseases* **36**, 1,245.
Spain, D. M., Molomut, N., and Warshaw, L. J. (1948). *Proc. Soc. Exptl. Biol. Med.* **69**, 134.
Staehelin, R. (1904). *Arch. Hyg.* **50**, 77.
Stein, L., and Wertheimer, E. (1942). *Ann. Trop. Med. Parasitol.* **36**, 17.
Strangeways, W. I. (1937a). *Ann. Trop. Med. Parasitol.* **31**, 387.
Strangeways, W. I. (1937b). *Ann. Trop. Med. Parasitol.* **31**, 405.
Tainter, M. L., and Cutting, W. C. (1933). *J. Pharmacol. Exptl. Therap.* **49**, 187.
Tasaka, M. (1935). *Fukuoka Ikwadaigaku Zasshi* **28**, 27.
Thatcher, J. S., and Radike, A. W. (1947). *Am. J. Physiol.* **151**, 138.
Thomas, J. O. (1942). Unpublished Ph. D. dissertation. Stanford University.
Tobie, E. J., von Brand, T., and Mehlman, B. (1950). *J. Parasitol.* **36**, 48.
Town, B. W., Wills, E. D., and Wormall, A. (1949). *Nature* **163**, 735.

Trensz, F., and Jardon, M. (1933). *Bull. soc. path. exotique* **26**, 442.
Tubangui, M. A., and Yutuc, L. M. (1931). *Philippine J. Sci.* **45**, 93.
Voegtlin, C. (1925). *Physiol. Revs.* **5**, 63.
Voegtlin, C., Dyer, H. A., and Leonard, C. S. (1923). *U. S. Pub. Health Service Rept.* **38**, 1882.
Voegtlin, C., Dyer, H. A., and Miller, D. W. (1924). *J. Pharmacol. Exptl. Therap.* **23**, 55.
Voegtlin, C., Rosenthal, S. M., and Johnson, J. M. (1931). *U. S. Pub. Health Service Rept.* **46**, 339.
Voegtlin, C., and Smith, H. W. (1920). *U. S. Pub. Health Service Rept.* **35**, 2264.
von Brand, T. (1933). *Z. vergleich. Physiol.* **19**, 587.
von Brand, T. (1935). *Ergeb. Biol.* **12**, 161.
von Brand, T. (1938). *Quart. Rev. Biol.* **13**, 41.
von Brand, T. (1946). Anaerobiosis in invertebrates. Biodynamica Monographs No. 4. Biodynamica. Normandy, Missouri.
von Brand (1951). In Most, Parasitic Infections in Man. Columbia Univ. Press, **90**.
von Brand, T., and Johnson, E. M. (1947). *J. Cellular Comp. Physiol.* **29**, 33.
von Brand, T., Johnson, E. M., and Rees, C. W. (1946). *J. Gen. Physiol.* **30**, 163.
von Brand, T., and Krogh, A. (1935). *Skand. Arch. Physiol.* **72**, 1.
von Brand, T., and Regendanz, P. (1931). *Biochem. Z.* **242**, 451.
von Brand, T., Regendanz, P., and Weise, W. (1932). *Zentr. Bakt. Parasitenk. Abt. I.*, Orig. **125**, 461.
von Brand, T., and Tobie, E. J. (1948a). *J. Cellular Comp. Physiol.* **31**, 49.
von Brand, T., and Tobie, E. J. (1948b). *J. Parasitol.* **34** (suppl.), 19.
von Brand, T. Tobie, E. J., Kissling, R. E., and Adams, G. (1949). *J. Infectious Diseases.* **85**, 5.
von Brand, T., Tobie, E. J., and Mehlman, B. (1950). *J. Cellular Comp. Physiol.* **35**, 273.
Walravens (1931). *Ann. soc. belg. méd. trop.* **11**, 213.
Waters, L. A., and Stock, C. (1945). *Science* **102**, 601.
Wiechmann, E., and Horsters, H. (1927). *Deut. Arch. klin. Med.* **155**, 177.
Wien, R. (1946). *Brit. J. Pharmacol.* **1**, 65.
Wilde, J. K. H., and French, M. H. (1945). *J. Comp. Path. Therap.* **55**, 206.
Williamson, J., and Lourie, E. M. (1946). *Ann. Trop. Med. Parasitol.* **40**, 255.
Williamson, J., and Lourie, E. M. (1948). *Nature* **161**, 103.
Work, T. S., and Work, E. (1948). The basis of chemotherapy. Interscience Publishers, New York.
Wormall, A. (1932). *Biochem. J.* **26**, 1777.
Wright, W. H. (1946). In Powers, Advancing fronts in chemistry **2**, 105. Reinhold Publishing Corp., New York.
Wu, H. (1922). *J. Biol. Chem.* **51**, 33.
Yorke, W. (1940). *Trans. Roy. Soc. Trop. Med. Hyg.* **33**, 463.
Yorke, W. (1944). *Brit. Med. Bull.* **2**, 60.
Yorke, W., Adams, A. R. D., and Murgatroyd, F. (1929). *Ann. Trop. Med. Parasitol.* **23**, 601.
Yorke, W., and Murgatroyd, F. (1930). *Ann. Trop. Med. Parasitol.* **24**, 449.
Yorke, W., Murgatroyd, F., and Hawking, F. (1931). *Ann. Trop. Med. Parasitol.* **25**, 521.

Zeuthen, E. (1947). *Compt. rend. trav. lab. Carlsberg. Ser. chim.* **26**, 17.
Zotta, G., and Radacovici, E. (1929a). *Arch. roumaines path. exptl. microbiol.* **2**, 55.
Zotta, G., and Radacovici, E. (1929b). *Compt. rend. soc. biol.* **102**, 129.
Zschucke, J. (1932). *Z. Hyg. Infektionskrankh.* **114**, 464.
Zwemer, R. L., and Culbertson, J. T. (1939). *Am. J. Hyg.* **29**, Section C, 7.

Nutrition of Parasitic Amebae

MARGUERITE LWOFF

Pasteur Institute, Paris

CONTENTS

I. Introduction

The problem of the nutrition of the free-living amebae is analogous to that of the nutrition of the free-living ciliates (see p. 323). The problem of the nutrition of the parasitic amebae appears to be quite different. It is complicated at the start by the fact that none of these amebae have yet been isolated in pure culture. This review is for the most part a description of the many attempts, no doubt soon to be crowned with success, to liberate these parasitic amebae from their satellite bacterial flora. In fact, such efforts have mostly been devoted to the ameba causing amebic dysentery, which, because of its pathogenic role and because of certain other obscure aspects of amebic dysentery, seems to have fascinated investigators. Almost all the parasitic amebae have been cultured more or less successfully. When the characteristics and requirements of the amebic dysentery organism are better known, a great step will have been taken toward a knowledge of the others. But it is not certain that *Entamoeba histolytica* constitutes the material of choice for these studies and that other amebae were not shown to be more favorable. This is the point of view that Lamy seems to have adopted when, after studies on the dysentery ameba and after various attempts at purification of cultures, notably with the aid of sulfonamides (1944), he took this question up again by experimenting with a parasite of reptiles, *E. invadens*. He obtained a pure culture of this organism (1948, 1949) with the use of chick embryo, and, if these researches proceed favorably, they will be of great interest.

To return to the dysentery ameba, its ability to multiply in the absence of bacteria cannot be placed in doubt. The existence of sterile hepatic amebic abscesses, both spontaneous (in man) or induced (in the cat), is well known. And it will be seen (p. 237) that these pure *in vivo* cultures have been used as inoculation material for *in vitro* cultures, without success, we hasten to add. But to reproduce in one step *in vitro* all the conditions which allow the amebae to multiply indefinitely without the aid of the associated flora is not easy. Bacteria certainly have a part to play in these multibacterial cultures, and if one suppresses them, one must know how to reproduce the medium which they were building up. The difficulty is obvious.

Nevertheless, if the biochemical characteristics of parasitic amebae have not yet been defined in pure culture, those which have been identified in monobacterial mixed cultures allow us to arrive more rapidly at total purification. Such information consists of, for example, data on the most favorable oxidation-reduction potential; and also the need for certain substances such as cholesterol.

Many methods have been used to eliminate the bacterial flora: isolation of cysts by means of micromanipulation and washing; isolation of the organism from sterile hepatic abscesses; but it would seem that in this instance, as in many others, the best chance of success may be found in a systematic use of antibiotics, among which the sulfonamides and penicillin would appear to be the most effective.

Before beginning the study of the principal aspects of amebic physiology to which attention has been drawn the last few years, it is necessary to emphasize certain points. At the present time there exist no pure cultures of the dysentery ameba whose study had led to any definite conclusions. Synthetic culture media exist neither for the culture of this ameba nor for other parasitic amebae. The expression "synthetic medium" has perhaps been seriously abused: egg (whole or in part), serum, rice starch, and bacteria—few or otherwise—are still the basis of media; nevertheless, the progress made is such as to justify a certain optimism.

II. Nutrition of the Dysentery Ameba, *Entamoeba histolytica*

Schaudinn, 1903 (emended by Walker, 1911)*

In 1924 Boeck and Drbohlav obtained for the first time a multiplication of the dysentery ameba in an egg medium based on that of Dorset, in the presence of a highly diverse bacterial flora (1925a,b). This cultural technique has since become commonplace, and many attempts at purification have been made since then, such as either isolations from spontaneous or from experimentally produced amebic abscesses, or from occasional initially pure cultures (Friedrichs and Harris, 1929; Cleveland and Sanders, 1930), or purifications by washing in sterile water or antiseptic solutions (Cleveland and Sanders, 1930; Reardon and Bartgis, 1949); or isolations with a micromanipulator (Reardon and Rees, 1939; Rees *et al.*, 1941), or, finally, more or less successful attempts to suppress the associated flora with the help of antibiotics (Lamy, 1944, 1948; Jacobs, 1947; Schaffer *et al.*, 1948; Fuller and Faust, 1949 and others). A culture at least temporarily pure has been described by Jacobs (1947) who, after eliminating with the help of penicillin the sole remaining contaminant, *Clostridium welchii*, was able to maintain the ameba with transplants every 3½ months. Antedating so many failures, but before, it is true, the era of antibiotics, Snyder and Meleney (1942) attacked the problem in an indirect manner: they tried to take into account the role played by the more or less well-defined bacterial flora and from this to draw conclusions as to the metabolism of the ameba itself. In this manner they succeeded in distinguishing two important aspects of its metabolism which have been confirmed by subsequent workers: on one hand, the importance of the oxidation-reduction potential of the culture; on the other hand, the probable need of cholesterol in the nutrition of *Entamoeba histolytica.*

A. The Oxidation-Reduction Potential

Oxidation-reduction potential is a very important factor when dealing with protozoa whose natural habitat is heavily laden with bacteria. This was first expressed clearly by Jahn (1934). According to him, the E_h of the interior of the rat cecum, for example, is approximately -200 mv.; values of the same order are attained *in vitro,* because, for the most part, of the bacterial flora. In fact, Jacobs (1941) observed that cultures of the dysentery ameba survived longest when they contained bacteria which

* After C. Dobell, 1919. The amebae living in man.

did not induce a rapid decline of the oxidation-reduction potential (*Streptococcus hemolyticus, Bacillus subtilis, Bacterium coronafaciens*), whereas cultures had to be reinoculated every 3 or 4 days when the satellite bacteria caused a rapid lowering of this potential (*Leptotrichia buccalis, Clostridium welchii*). In spite of the fact that the ameba can multiply when the potential is −114 to −150 mv., cultures are much heavier at −300 to −500 mv.; results which led Jacobs to conclude that to obtain a

TABLE I. Oxygen and Carbon Dioxide Content of Air above Egg White Medium and Locke's Solution in Florence Flasks; and of Air-Filled Flasks after Specified Periods of Storage at 10° C.
(After von Brand *et al.*, 1946)

Series number	Kind of material in flasks	Storage period days	Oxygen per cent	Carbon dioxide per cent
1	25 ml. egg white base, 100 ml. overlay	1	20.35	0.50
		8	19.85	0.72
		15	19.85	0.75
		22	19.87	0.69
2	50 ml. egg white base, 3 g. egg shell	1	20.35	0.27
		8	19.75	0.27
		15	19.76	0.28
		22	19.83	0.23
3	50 ml. egg white base, without egg shell	1	20.36	0.33
		8	20.11	0.16
		15	20.02	0.23
		22	20.21	0.17
4	125 ml. Locke's solution	1	20.59	0.13
		8	20.53	0.14
		15	20.76	0.21
		22	20.63	0.24
5	Air-filled flask	1	20.66	0.23
		8	20.75	0.14
		15	20.62	0.11
		22	20.73	0.20

bacteria-free culture an exact control of the oxidation-reduction of the medium was essential. Snyder and Meleney (1942, 1943), having confirmed the extreme sensitivity of the ameba to oxygen, made the following observations. The necessary anaerobiosis must be produced, in cultures, by the bacteria. Neither washed bacteria alone, nor culture filtrates sufficed to create a suitable oxidation-reduction potential. Furthermore, washed live bacteria could not be replaced by dead bacteria, but reducers such as cysteine could be substituted for the living bacteria with good results. Jacobs noted that at the concentrations used, cysteine hydrochloride gave a potential of −160 to −200 mv.—values compatible with a moderate multiplication of the amebae. Besides, even in the presence of bacteria, reducing substances decidedly favored their growth; thus reductose, which is rich in reductone, speeded the initiation of growth, and allowed richer and longer lasting cultures (Pautrizel *et al.*, 1949).

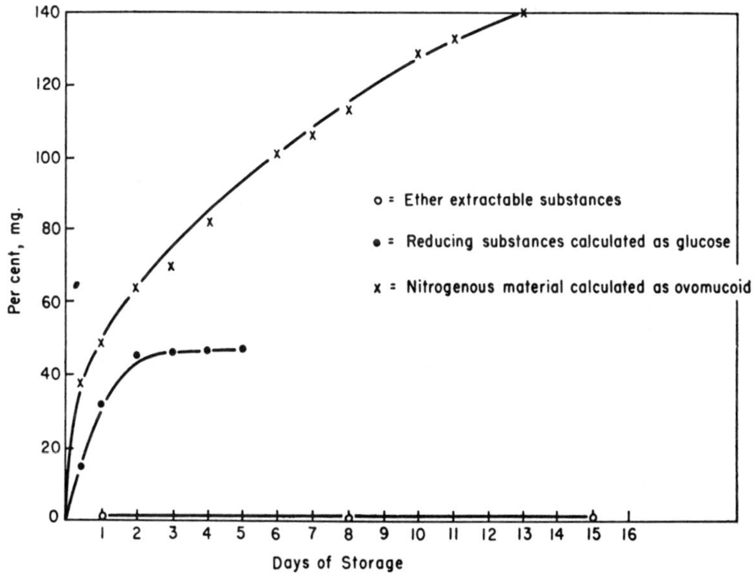

FIG. 1. Occurrence in milligram per cent of reducing substances (calculated as glucose), of nitrogenous material (calculated as ovomucoid) and of ether-extractable substances in the fluid phase of egg white medium during specified periods of storage at 37° C. Each point on the reducing substances curve is the mean value of 4 to 6 flasks, on the ovomucoid curve of 2 flasks, and on the ether-extractable substances curve of 4 flasks. (After von Brand *et al.*, 1946.)

Likewise, Rees and Reardon (1945), in the course of their investigations on the nutrition of *E. histolytica* (see p. 246) grown in the presence of a single bacterial strain, observed the favorable effect of a rather low oxidation-reduction potential: an egg white medium kept in cotton-plugged containers became unfavorable for the multiplication of the ameba, while if it was distributed in rubber-stoppered tubes in the presence of pyrogallic acid, it retained all its properties (see Table II). The question has been raised as to whether certain growth substances neces-

TABLE II. Yields of *Entamoeba histolytica*—Organism *t* in Stored Egg White Medium Enriched with the Vitamin-Cholesterol Formula*
(After Rees and Reardon, 1945)

Kind of medium	Temperature during storage °C.	Period of storage days	Number of tests	Average Yields		Average Control	
				E. histolytica	organism *t*	*E. histolytica*	organism *t*
C	37	10–20	7	20	46	139	86
C	37	30–60	2	70	60	266	
C	10	10–20	4	82	43	196	68
C	10	30–60	2	5	48	249	70
Cb	37	14–30	2	142	84	171	
Cb	10	10–20	2	96	98	181	
Cb	10	30–60	2	74	51	172	40
C fO/B	37	10–20	7	35	16	174	64
C fO/B	37	31	1	86	38		
C fO/B	10	10–20	3	106	51	215	64
C O/fB	37	15	1	186		266	
C O/fB	10	16	1	149		171	
CO	37	10–20	6	21		134	
CO	10	10–30	3	44	71	181	
COb	37	16	1	132		171	
COb	10	10–30	4	94	89	181	
R	10	10–20	4	173	60	235	
R	10	30–60	3	174	60	317	90
Rb	10	10–20	2	289	104	295	64
Rb	10	30–60	2	172	81		
R fO/B	10	10–20	2	155	48	190	
R fO/B	10	30–60	2	81	24		
R O/fB	10	20–40	2	244	91	190	
RO	10	20–40	2	113	87		
Rpy	10	20–30	2	267	93	135	92
Rpy fO/B	10	20–30	2	196	85	135	92
Rpy O/fB	10	30	1	135	52		
RpyO	10	20–30	4	148	73		

* The figures by 10,000 represent the number of amebae per Florence flask culture. The figures for organism *t* represent milliliters of gas per flask produced during the 72-hour period of incubation.

Key. C = cotton-stoppered stored medium.
Cb = same, boiled to drive out oxygen.
C fO/B = cotton-stoppered stored medium, overlay poured off, fresh overlay added to base.
C O/fB = cotton-stoppered stored overlay on fresh base.
CO = cotton-stoppered stored overlay as a wholly liquid medium.
COb = same, boiled.
R = rubber-stoppered stored medium.
Rb = same, boiled.
R fO/B = rubber-stoppered stored medium, overlay poured off, fresh overlay added to base.
R O/fB = rubber-stoppered stored overlay on fresh base.
RO = rubber-stoppered stored overlay as wholly liquid medium.
Rpy = rubber-stoppered stored medium with residual oxygen adsorbed by pyrogallic acid.
Rpy fO/B = same, overlay poured off, fresh overlay added to base.
Rpy O/fB = stored overlay on fresh base.
RpyO = stored overlay as wholly liquid medium.

sary for the growth of the ameba are destroyed by oxidation. Taking up this question, with the observations of Rees and Reardon as a point of departure, von Brand *et al.* (1946) studied the diffusion of reducing and nitrogenous substances in an egg medium (p. 239 and Table II) in the presence of a single bacterium (organism *t*). It could well be that the oxidation of the medium does not affect the ameba directly, but rather a substance needed by it. The diffusion of reducing substances, mostly glucose, from the solid to the liquid phase of the medium (Figs. 1 and 2)

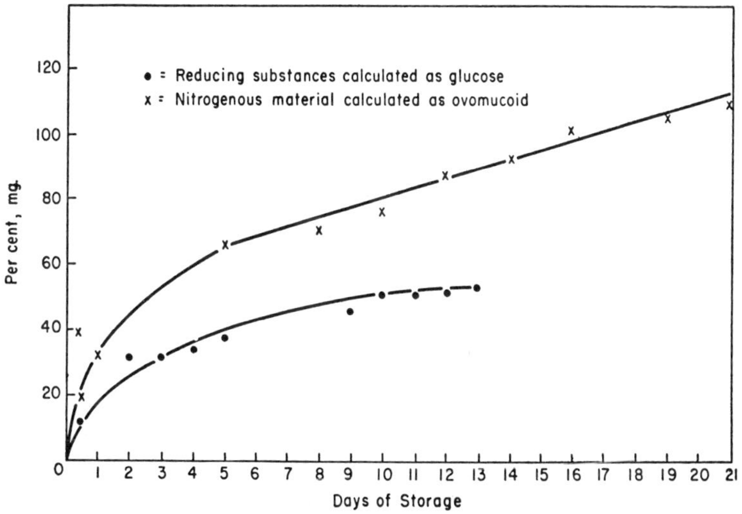

FIG. 2. Occurrence in milligram per cent of reducing substances (calculated as glucose), and of nitrogenous material (calculated as ovomucoid) in the fluid phase of egg white medium during specified periods of storage at 10° C. Each point on the reducing substances curve is the mean value of 5 flasks, on the ovomucoid curve of 2 flasks. (After von Brand *et al.*, 1946.)

is readily observable. The rate of diffusion varies with the temperature: it is greater at 37° than at 10° C. Nitrogenous substances also diffuse, but more slowly (Figs. 1 and 2). Moreover, spontaneous oxidations occur gradually which might inactivate substances indispensable for growth. In any case, all these factors have to be taken into account in creating a medium suitable for *E. histolytica.*

Other results point to similar conclusions. Chang (1946), using several strains of *E. histolytica* and a potentiometer specially adapted for the purpose, made some direct measurements. The best growth occurred at potentials of −350 to −425 mv., less growth between −275 and −350 mv., becoming scantier with increase of potential, then becoming extinct at −50 to −150 mv. The longevity of cultures is also related to the E_h, as already shown by Jacobs. As the lowering of the oxidation-reduction potential is carried out mostly by bacteria, this explains at least in part why they are indispensable (Fig. 3).

Schaffer and Frye (1948) and Schaffer *et al.* (1948, 1949) came to an analogous conclusion by a different route. In using a highly complex medium (to which later reference will be made, see p. 245) in which bac-

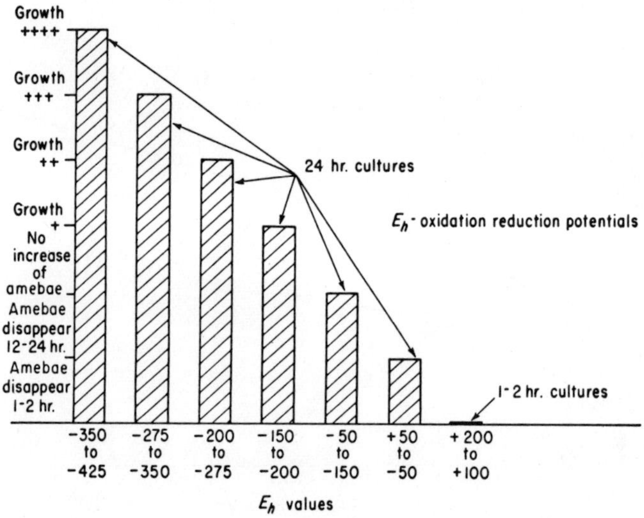

FIG. 3. Relation of oxidation-reduction potentials to growth of *E. histolytica* in culture. pH = 6.6–6.8. (After Chang, 1946.)

teria were very rare, these authors noted the favorable action of anaerobiosis on multiplication: "after 48 hours, some evidence of multiplication of the amebae was observed in tubes which were incubated anaerobically . . ." (Schaffer *et al.,* 1948) and, further on: "No evidence of multiplication of the amebae was observed in the tubes incubated aerobically." Also, the presence of oxygen is fatal to cultures: "The third passage failed, however, when the anaerobic jar developed a leak permitting oxidation of the substrates of the culture." Even though oxygenation in this case was accidental, the fact remains that in the aggregate the examples cited show that *E. histolytica* lives only in a very poorly oxygenated environment. This important question of oxido-reduction potential has just been carefully discussed in a recent paper by Jacobs (1950b).

It is seen therefore that one of the prerequisites for the purification of the dysentery ameba should be provision of a medium which had an oxidation-reduction potential such that bacteria as well as reducers could be excluded without harm.

B. The Need for Cholesterol

As mentioned before (p. 237), Snyder and Meleney (1942, 1943) raised the question of the requirement of a specific substance of a cholesterol-like nature, although Dopter and Deschiens (1938) had not been able to demonstrate a clear-cut effect. After obtaining fairly regular cultures with a medium based on an infusion of coagulated egg with rice starch added, Snyder *et al.* stated that neither washed bacterial cells nor culture filtrates were active alone. The living washed bacteria, which could not be replaced by killed bacteria, could be replaced by cysteine or other reducing agents (see p. 238). As for the culture filtrate, it could only be replaced by Ringer peptone in combination with cholesterol. The table in Snyder and Meleney's paper (1943, p. 283), in which the necessary elements for the multiplication of the ameba are listed, includes, besides a reduced oxygen tension obtained in mixed cultures by means of the bacterial population* and, in their experiments, by means of reducers (cysteine or others), cholesterol also. The need for other factors is suggested but only as a hypothesis.

More recently, experiments carried out with monobacterial cultures have confirmed the results of Snyder and Meleney. In the course of several years of methodical research, Rees and his collaborators have obtained mixed and single bacterial cultures of amebae and have begun the study of the nutritional needs of the ameba. In 1941 Rees succeeded in separating the cysts from the accompanying flora by means of micromanipulation and grew them in the presence of an unidentified bacterium which was designated as "organism *t*." Among the bacteria capable of filling this function it seemed to the authors to be the most suitable. It is a small rod-like microaerophile bacterium which is extremely pleomorphic when cultured in the egg media currently used for amebae. It is gram-negative and can survive heating for 15 minutes at 80° C. Organism *t* produces hydrogen in L.E.S. (Locke-Egg-Serum) medium, thus creating an oxidation-reduction potential favorable for the ameba. By using this

* According to a personal communication from Meleney (in Schaffer *et al.*, 1948), the bacterial flora seems to be mainly composed of a diplobacillus often appearing in chains.

organism in their media, Rees *et al.* (1944) showed the need of cholesterol for *E. histolytica.*

The medium consisted essentially of a solid phase, made with either coagulated egg (whole or either white or yolk) or with commercial coagulated albumin, and a liquid phase which consisted of glucose-Locke. A small amount of rice starch completed the medium. The cholesterol (Merck) was suspended in distilled water, sterilized, shaken with glass beads, and added to obtain a final concentration of 0.1 mg./ml. An autoclaved solution of Na cholesteryl sulfate in distilled water was also used (1µg./ml.).

In spite of the complexity of the medium and the presence of organism *t,* certain conclusions were drawn which deserve confirmation. These are shown in Table III. The B vitamins are inactive in the absence of cholesterol, and cholesterol is inactive without these same vitamins. When the basal medium contained whole egg, the culture was very heavy. When it contained egg white, a certain number of vitamins had to be added (thiamine, riboflavin, nicotinic acid, pyridoxine, pantothenic acid, *p*-aminobenzoic acid, inositol, choline, and cholesterol) to obtain a culture comparable to the one on whole egg medium. Other substances such as amino acids and purine bases did not improve the medium. Finally, the associated bacteria had no need for cholesterol, but they did require vitamins. Cholesterol is therefore a growth factor for the ameba only. As for the bacterium, it did not supply all the factors needed by the ameba.

A medium of egg yolk alone—with the addition of the above vitamins and cholesterol—did not permit appreciable growth. The egg white must also contain an indispensable element. In summary, whole egg constituted a satisfactory medium, and cholesterol in an egg white medium with the above-mentioned vitamins allowed multiplication of the ameba.

It would appear likely therefore that cholesterol plays a significant part in the nutrition of the dysentery ameba. The requirement for cholesterol is therefore common to it and *Trichomonas* (see p. 162).

Other factors found in egg besides cholesterol are equally important since the whole egg, adequate by itself, could be replaced by the white enriched with cholesterol and vitamins, but not by the yolk enriched with

---

* Based on the combined counts of 2 to 3 workers on the pooled harvests from 5 flasks.

Components of solutions:

Solution I. Thiamine hydrochloride, riboflavin, nicotinic acid, pyridoxine hydrochloride. calcium pantothenate, choline, inositol, and *p*-aminobenzoic acid.
Solution II. Biotin, β-alanine, and casein hydrolyzate.
Solution III. Adenine, guanine, thymine, uracil, and salt solution B.
Solution IV. Thiamine hydrochloride, riboflavin, and nicotinic acid.
Solution V. Pyridoxine hydrochloride, and calcium pantothenate.
Solution VI. Choline, inositol, and *p*-aminobenzoic acid.

TABLE III. Comparative Yields of *Entamoeba histolytica* from Various Media in Florence Flask Cultures
(After Rees *et al.*, 1944)

Composition of medium	Designation	Number of tests	Harvest of amebae in tens of thousands per flask*		
			Minimum	Maximum	Average
Whole egg	WE	20	24	274	153.0
Egg white	A	20	1	45	20.5
Egg white enriched with the following:					
Cholesterol	AC	4	11	55	25.0
Lecithin	AL	1			12.0
Cholesterol and lecithin	ACL	2	10	23	16.5
Vitamins, amino acids, and purine bases (Solutions I, II, and III)	AF_1	3	10	24	19.0
Vitamins and purine bases (Solutions I and III)	AF_2	2	28	84	56.0
Vitamins, amino acids, and purine bases (Solutions II and III)	AF_3	2	16	52	34.0
Vitamins and amino acids (Solutions I and II)	AF_4	1			28.0
Vitamins (Solution I)	AF_5	2	28	54	41.0
Vitamins, amino acids, purine bases, and cholesterol (Solutions I, II, and III)	AF_1C	3	99	171	128.0
As above plus lecithin	AF_1CL	6	53	169	96.0
Vitamins, purine bases and cholesterol (Solutions I and III)	AF_2C	2	77	197	137.0
As above plus lecithin	AF_2CL	1			110.0
Vitamins, amino acids, purine bases, and cholesterol (Solutions II and III)	AF_3C	2	52	57	54.5
As above plus lecithin	AF_3CL	1			64.0
Vitamins, amino acids, and cholesterol (Solutions I and II)	AF_4C	3	81	133	109.0
As above plus lecithin	AF_4CL	2	25	50	37.5
Vitamins and cholesterol (Solution I)	AF_5C	9	90	160	125.0
Vitamins and cholesterol (Solution IV)	AF_6C	4	65	115	93.0
Vitamins and cholesterol (Solution V)	AF_7C	1			17.0
Vitamins and cholesterol (Solution VI)	AF_8C	1			19.0
Commercial "glassy" albumin enriched with vitamins and cholesterol (Solution I)	A_2F_5C	1			9.0
Egg yolk	Y	12	7	51	23.0
Egg yolk enriched with the following:					
Vitamins, amino acids, purine bases, and cholesterol (Solutions I, II, and III)	YF_1C	1			14.0
Vitamins and cholesterol (Solution I)	YF_5C	4	2	21	11.0
Ovalbumin	YO_1	1			7.0
Ovomucin	YO_2	1			56.0
Ovomucoid	YO_3	1			43.0
Ovalbumin, ovomucin, and ovomucoid	$YO_1O_2O_3$	2	27	73	72.5
Commercial powdered albumin	YA_1	1			18.0
Commercial "glassy" albumin	YA_2	1			53.0
Difco Bacto-Peptone	YP	1			18.0
Wholly liquid infusate of egg albumin, vitamins, and cholesterol (Solution I)	IF_5C	1			70.0

these same substances (see Table III). Therefore other necessary substances must be present in the egg white which organism *t*, always present, could not furnish to the ameba. Rees and Reardon (1945) have tried without success to elucidate the nature of these substances. In view of the importance of oxygen tension revealed in the course of these studies, they concluded that it was likely that the white of egg "probably acting in conjunction with organism *t* is concerned in the oxygen potential of the medium" which is perfectly plausible.

Schaffer *et al.* (1948, 1949) took up this problem again, trying first to purify some cultures with the help of antibiotics. Using a culture isolated by Dobell from a macaque and maintained since 1934, they first defined the associated flora: three aerobic gram-negative bacteria, one gram-positive bacterium, and an anaerobic gram-negative streptobacillus. Penicillin and streptomycin inhibited the bacteria without affecting the amebae. They did find, however, fewer amebae in the presence of antibiotics without their multiplication being inhibited: "there were consistently fewer amebae in the antibiotic treated tubes, but no marked inhibition of multiplication was indicated" (Schaffer and Frye, 1948). This decrease in the number of amebae does not necessarily imply a sensitivity to antibiotics; it could merely be due to a decrease in the associated flora. In any case, one can eliminate most of the bacteria with the combined use of antibiotics and heat, with the exception of the streptobacillus, which apparently plays an important role.

Finally, the dysentery ameba can be maintained in a medium as yet very complex, but not containing any actively multiplying bacteria. This medium consists essentially of:

1. The supernatant liquid from a culture of the streptobacillus in thioglycolate medium (2.5 ml.).
2. Physiological saline (2.0 ml. of 0.85 per cent NaCl).
3. Normal horse serum (0.25 ml.).
4. Five hundred units of Na penicillin G.

The thioglycolate medium used to grow the bacteria must contain glucose, which apparently is indispensable for the ameba. But the study of its nutrition has hardly been touched; it still presents many difficulties.

Hansen and Anderson (1948) define the needs for growth of *E. histolytica* in association with organism *t* according to an outline reproduced here (p. 247). Their medium as given in Table IV contains essentially inorganic salts, twelve amino acids, ten synthetic vitamins, nucleic acid, cholesterol, and rice starch.

Nutritional Requirements of *Entamoeba histolytica**
(After Hansen and Anderson, 1948)

Protein	Nitrogenous bases	Vitamins	Carbohydrate and sterol
Solid/Liquid Media			
	Egg slope, or agar containing liver, overlaid by saline containing serum or egg albumin		Rice starch
Liquid Media from Natural Products			
	Dehydrated proteose peptone	Enzyme digest of liver	Rice starch cholesterol
Essentially Synthetic Liquid Media			
Twelve amino acids	Nucleic acid	Ten synthetic vitamins of B complex	Rice starch cholesterol

* In association with organism *t*, isolated by Dr. C. W. Rees, National Institute of Health, Bethesda, Md.

TABLE IV. Composition of an Essentially Synthetic Medium (quantities for 100 ml. Total)
(After Hansen and Anderson, 1948)

Trace minerals		
$MgSO_4 \cdot 7H_2O$	10	μg.
$CaCl_2 \cdot 2H_2O$	10	
$FeCl_3 \cdot 6H_2O$	125	
$MnCl_2 \cdot 4H_2O$	5	
$ZnCl_2$	5	
Amino acids		
L-(+)-Lysine HCl	5	mg.
L-(−)-Histidine $HClH_2O$	5	
L-(−)-Leucine	5	
L-(−)-Tryptophan	5	
L-(+)-Arginine HCl	5	
L-(+)-Cysteine $HClH_2O$	5	
D,L-Isoleucine	10	
D,L-Methionine	10	
D,L-Phenylalanine	10	
D,L-Valine	10	
D,L-Serine	10	
Glycine	10	

Buffered saline		
NaCl	0.5	g.
$Na_2HPO_4 \cdot 7H_2O$	1.0	
KH_2PO_4	0.1	
Vitamins		
Thiamine*	20	μg.
Riboflavin*	40	
Calcium pantothenate	40	
Nicotinic acid	40	
Pyridoxine HCl	80	
p-Aminobenzoic acid	40	
i-Inositol	40	
Choline	80	
Lactobacillus casei factor†	54	
Biotin	2	
Other nutrients		
Nucleic acid‡	5.0	mg.
Cholesterol sodium sulfate§	0.1	
Cholesterol	4.0	
Rice powder-Difco (approx.)	100.0	

* Thiamine and riboflavin may be completely inactivated during preparation.
† Crystalline fraction; obtained from the Lederle Laboratories.
‡ From yeast; obtained from the Schwarz Laboratories.
§ Recommended by Dr. C. W. Rees.

As mentioned before (p. 239), these results show above all the importance of control of the oxygen tension in these cultures.

In conclusion, one feature stands out clearly from most of the work that has been done: the importance of the oxidation-reduction potential of the medium. Although it would seem that the ameba apparently is better fitted for life at reduced oxygen tensions (Snyder and Meleney, Jacobs, Chang, Schaffer and Frye), it is also possible that oxidation of the medium acts only indirectly by changing some substances indispensable to the ameba (Rees *el al.*, von Brand *et al.*). This is a question of

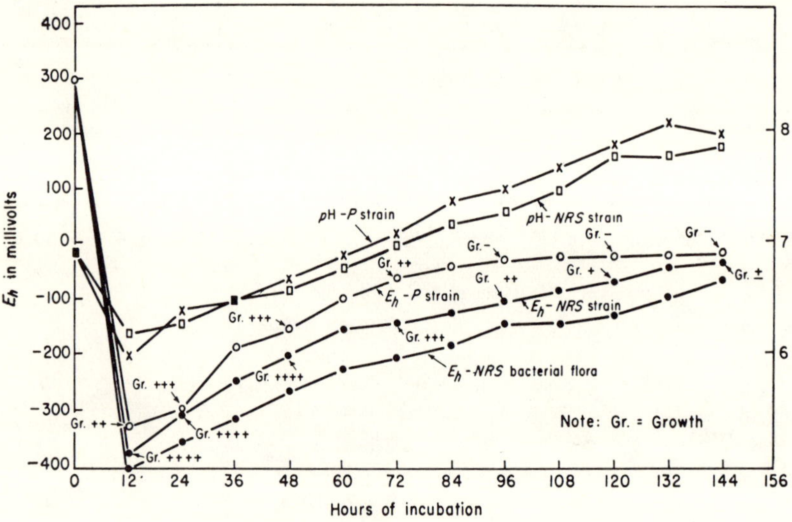

FIG. 4. Relation of oxidation-reduction potentials to the life span of a culture of *E. histolytica*. (After Chang, 1946.)

primary importance in obtaining a bacteriologically pure culture and, as such, should not be slighted.

From the point of view of its nutrition, as such, the need of cholesterol for *E. histolytica* seems to have been proved. Obviously this will have to be confirmed under various conditions (absence of bacteria, less complex media), and certain aspects of the data acquired will have to be extended through the study of the activity of various sterols.

It is known that serum, which permits *in vitro* growth of certain bacteria, acts sometimes by neutralizing the fatty acids in the medium. It can then be replaced by cholesterol, whose ability to fix fatty acids is well known. Certain organisms, such as the whooping cough bacterium *He-*

mophilus pertussis, are very sensitive to fatty acids. It is also known that in certain instances the concentrations which cause inhibition are very close to those in which the fatty acid acts as a growth substance (see on this subject papers of Kodicek, 1948; Pollock, 1947, 1949; A. Lwoff, 1947).

It happens too that certain bacteria, *Hemophilus pertussis,* for example, produce fatty acids which inhibit their own growth as well as that of other bacteria. One might therefore ask if cholesterol did not neutralize in the ameba cultures the fatty acids resulting from the metabolism of organism *t.* Indeed, Griffin and McCarten (1949), after having confirmed the indispensability of cholesterol, stated clearly that in a medium based on serum and liver (LSB medium), serum could be replaced by cholesteryl oleate or palmitate. Very low concentrations of oleic acid (0.01–0.03 mg./ml.) have a stimulating effect on an ameba of reptiles, *Endameba terrapinae,* and a toxic effect at higher concentrations. A mixture of cholesterol + oleic acid nevertheless would not permit multiplication.

Whatever may be the true role of cholesterol in the metabolism of microorganisms, it is interesting to compare the case of the dysentery ameba to that of *Trichomonas* which has previously been discussed (p. 162) and to which we refer the reader.

REFERENCES

Boeck, W. C., and Drbohlav, J. (1925a). *Am. J. Hyg.* **5**, 371.

Boeck, W. C., and Drbohlav, J. (1925b). *Trans. Roy. Soc. Trop. Med. Hyg.* **18**, 238.

Bradin, J. L., Jr., and Hansen, E. L. (1948). *J. Parasitol.* **34** (suppl.), 11.

Chang, S. L. (1946). *Parasitology* **37**, 101.

Cleveland, L. R., and Sanders, E. P. (1930). *Science* **72**, 149.

DeLamater, J. N., and Hallman, F. A. (1947). *Proc. Soc. Exptl. Biol. Med.,* **65**, 26.

Dobell, C. (1919). The Amoebae Living in Man. John Bale Sons & Danielsson Ltd., London.

Dobell, C., and Laidlaw, P. P. (1926). *Parasitology* **18**, 283.

Dopter, C., and Deschiens, R. (1938). *Compt. rend. soc. biol.,* **129**, 626, 628.

Friedrichs, A. V., and Harris, W. H. (1929). *Proc. Soc. Exptl. Biol. Med.* **27**, 90.

Fuller, F. W., and Faust, E. C. (1949). *Science* **110**, 509.

Griffin, A. M., and McCarten, W. G. (1949). *Proc. Soc. Exptl. Biol. Med.* **72**, 645.

Griffin, A. M., and McCarten, W. G. (1950). *J. Parasitol.* **36**, 238, 253.

Griffin, A. M., and Michini, L. J. (1950). *J. Parasitol.* **36**, 247.

Hallman, F. A., Michaelson, J. B., and DeLamater, J. N. (1950). *Am. J. Trop. Med.,* **30**, 363.

Hansen, E. L. (1950), *J. Lab. Clin. Med.,* **35**, 308.

Hansen, E. L., and Anderson, H. H. (1948). *Parasitology* **39**, 69.
Jacobs, L. (1941). *J. Parasitol.* **27** (suppl.), 31.
Jacobs, L. (1947). *Am. J. Hyg.* **46**, 172.
Jacobs, L. (1950a). *J. Parasitol.*, **36**, 128.
Jacobs, L. (1950b). *Am. J. Trop. Med.* **30**, 803.
Jahn, T. L. (1934). *Cold Spring Harbor Symposia Quant. Biol.* **2**, 167.
Kodicek, E. (1948). *Bull. soc. chim. biol.* **30**, 946.
Lamy, L. (1944). *Ann. inst. Pasteur* **70**, 318.
Lamy, L. (1948). *Compt. rend.* **226**, 401, 2021.
Lamy, L. (1949). *Thèse Faculté des Sciences,* University of Paris.
Lwoff, A. (1947). *Ann. Inst. Pasteur* **73**, 735.
Pautrizel, R., Pintaud, C., and Masquelier, J. (1949). *Bull. trav. soc. pharm. Bordeaux* **87**, 83.
Pollock, M. R. (1947). *Brit. J. Exptl. Path.* **28**, 295.
Pollock, M. R., Wainwright, S. D., and Manson, E. E. D. (1949). *J. Path. Bact.* **61**, 274.
Reardon, L. V., and Bartgis, I. L. (1949). *J. Parasitol.* **25** (suppl.), 13.
Rees, C. W., Bozichevich, J., Reardon, L. V., and Daft, F. S. (1944). *Am. J. Trop. Med.* **24**, 189.
Rees, C. W., and Reardon, L. V. (1945). *Am. J. Trop. Med.* **25**, 109.
Rees, C. W., Reardon, L. V., and Jacobs, L. (1941). *Am. J. Trop. Med.* **21**, 695.
Rees, C. W., Reardon, L. V., Jacobs, L., and Jones, F. (1941). *Am. J. Trop. Med.* **21**, 567.
Schaffer, J. C., and Frye, W. W. (1948). *Am. J. Hyg.* **47**, 214.
Schaffer, J. C., Ryden, F. W., and Frye, W. W. (1948). *Am. J. Hyg.* **47**, 345.
Schaffer, J. C., Ryden, F. W., and Frye, W. W. (1949). *Am. J. Hyg.* **49**, 127.
Schaffer, J. C., Walton, J. G., and Frye, W. W. (1948). *Am. J. Hyg.* **47**, 222.
Snyder, T. L., and Meleney, H. E. (1942). *J. Parasitol.* **28** (suppl.), II.
Snyder, T. L., and Meleney, H. E. (1943). *J. Parasitol.* **29**, 278.
von Brand, T., Rees, C. W., Jacobs, L., and Reardon, L. V. (1942). *J. Parasitol.* **28**, 11.
von Brand, T., Rees, Ch. W., Jacobs, L., and Reardon, L. V. (1943). *Am. J. Hyg.* **37**, 310.
von Brand, T., Rees, C. W., Reardon, L. V., and Simpson, W. F. (1946). *J. Parasitol.* **32**, 190.

Biochemistry of *Plasmodium* and the Influence of Antimalarials

RALPH W. McKEE

Department of Biological Chemistry, Harvard Medical School, Boston, Massachusetts

CONTENTS

I. Introduction

A. Boundaries of Subject

Blood is the life-giving fluid which determines the type of existence of all its environmental cells, the organ and structural tissue cells as well as the circulating erythrocytes. Blood is then of paramount importance in a discussion of the intricate mode of life of the blood protozoa, *Plasmodium,* or malarial parasites which infest red blood cells. These organisms, which belong to the phylum, *Protozoa,* class *Sporozoa,* order *Haemosporodia* and genus *Plasmodium,* include human, simian, avian, saurian, and rodent species. Biochemical investigators have concentrated the major portion of their studies on the asexual blood forms of only a few of these *Plasmodium,* so too these same few will be the ones of main consideration in this treatise. The species to be discussed are the human malarial parasites, *Plasmodium falciparum, P. malariae* and *P. vivax,* the simian species, *P. cynomolgi* and *P. knowlesi,* the avian species, *P. cathemerium, P. elongatum, P. gallinaceum, P. lophuruae,* and *P. praecox* (relictum) and the rodent parasite, *P. berghei.*

It is an integration of our present knowledge regarding the biological processes of the malarial parasites which will be presented in this chapter. The author in so doing has drawn considerable information from the reviews of this subject by Geiman (1943, 1948a), Geiman and McKee (1948), Maegraith (1948), and Moulder (1948), and from the detailed experimental results of a host of other workers to whom references will be made.

Concerning the action of chemotherapeutic agents on the biochemistry of *Plasmodium,* space will allow only a review of our knowledge of the influence of antimalarials on the metabolic processes of the parasites. No attempt will be made to deal with the diagnosis or the prophylactic and therapeutic treatments of malaria. These aspects of the malaria problem have been dealt with in detail elsewhere by Geiman (1944), Russell, West, and Manwell (1946), Boyd (1949), and by many others.

During the past seven years, due principally to the stimulus given by World War II, rapid strides have been made in our knowledge of plasmodia. However, the need for even more basic knowledge regarding the complicated life processes of malarial parasites has been realized from the empirical testing of some 15,000–20,000 antimalarial compounds which yielded only a few satisfactorily effective drugs. This lack of adequate data regarding the metabolic processes of the *Plasmodium* is particularly apparent for the mosquito and exoerythrocytic forms. The gaining of further basic biological information will lead eventually, not only to a more complete explanation of the mechanism of antimalarial action, but also to a more rational consideration of the biochemical and physiological processes and a more effective chemotherapeutic treatment of other parasitic organisms.

B. Life Cycle of Malarial Parasites

So that one can gain a better understanding of the biochemical processes of the parasites of malaria, it is necessary first to consider the developmental stages through which the organisms pass during their life cycle. The parasite, to survive and propagate itself, must traverse a series of transitory stages, periods during which not only morphological changes but also metabolic adaptations take place. Many of the malarial parasites are known to pass through three such stages: one in the *Anopheles* or *Culex* mosquitoes, the second in the tissue cells (exoerythrocytic) of the host animal, and lastly the asexual erythrocytic or blood stages of the host. Although these stages in the life cycle of malarial parasites have been well described by Wenyon (1926) and Huff (1947), a brief review will be given here in order to present a rounded picture of the problems involved.

Mosquitoes, the only insect vectors known for plasmodia, may become infected when they ingest blood from the animal host which carry the sexual forms of the parasites (gametocytes). In the stomach of the mosquito the microgametocyte produces the microgamete which penetrates the macrogametocyte and fuses with its nucleus. The resulting fertilized cell (zygote) elongates and becomes a mobile ookinete. This form then penetrates the lining of the mosquito's stomach where the oocyst develops. The oocyst, which undergoes nuclear division and the differentiation of sporozoites, ruptures freeing the sporozoites into the body cavity of the mosquito. The sporozoites then penetrate the salivary glands of the insect from which they are ejected at the next feeding. This cycle in the *Anopheles* mosquito requires from 16 to 35 days, depending upon the species of *Plasmodium* and the body temperature of the mosquito, the rates of biochemical processes being influenced by temperature.

Within the animal host, certain species of parasites are known to undergo preerythocytic changes in preparation for the erythrocytic life under a different chemical and physical environment. James and Tate (1938) first discovered the exoerythrocytic development of *Plasmodium gallinaceum*. Huff and Coulston (1944) observed the complete exoerythrocytic development of *P. gallinaceum* in the skin of chickens near the site of sporozoite inoculation. The sporozoites undergo rapid growth and development through stages called cryptozoites and metacryptozoites. This process is completed in 36 to 48 hours with the formation of 75 to 150 merozoites per segmenter. The merozoites then undergo several subsequent generations in the lymphoid—macrophage cells and endothelial cells of the blood vessels. After 5 to 10 days the erythrocytic parasites gradually begin to appear in the circulating blood. Also preerythrocytic stages have been shown for *P. relictum* by Coulston, and Huff (1947), for *P. lophurae* by Huff, Coulston, Laird, and Porter (1947) and for *P. cathemerium* by Huff and Coulston (1948). Similar preerythrocytic forms of *P. cynomolgi* have been reported by Shortt and Garnham (1948) and by Hawking and Perry (1948) to be in the liver parenchymal cells of the monkey. More recently Shortt, Garnham, and Malamos (1948) have observed comparable forms for *P. vivax*. The time required for the development of the tissue forms varies with the species, due to differences in the biological processes involved. This time interval explains the fact that parasites do not appear in the circulating blood for a few days following the bite of the mosquito. This reservoir of tissue stages also furnishes an explanation for the relapses which occur with malarial infections.

The erythrocytic forms of the malarial parasites in the circulating blood undergo a series of changes within the red blood cell. The youngest form of the parasite is called a "ring." This is so named because of a central vacuole surrounded by a ring of cytoplasm containing a chromatin granule. The parasite then grows and develops though trophozoite and schizont stages followed by division of the chromatin to produce the fully grown segmenter. At a precise time, which varies between 24 and 72 hours with the different species, the segmenter ruptures, and the cell contents and new progeny (merozoites) are freed. These merozoites immediately and rapidly invade other red blood cells to start a new asexual erythrocytic cycle. During this period of growth within the erythrocytes a few sexual forms (gametocytes) differentiate. It is this form of the parasite which perpetuates the next mosquito cycle and makes possible the transmission of the disease to other animals.

C. Biochemical Correlations

It is concerning the period of growth and multiplication within red cells that most of our biochemical knowledge has been gained and toward which the discussions of this chapter are directed. It is indeed unfortunate that our knowledge regarding the metabolic processes of the sporozoite and exoerythrocytic stages is not more extensive, because control of the organism for the prevention of the disease should be directed at these earlier stages.

During this erythrocytic cycle numerous biochemical reactions take place. Some of the more important life processes which have been investigated and which will be presented in detail in this chapter are summarized here. Glucose, the principal nutrient for the parasitized red cell, undergoes phosphorolative degradation with the production of pyruvate and lactate. This process together with subsequent oxidation of a portion of the lactate produces the required energy for the parasitic cell. Excess lactate formation by the parasite requires the host animal's body for its metabolism. The oxygen for the oxidative processes of the organism is supplied by the oxyhemoglobin present in the host red cells. The quantities of this gas which are utilized vary with the species of parasite and increase as the parasite grows. The hemoglobin of the erythrocyte is split and the globin portion degraded by enzymes of the contained parasite. A portion of the formed amino acids and peptides are utilized for the synthesis of parasite protein. Hematin is deposited as granules in the cytoplasm of the parasite and utilized for pigment production. Lipids are synthesized by the parasitized cell in large quantities. All these processes are essential for the growth and multiplication of the organism.

The parasite not only consumes components of the red cell but draws on the surrounding plasma for many of its nutrients and metabolites—glucose, methionine, *p*-aminobenzoic acid, riboflavin, pantothenate, biotin, certain purines and pyrimidines, phosphate, etc. The extent of this list is determined by the requirements of the particular species of parasite, which have not been determined completely for any *Plasmodium*.

The integrated metabolic processes of the parasite are of the utmost importance to the synthetic processes of the cell. For example, the synthesis of protein depends on oxidative carbohydrate metabolism, and the

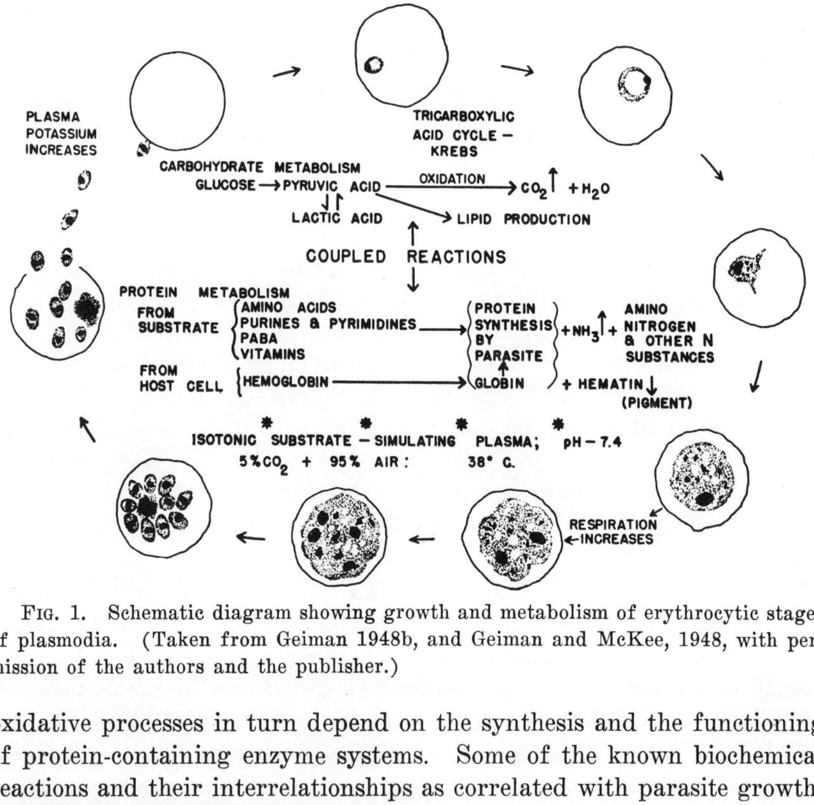

FIG. 1. Schematic diagram showing growth and metabolism of erythrocytic stages of plasmodia. (Taken from Geiman 1948b, and Geiman and McKee, 1948, with permission of the authors and the publisher.)

oxidative processes in turn depend on the synthesis and the functioning of protein-containing enzyme systems. Some of the known biochemical reactions and their interrelationships as correlated with parasite growth, are pictured diagrammatically in Fig. 1.

In addition to these intimate chemical agents there are other factors which influence the growth and control of malarial parasites. The host animal is able to synthesize humoral materials (hormones, antibodies, etc.) which markedly alter the course of malarial infections *in vivo*.

It is because of this close interplay of the life processes that it is possible for a variety of basically different chemicals to be effective as antimalarials (i.e., quinine, atabrine, paludrine, pentaquine, naphthoquinones, and sulfanilamide). Although each compound produces the same end result, namely death of the parasite, this does not mean that all the drugs act initially in the same manner. This merely means that an attack at any point in the highly integrated biological processes results in a chainlike series of detrimental effects, ultimately resulting in death to the parasite. As will be indicated later there are many different metabolic processes and life stages which antimalarial compounds attack.

One must not draw the inference from this brief attempt at a correlation of the biochemical processes of the parasite that all species of *Plasmodium* possess identical or even similar metabolic processes. Differences do exist as indicated by precise requirements for metabolites and specific host-parasite relationships. It is these variations in biochemical reactions among the parasite species that make for differences in morphology, pathogenicity, and antigenic reactions.

II. Chemical and Metabolic Constituents of Parasitized Erythrocytes and of Plasma

A. Inorganic Composition of Normal and Parasitized Blood

Since the malarial parasite inhabits the mammalian host red blood cell during most of its asexual life cycle and draws on the host cell and surrounding plasma for its building blocks, a study of the chemical and metabolic properties of the environmental erythrocytes and plasma should aid in the obtaining of an understanding of the metabolic processes of the parasite itself.

The early major investigations of the inorganic components of blood containing malarial parasites were concerned principally with potassium. Pinelli (1929) detected a marked increase in serum potassium during the fever period of tertian and subtertian malaria. Zwemer, Sims, and Coggeshall (1940) determined that the paroxysm and fever in human malaria coincide with sporulation of the parasite, red cell rupture, and the simultaneous liberation into the plasma of the contents of the parasitized erythrocytes. The resulting serum or plasma potassium increases are accompanied by the chill and precede the rise in body temperature. It is not generally believed, however, that the potassium increases are responsible for the fever. This is because at low infections too few erythrocytes are destroyed to cause large plasma increases of potassium.

In Rhesus monkeys infected with *Plasmodium knowlesi* much greater plasma potassium increases may be encountered because of the higher infections obtained. Zwemer *et al.* (1940) obtained increases in the serum potassium values from 5.0 to 8.9 millimoles per liter. McKee, Ormsbee, Anfinsen, Geiman, and Ball (1946) using *P. knowlesi* in monkeys found a similar marked increase in one monkey at the time of parasite segmentation. However, the average plasma potassium increase for seven infected monkeys was only from 5.1 to 6.4 millimoles per liter, the blood having been drawn from the animals 6 to 8 hours before the start of parasite segmentation. These and other inorganic constituents of normal and parasitic blood are shown in Table I. Velick and Scudder (1940)

TABLE I. Inorganic Constituents of Normal and Parasitized Monkey Bloods

(Taken from McKee *et al.*, 1946, with the permission of the authors and the publisher)

Inorganic constituent*	Whole blood		Plasma		Red cell†	
	Normal	Parasitized	Normal	Parasitized	Normal	Parasitized
Sodium	93.9(5)	127.8(3)	157.7(5)	165.2(3)	8.3(5)	11.8(3)
Potassium	52.3(8)	32.2(7)	5.1(8)	6.4(7)	113.1(8)	121.4(7)
Chlorides	82.5(4)	88.9(3)	102.2(4)	108.6(3)	55.5(4)	54.2(3)
Inorganic phosphate	1.5(16)	1.2(15)	2.1(13)	1.4(15)	0.9(13)	0.5(15)‡
Bicarbonate§	23.6(2)		28.0(2)			
pH	7.47(2)					

* Values are expressed as millimoles per liter. The number of different animals furnishing blood samples is given in parentheses.

† Calculated from whole blood and plasma data with the aid of hematocrit values.

‡ For the fifteen parasitized blood samples, the average hematocrit was 23, while for the thirteen normal bloods, the average was 43.

§ These determinations were kindly carried out by Dr. M. E. Krahl.

working with *P. cathemerium* infections found increases from 5.9 to 8.7 millimoles in the plasma of infected canaries. In ducks where there is less synchronicity of the parasitized cells there was only a slight increase, from 4.9 to 5.9 millimoles. With *P. lophurae* infections in chickens they found no increase in plasma potassium. The potassium values for plasma depend largely on the time when the blood is drawn for analysis and the degree of parasite synchronicity. The greatest potassium increases will be during the period when the segmentation of the parasites and rupture of the red cells is maximum. Due to the rapid rate of potassium excretion the blood values fall precipitously after cell rupture.

Flosi (1944) investigated the plasma potassium and sodium levels in

12 humans with *P. vivax, P. falciparum,* and mixed infections. In general there was observed an increase of potassium and a decrease in the sodium during the attack and immediately after treatment with quinine. The plasma potassium changes were similar to those observed by Zwemer *et al.* (1940). The plasma sodium values simultaneously dropped from 140.9 to 127.4 millimoles per liter. The plasma sodium values obtained by McKee *et al.* (1946) on normal and *P. knowlesi* infected monkeys showed a slight increase in the infected animals, from 157.7 to 165.2 millimoles. The blood, however, was drawn from 6 to 8 hours before segmentation took place.

Unlike sodium and potassium, the calcium content of serum appears to remain constant in animals with malarial infections. Serum calcium values have been determined by Wats and Das Gupta (1934) for normal and *P. knowlesi* infected Rhesus monkeys. They report serum values of 2.83 millimoles per liter for three normal monkeys and a similar value for three parasitized animals.

The chloride content of the blood and plasma of malarious patients has been variously reported as normal, reduced, and elevated. The truth can be gleaned only by careful consideration of the species of malaria and the time at which the blood is drawn for chemical analysis. Miyahara (1936) reported a reduction in the plasma chlorides of 100 patients with *P. vivax* infections. Wakeman and Morrell (1929) obtained normal serum chloride values. McKee *et al.* (1946) found elevated plasma values (102.2 to 108.6 millimoles per liter) in monkeys whose blood contained presegmenting *P. knowlesi* parasites.

One of the most interesting changes in inorganic composition of blood, from a metabolic standpoint, is the decrease in phosphorus. This has been reported by McKee *et al.* (1946) for *P. knowlesi* infections in *Macaca mulatta* monkeys (Table I). Fairley and Bromfield (1934), however, observed an increase in plasma inorganic phosphate of *P. knowlesi* infections in *M. mulatta.* Although these differences are difficult to reconcile, the time during the parasite cycle at which blood was drawn for phosphorus analysis may explain the apparent discrepancy. The need for phosphorus in the production of adenosinetriphosphate, nucleic acid, lipid and protein materials and the inability of *P. knowlesi* to grow in culture without phosphorus, all of which will be discussed in more detail later, are indicative of the utilization of phosphorus by parasitized cells both *in vitro* and *in vivo.*

Fairley and Bromfield (1934) report decreases in the carbon dioxide content of the plasma of humans infected with *P. vivax*. This change is probably due to the reduction in plasma bicarbonate caused by the excretion of acid metabolites produced by the parasites.

Kehar (1936) observed a distinct drop in the pH values of the blood of *M. mulatta* in the later stages of *P. knowlesi* infections. This drop to a pH value of 7.05 is understandable with the high infections obtained with *P. knowlesi* and the excessive quantities of lactic acid produced by the parasites.

By far the most extensive investigations on the ionic composition and

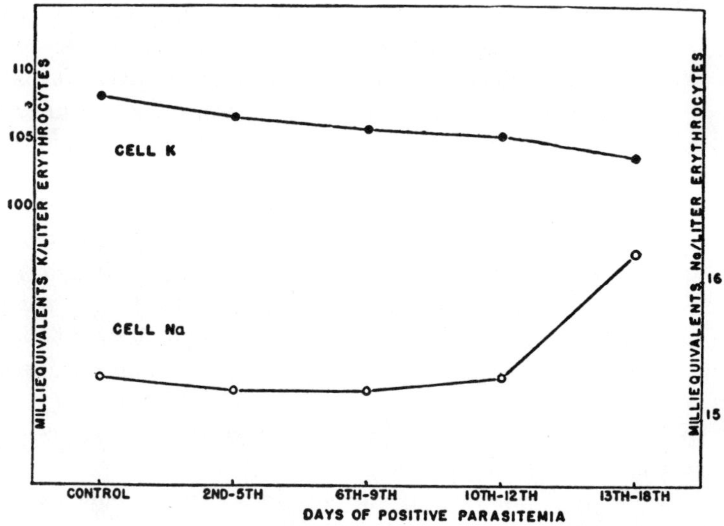

FIG. 2. Average Na and K concentration changes in erythrocytes. (Taken from Overman *et al.*, 1949, with the permission of the authors and the publisher.)

body fluid changes in man with infections of *P. vivax* and *P. falciparum* are those of Overman, Hill, and Wong (1949). Their findings parallel earlier observations in monkeys with *P. knowlesi* (Overman, 1946, Overman and Feldman, 1947; and Overman *et al.*, 1949). As stated by these workers, the general physiological picture exhibited by both monkey and man infected with malarial parasites is one of red cell destruction with in turn initiates a compensatory passage of tissue fluids and of protein into the blood. Toxic products of parasite metabolism or the effects of parasite metabolism on the host response cause changes which lead to abnormal cellular permeability of sodium and chloride. Cell permeability, as well as sweating in man, tend to cause an excessive loss of Na and Cl, which is partially prevented by increased renal tubular reabsorption. The authors postulate that these alterations in the intracellular ionic pattern may produce drastic changes in the functioning of cellular enzyme systems. These cellular changes persist for days or weeks after the parasitemia has been stopped by therapeutic treatment or by spontaneous remission. Overman *et al.* (1949) have observed a loss of erythrocyte K and an increase of red cell Na for the non-parasitic cells of patients infected with *P. vivax* and *P. falciparum* (Fig. 2). More recent work by Overman, Bass, Davis, and Golden (1949) indicates that exhaustion of the adrenal cortex, or an imbalance among the steroid hormones of the gland, allows the electrolyte shift. Overman, Hill, and Wong (1949) caution against attributing the fatal outcome of a malarial infection to a single phenomenon such as anemia or peripheral anoxia or even to alterations in cellular permeability to Na, Cl, or K. It is important to consider the total physiological response of the host and the accumulated effects.

B. Organic Components and Metabolic Constituents of Parasitized Red Cells

A consideration of the organic, as well as the inorganic constituents of parasitized blood, particularly of the infected erythrocytes, is of importance for piecing together our knowledge of malarial parasites into a picture of the overall metabolic functionings.

A decrease in the quantity of hemoglobin, as first shown by Chris-

TABLE II. Phospholipid Phosphorus and Fatty Acid Content of Normal and Parasitized (*Plasmodium knowlesi*) Monkey Red Blood Cells

(Adapted from Ball *et al.*, 1948, with the permission of the authors and the publisher)

	Blood sample number (1)	Parasites (2) %	Phospholipid P (3) mM per 5 × 10^{12} cells*	Fatty acids (4) mM per 5 × 10^{12} cells	Ratio (4) to (3)
Normal	1–7		1.78–2.08	3.9–5.2	2.17–2.75
Average	7		1.91	4.7	2.46
Parasitized	1	7.4	2.59	—	—
	3	17.3	2.70	7.3	2.70
	6	28.0	2.80	7.2	2.57
	9	40.6	4.0	9.4	2.35
	12	50.0	4.5	11.1	2.47
Average	5		3.5	8.75	2.52

* The choice of 5 × 10^{12} as the unit number of cells in this and the subsequent tables is an arbitrary one. It is, however, about the number resent in 1 liter of blood.

tophers and Fulton (1938), and the appearance of three new parasite proteins in the parasite *Plasmodium knowlesi* (Morrison and Jeskey, 1947) is of considerable signficance in the protein metabolism of the parasite. The released hematin is the precursor of the parasite's cytoplasmic pigment and a component of methaemalbumin present in the plasma of infected humans (Fairley and Bromfield, 1934).

Morrison and Jeskey (1948) report that *P. knowlesi* parasites contain, in addition to the three proteins, about 20 per cent of lipid on a dry weight basis. The lipid material consists principally of cholesterol and fatty acid-containing glycerides. A large part of the fatty acid they

TABLE III. Chemical and Metabolic Changes Occurring in Red Blood Cells as Number of Parasites increases *in Vivo*

(Taken from Ball *et al.*, 1948, with the permission of the authors and the publisher)

Monkey 71; 5.6 kilos; infected with *Plasmodium knowlesi.*

	11.30 A.M., Jan. 2, 1945		10.00 A.M., Jan. 8, 1945		12.00 noon, Jan. 9, 1945	
Blood sample, *cc.*	13		13		120*	
Red cells, per c.mm.	5.26×10^6		4.84×10^6		2.70×10^6	
White cells, per c.mm.	9800				11,000	
Hematocrit, %	39.8		37.4		21.8	
Parasites, %	0		17.3		45.0	
Rings			1.5		1.5	
Trophozoites, early			49.0		41.5	
Trophozoites, late			30.5		27.5	
Schizonts			15.0		26.0	
Segmenters			4.0		3.5	
	Total	Parasitized	Total	Parasitized	Total	Parasitized
	mM per hr. per 5×10^{12} cells					
O_2 consumption	0.60		5.65	29.8	17.0	37.1
Glucose consumption	0.53		4.87	25.6	15.43	33.7
Lactate consumption	0.00		3.71	21.4	13.80	30.7
Lactate accumulation	1.06		6.03		17.06	
	mM per 5×10^{12} cells					
Fatty acids	4.9		7.3	19.7	11.8	20.4
Total P	9.7		12.9	28.3	20.1	32.9
Acid-soluble P	6.50		5.9		6.6	
15 min.-hydrolyzable P	1.30		1.45	2.0	2.1	3.1
Phospholipid P	1.83		2.7	6.5	4.2	7.0
Nucleic acid P	1.4		4.3	18.2	9.3	19.0
	Mg. per 5×10^{12} cells					
Flavin-adenine dinucleotide	0.21		0.45	1.62	0.98	1.91

* Terminal.

believe to be oleic acid. From the work of Ball, McKee, Anfinsen, Cruz, and Geiman (1948) it is apparent that a fair proportion of this lipid is phospholipid, with a fatty acid to phosphorus ratio of 2.64. These data are summarized in Table II. These same investigators also analyzed normal and *P. knowlesi* parasitized cells for total, acid-soluble, 15-minute hydrolyzable, and nucleic acid phosphorus components the values of which are shown in Table III. A comparison of these values with those for normal, non-parasitized erythrocytes indicates a remarkable ability of the parasite to synthesize phosphorus-containing organic compounds, fatty acids and lipid materials, and nucleoproteins all of which are essential for the synthesis of protoplasm and the structural material for parasite substance. Unfortunately similar data are not available for other species of malarial parasites.

Although, as will be discussed in detail later, in terminal stages of *P. knowlesi* infections blood glucose and liver glycogen drop (Christophers and Fulton, 1938), no glycogen is produced by the parasites. Simultaneous with the utilization of the carbohydrate reserves of the body there is the appearance of lactic acid which although completely metabolized by the host in the early stages of the infection, accumulates in the terminal stages causing an increase in the acidity of the blood. Not only is glucose converted to lactate but also lactate and pyruvate are oxidized, the rate of which is increased as the parasites grow (Table IV). This is particularly true for *P. knowlesi* and *P. falciparum* (McKee, 1951). One component of certain oxidative enzyme systems, namely flavin-adenine dinucleotide, has been shown to increase in *P. knowlesi* infected monkey blood (Table III). Likewise the enzyme, lactic dehydrogenase, has been shown to increase in the parasitic erythrocyte (McKee *et al.*, 1946). Further discussion of oxidative systems and the importance of flavin-adenine dinucleotide will be given later.

Certain changes in the vitamin content of the blood are also known to take place in malarial infections. Trager (1943b) has shown an increase in the biotin activity of both red cells and plasma of chickens and ducks infected with *P. lophurae* and *P. cathemerium*. Marked decreases have been observed in the plasma ascorbic acid of patients infected with *P. vivax* (McKee, 1951) and in monkeys infected with *P. knowlesi* (McKee and Geiman, 1946) although slight increases were observed by Josephson *et al.* (1949) in chickens infected with *P. gallinaceum*. The importance of these changes to the growth of malarial parasites will be considered in detail later. Large increases in plasma and red cell *p*-aminobenzoic acid (PAB) have been obtained with *P. knowlesi* infections in the monkey

TABLE IV. Chemical and Metabolic Changes Occurring in Red Blood Cells as Parasite Size Increases *in Vivo*

(Taken from Ball *et al.*, 1948, with the permission of the authors and the publisher)

	Monkey 86; 6.1 kilos				Monkey 98; 4.4 kilos			
	4.20 P.M., Apr. 11, 1945		1.00 P.M., Apr. 12, 1945		10.00 A.M., June 12, 1945		9.00 P.M., June 12, 1945	
Blood sample, *cc.*			110*		17		17	
Red cells, per c.mm.	3.45×10^6		2.08×10^6		3.81×10^6		2.37×10^6	
White cells, per c.mm.	13,200		14,400		34,400		38,700	
Hematocrit, %	28.0		15.5		32.3		21.3	
Parasites, %	33.7		47.8		47.4		34.4	
Rings	41		6		4		2	
Trophozoites, early	52		27		88		16	
Trophozoites, late	3		40		1		71	
Schizonts	2		25		2		4	
Segmenters	2		2		1		0.5	
	Total	Parasitized†	Total	Parasitized†	Total	Parasitized†	Total	Parasitized†
	mM per hr. per 5×10^{12} cells							
O_2 consumption	6.48	17.9	16.70	34.2	8.78	17.8	17.95	51.0
Glucose consumption	6.44	17.9	14.14	28.9	10.77	22.1	18.27	52.0
Lactate consumption	4.83	14.3	13.07	27.3	5.83	12.3	9.37	27.2
Lactate accumulation	8.05	21.5	15.21	30.5	15.71	31.9	27.71	76.8
	mM per 5×10^{12} cells							
Fatty acids					11.8	19.7	10.0	20.1
Total P	12.77	19.0	22.33	36.3	15.43	21.9	17.45	32.4
Acid-soluble P	7.18	8.35	8.48	10.5	7.10	7.67	7.61	9.57
15-min.-hydrolyzable P	1.78	2.7	2.46	3.7	1.68	2.09	2.28	4.1
Phospholipid P	3.1	5.4	4.4	7.1	3.7	5.7	4.1	8.3
Nucleic acid P	2.5	5.3	9.5	18.7	4.6	8.5	5.8	14.5
	Mg. per 5×10^{12} cells							
Flavin-adenine dinucleotide	0.54	1.28	0.84	1.57	0.53	0.93	0.85	2.15

* Terminal.

† The values for parasitized cells were calculated according to the method given in the text, with the following values (mM) for 5×10^{12} normal red cells: O_2 consumption 0.70; glucose consumption 0.60; lactate consumption 1.20 mM per hour; fatty acids 4.7, total P 9.61, acid-soluble P 6.60, 15-minute-hydrolyzable P 1.31, phospholipid P 1.91 mM; flavin-adenine dinucleotide 0.17 mg.

Macaca mulatta (McKee and Geiman, 1950). The significance of these changes are not understood but may be due to increased mobilization of biotin and PAB from other parts of the body for use by the parasites.

As stated above, many organic and metabolic components of the blood of monkeys infected with malarial parasites increase rapidly. These increases occur not just as there is a multiplication of parasites (Ball, 1946, Fig. 3), but also as the parasites grow and increase in size (Ball *et al.,* 1948, Table IV). Velick (1942) from a study of *P. cathemerium* infections in canaries demonstrated increases of respiration, respiratory quotient, and cytochrome oxidase activity as the parasites grew in size. Silverman, Ceithaml, Taliaferro, and Evans (1944) working with *P.*

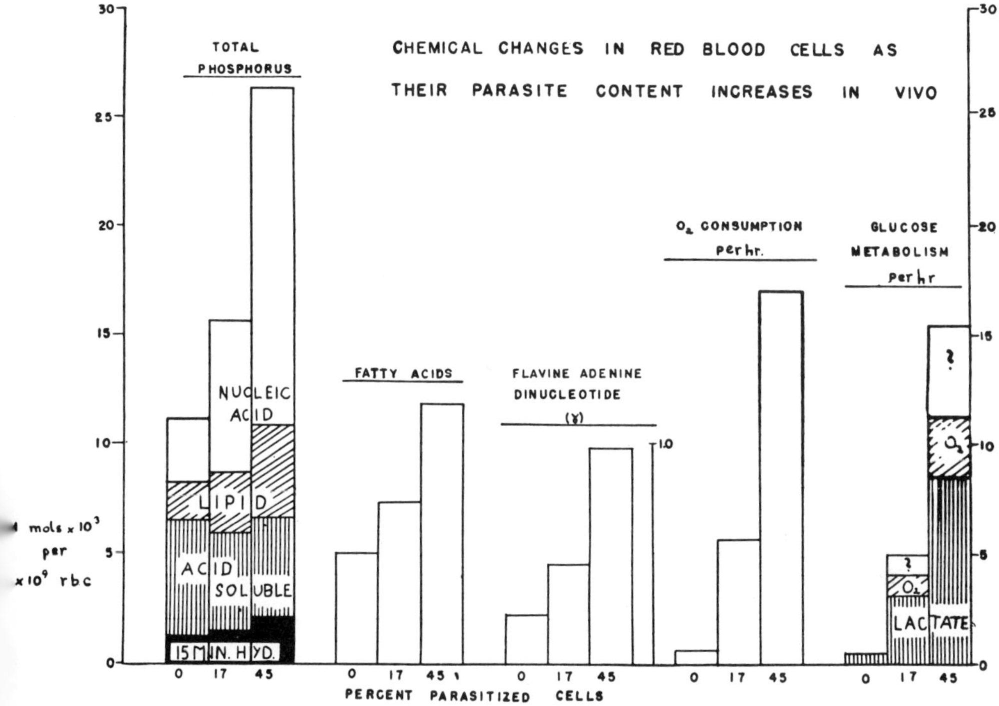

FIG. 3. Chemical changes in red blood cells as their parasite content increases *in vivo.* (Taken from Ball, 1946, with the permission of the author and the publisher.)

gallinaceum were able to correlate parasite surface area (parasite mass) with respiration and with glucose and lactate utilizations. Manwell and Feigelson (1949) working with *P. gallinaceum* found the glycolytic rate to be proportional to the surface area of the parasites. Increases in the quantities of these organic components and metabolic products are undoubtedly produced by all species of malarial parasites; however, specific data are lacking for the other *Plasmodium.*

III. Metabolism of Malarial Parasites

A. Historical

From the time of the discovery of malarial parasites by Laveran (1880) until the biochemical investigations of Christophers and Fulton (1938) few studies were made on the metabolism of the parasites of malaria. During the intervening years the main contributions to our knowledge of malarial parasites were: (1) the use of polychrome stains in the establishment of the mosquito cycle of the parasite *Plasmodium vivax* (Ross, 1897); (2) the *in vitro* survival and maintenance of *P. falciparum* and *P. vivax* by Bass and Johns (1912) and *P. malariae* by Row (1917); (3) the investigations by Brown (1911) on the production of hematin from hemoglobin and the isolation and identification of the hematin by Sinton and Ghosh (1934); (4) preliminary studies on some physical and chemical changes of the blood of infected monkeys and man by Fairley and Bromfield (1934), Chopra, Mukherjee, and Sen (1935), Krishnan, Ghosh, and Bose (1936), and by Kehar (1936); and (5) observations by Johns (1930) on the influence of dextrose and low temperature on the preservation of parasites.

Since 1938 the progress of research on the metabolic aspects of malarial parasites has been rapid and fruitful. The work of Christophers and Fulton was followed by a rapid progression of investigations by English, Australian, Indian, and American workers. Perhaps the most notable achievement from the biochemical standpoint is the elucidation of the carbohydrate metabolism and an attempted explanation of the mode of inhibition of certain antimalarial drugs on some of the enzyme systems involved. The principal investigators and their findings will be given under what the author feels to be the most appropriate metabolism subtopic. Space will not permit a complete listing of all the available references nor a detailed accounting of all the findings; however, it is the purpose of this review to give an integrated picture of the more important aspects of the life processes of the *Plasmodium*. Although considerable progress has been made in the elucidation of the metabolic processes of the malarial parasites, many questions remain unanswered. A striking example of this is the lack of adequate information to formulate a nutrient medium that will give maximum growth and multiplication of a single species of parasite in the absence of plasma, although varying amounts of growth and multiplication have been obtained with a number of different parasites. These data will be presented in detail in a later section.

B. Enzyme Systems and Oxidative Reactions

The concept of enzymes as functional components of the cell was introduced by Traube about 1865 after his studies on the fermentation of sugar. He considered that all fermentation, putrefaction, and cell oxidation processes were due not to the cell *per se* but to the cellular substances which could take up oxygen and transfer it to the oxidizable groups of atoms or molecules. Büchner in 1897 substantiated this theory by discovering that a cell-free extract of yeast could ferment sugar to alcohol and carbon dioxide. In this manner the chemical processes of the cell were linked to the contained enzymes.

Thus it is apparent that the activity of the cell's enzyme systems must be the prime consideration for an understanding of the metabolism of the cell. When one sets out to study a cellular material which has not been previously investigated, he would like to know immediately whether or not the organisms require oxygen, then what oxidative enzymes are involved—whether the flavoproteins and the cytochrome systems are present. The possible presence of anaerobic processes should be considered. Next one would like to determine the type of nutrients that are required, whether carbohydrate, fat or protein are utilized by the cells. Knowledge of the protein metabolism, particularly of its synthesis, is of great importance to an understanding of the cellular economy. This exemplifies only a few of the many problems involved but typifies the importance of enzymes in the metabolism of the cell.

A considerable amount of precise information has been obtained regarding some of the enzymes involved in the metabolic processes of a few of the plasmodia. Speck and Evans (1945a) prepared cell-free enzyme extracts of *Plasmodium gallinaceum*. Their preparations were capable of converting glucose to lactic acid via phosphorolative, dismutative, and oxidative processes similar to the classical glycolytic reactions produced by muscle preparations. The enzymes involved are hexokinase, phosphohexokinase, aldolase, glyceraldehyde dehydrogenase, and lactic dehydrogenase. Evidence for the functioning of these enzymes in water hemolysates and extracts of *P. gallinaceum* was obtained by showing (1) the disappearance of adenosine triphosphate (ATP) phosphorus, (2) the formation of phosphorolated carbohydrate intermediates, and (3) the action of specific enzyme inhibitors. Analyses for the phosphorus of ATP and for fructose-1,6-diphosphate indicate the formation of the latter simultaneously with the disappearance of ATP phosphorus. The transfer of phosphate from ATP to a hydroxyl group of glucose results in the production of acid, which was measured manometrically by the evolution of carbon dioxide in a bicarbonate medium. The functioning of the enzyme, aldolase, which splits fructose-1,6-diphosphate, was demonstrated by binding the product of its action, triose phosphate, by cyanide and by the ready hydrolyzability of the phosphate groups of the trioses, 3-phosphoglyceraldehyde and dihydroxyacetone phosphate, which are formed. The enzymatic oxidation of 3-phosphoglyceraldehyde is characteristically accelerated by arsenate and 90 per cent inhibited by iodoacetate. Thus the work of Speck and Evans (1945) indicates that *P. gallinaceum* is capable of converting glucose to fructose-1,6-diphosphate, this compound in turn to 3-phosphoglyceraldehyde and then of oxidizing this latter compound by a series of reactions similar to those functioning in muscle extracts. The precise steps will be given in detail later under carbohydrate metabolism. Although similar data have not been obtained, it is likely that most, if not all, of the other species of malarial parasites produce similar reactions.

Concerning the presence of oxidase and dehydrogenase enzyme systems in parasites, Speck. Moulder, and Evans (1946), working with *P. gallinaceum* freed of the host red cell, found certain enzymes of the Krebs tricarboxylic acid cycle to be operating. These enzymes, as shown later, are fumarase, malic dehydrogenase, aconitase, and succinic dehydrogenase. Evidence for the presence of these enzyme systems centers not on their actual isolation but around (1) the metabolism of all the seven substrates, (2) the catalyzing of the oxidation of pyruvate when these substrates are added in small amounts, (3) the formation of citrate and α-ketoglutarate from pyruvate and the dicarboxylic acids, and (4) the inhibition of pyruvate oxidation by the addition of the specific succinic dehydrogenase poison, malonate, with the accumulation of succinate and the subsequent removal of the inhibition with added fumarate. Additional evidence for the presence of the lactic and malic dehydrogenase enzyme systems in *P. gallinaceum* are the stimulation of pyruvate oxidation by the addition of diphosphopyridinenucleotide, which these enzymes require for their functioning.

Again comparable data are lacking for the other parasite species, but indirect evidence is available to suggest that similar oxidative systems are functional for many of the species of malarial parasites.

The supporting roles of the flavoprotein enzyme systems and of the cytochromes have been investigated. Christophers and Fulton (1938) determined that the oxygen consumption of *P. knowlesi* was almost completely inhibited by 0.001*M* cyanide. This has been confirmed by Wendel (1942) and by McKee *et al.* (1946). Likewise the respiration of *P. lophurae* was found by Bovarnick, Lindsay, and Hellerman (1946) to be cyanide-sensitive. Attempts to determine whether or not the cyanide-sensitive oxidative systems were iron-porphyrin or copper-containing systems have been only partially successful. McKee *et al.* (1946) found that carbon monoxide inhibited the uptake of oxygen by 64 per cent, but inconsistent reversals of the inhibition by irradiation with strong light were obtained. Strong light reverses the CO inhibition of iron porphyrin but not the inhibition of copper respiratory enzymes. Bovarnick and collaborators observed that the respiration of *P. lophurae* is less inhibited by azide than by cyanide. Velick found that *P. cathemerium* oxidizes *p*-phenylene diamine faster than do normal avian cells. All these findings point to the presence of iron-porphyrin respiratory enzymes in the malarial parasites. Concerning the presence of flavoprotein enzymes in parasites Bovarnick *et al.* (1946) observed that cresyl blue partially restores the respiration inhibited by cyanide, which is an indication of flavoprotein systems. Ball, McKee, Anfinsen, Cruz, and Geiman (1948) obtained increases in the flavine adenine dinucleotide content of *P. knowlesi* infected monkey red cells, both *in vitro* and *in vivo*.

The oxygen uptake by parasites, an overall measure of the functioning of the above discussed oxidative enzyme systems, has been measured for a number of parasite species. The parasites derive their main source of oxygen supply from the oxyhemoglobin and can draw on that source until most of the oxygen has been utilized and the blood is quite "blue." Oxygen utilization not only varies greatly with the species of parasite (Geiman, 1948b) but also with the size of the organism, as shown by Silverman *et al.* (1944) for *P. gallinaceum* and by Velick (1942) for *P. cathemerium.* That the uptake of oxygen is influenced by the oxygen tension in the atmosphere has been demonstrated by McKee *et al.* (1946). The maximum uptake was with a gas phase containing 5 per cent oxygen. Gas phases containing 0.34, 20.0, and 100 per cent oxygen were also studied. Not only is excessive oxygen detrimental to respiration but also to *in vitro* cultivation of *P. knowlesi,* as will be discussed later.

The observations of Moulder and Evans (1946) and of Morrison and Jeskey (1947, 1948) on the breakdown of hemoglobin, liberation of amino acids, and the production of new parasite proteins are good indirect evidence for the functioning of proteolytic enzymes in erythrocytes infected with *P. gallinaceum* and *P. knowlesi.* The precise enzymes which are involved have not been determined; however, details of the studies will be presented under protein metabolism. Another important aspect of enzymic reactions involved in parasite metabolism and economy of the cell is the coupling of oxidative reactions with protein breakdown and synthesis (Moulder and Evans, 1946).

The presence of nearly 30 per cent lipid in the dry weight of *P. knowlesi,* which represents nearly a sixfold increase over the amount present in the host red cell, is indicative of the action of lipid synthesizing enzyme systems. Again the exact enzymes present in the parasite have not been studied.

We see, then, a variety of enzymes present in the malarial parasite which are necessary for the breakdown of carbohydrate, protein, and lipid, and for the synthesis of new proteins (including nucleoproteins), lipids, and cellular components required by the organism for the synthesis of protoplasm.

C. Carbohydrate Metabolism

The importance of carbohydrate in the maintenance of malarial parasites was first realized by Bass and Johns (1912). They stated that glucose or maltose was necessary for the *in vitro* growth and division of *Plasmodium vivax* and *P. falciparum.* Johns (1930) also found that the addition of glucose to human blood containing *P. vivax* prolonged survival of the parasites stored at 0° C.

Evidence of the *in vivo* requirement of glucose by plasmodia has been advanced by a number of workers. Hegner and MacDougall (1926) observed that by increasing the blood sugar of birds infected with *P. gallinaceum* there were produced more favorable conditions for the parasites. Conversely, reducing the blood sugar with insulin appeared to produce a condition unfavorable for parasite multiplication. Christophers and Fulton (1938) were the first to observe a drop in the liver glycogen of monkeys infected with *P. knowlesi,* which becomes very pronounced after the infection has progressed and there are large numbers of parasites in the circulating blood. This situation is understandable since, as we see later, the malarial parasites consume large quantities of glucose.

Fulton (1939) was the first to make an extended *in vitro* study of the various carbohydrates that can be utilized by plasmodia. Working with *P. knowlesi* and using the conventional Warburg technique of measuring oxygen uptake he determined that glucose, fructose, maltose, mannose, and glycerol were oxidized. Maier and Coggeshall (1941) extended the carbohydrate studies to four other species of plasmodia and found that glucose, mannose, fructose, and glycerol can be used interchangeably to maintain for short periods of time the oxygen uptakes with *P. knowlesi, P. inui, P. cynomolgi. P. cathemerium,* and *P. lophurae.* Sodium D-lactate they found to be used to a lesser extent; sodium α- or β-glycerophosphate and maltose were used only to a slight degree; and malate, citrate, succinate, fumarate, raffinose, rhamnose, arabinose, and calcium hexose phosphate did not appear to be utilized at all. A demonstration of the utilization of the metabolic components of the citric acid cycle depends to a great extent on the completeness with which the preformed substrates are removed from the parasites. Bovarnick, Lindsay, and Hellerman (1946b) found that *P. lophurae,* separated from the red cells by means of saponin and washed free of glucose, gave oxygen uptakes with lactate and pyruvate comparable to that with glucose, whereas succinate and fumarate produced about 30 per cent of that with glucose. The oxygen consumption of the separated parasites was about 70 per cent of that of the original non-laked erythrocytes. Similar decreases have been observed for the respiration of *P. knowlesi* and *P. cynomolgi* isolated by means of saponin.

Wendel (1943) and others have found the respiratory activity of *P. knowlesi* to be equally good with glucose or sodium lactate. For glucose to yield as much oxygen uptake as sodium lactate it is necessary that the excessive lactic acid, produced by glycolysis and not oxidized, be neutralized. This phenomenon is explained by the fact that acid is detrimental to the respiratory activity of the parasites. Several workers have demonstrated this pronounced accumulation of lactic acid when glucose is metabolized. Fulton (1939) Wendel (1943) and McKee *et al.* (1946) observed an accumulation for *P. knowlesi* and McKee *et al.* (1949) for *P. falciparum, P. vivax, P. cynomolgi,* and *P. lophurae.* A better understanding of the magnitude of the acid production and accumulation can be obtained from the realization that the parasitic cell utilizes some 25 to 100 times the glucose that is consumed by a normal red cell. The value for normal human erythrocytes is about 0.8 mM per liter per hour. The major part of this glucose is converted, at least under anaerobic conditions, to lactic acid. With the plasmodia that have been studied, *P. knowlesi, P. cynomolgi, P. vivax, P. falciparum,* and *P. lophurae,* less than half of the lactic acid is oxidized further. In the animal organism this acid is converted mainly to glycogen. In cultivation experiments there is, however, the problem of dissipating the excess lactate. This subject will be discussed later under "Cultivation."

Glycerol was found by McKee *et al.* (1946) to give slightly, but probably significantly, greater oxygen utilization than either glucose or lactate. As we will see later, however, it is of considerable interest that, although glycerol is readily oxidized in short term Warburg experiments, this substrate will not maintain the parasites in 24-hour cultures. Glucose is necessary for the growth and multiplication of plasmodia in cultivation experiments.

The importance of glucose for the growth and multiplication of a number of parasite species led Geiman (1948b) and McKee (1951) to make a comparative study of the glucose requirements for two simian and two human species of parasites. In addition to glucose consumption the utilization of lactate was determined for the four parasites, and oxygen uptake for two species. These values are given in Table V. It is interesting to note that considerably larger values were obtained for *P. vivax* whereas the values for *P. cynomolgi* were the smallest. This striking difference may be due to the fact that the human species, *P. vivax,* is a much larger parasite and greatly swells and enlarges the red cell.

TABLE V. Comparative Rates of Glycolysis and Respiration for Four Species of Malarial Parasites (Altered from Geiman, 1948b)

Species of parasites	Utilizations* in mM/hr./5 × 10^{12} parasitized cells		
	Glucose	Lactate	Oxygen
P. knowlesi	19–52	12–31	18–51
P. cynomolgi	18–21	17–23	12–39
P. falciparum	37–47	16–30	—
P. vivax	103–238	48–85	—

* Age of parasite determines rate of glycolysis and respiration. Old parasites have 2–3 times the activity of young stages. 1 mM glucose forms 2 mM lactate.

Specific information concerning the enzyme systems involved in the metabolism of glucose by several malarial parasites has already been presented. The reactions involved and the intermediate products will now be considered.

Silverman, Ceithaml, Taliaferro, and Evans (1944) showed that glucose is quantitatively converted into lactic acid by *P. gallinaceum* under anaerobic conditions and suggested a typical muscle type of phosphorolative glycolysis. Speck and Evans (1945a) studying cell-free extracts of *P. gallinaceum* found that the preparations contained enzymes which catalyze (1) the phosphorolation of glucose by adenosine triphosphate (ATP), (2) the splitting of fructose-1,6-diphosphate to form 3-phosphoglyceraldehyde, and (3) dismutation between 3-phosphoglyceraldehyde and pyruvic acid. These three reactions are key ones in the glycolytic series which were first shown to be functional in yeast and muscle extracts. These systems are represented in Fig. 4.

Speck, Moulder, and Evans (1946) subsequently followed the oxidative removal of pyruvic acid by parasitized erythrocytes as well as by separated parasites. Although parasitic cells oxidize pyruvate almost completely, the free parasites also form appreciable quantities of acetate. This oxidative process according to these investigators undoubtedly

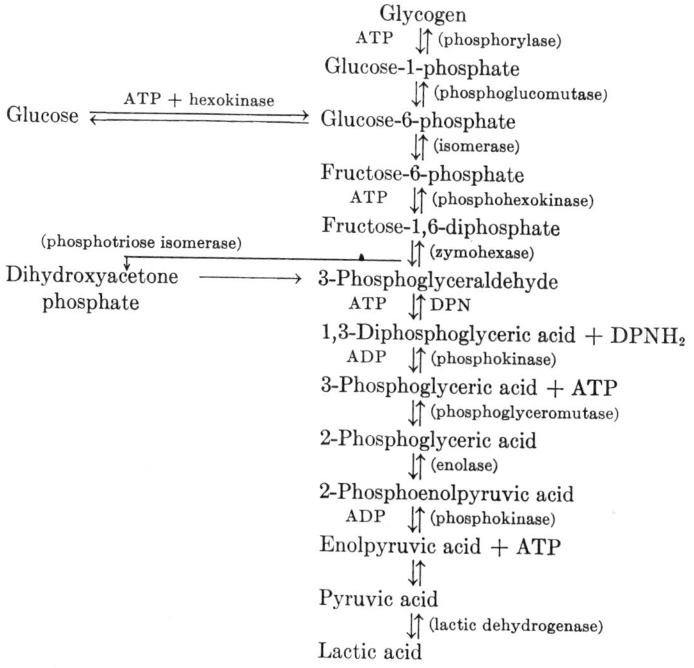

FIG. 4. Summary of glycolysis.

involves the tricarboxylic acid cycle of Krebs. In these cyclic reactions pyruvate or acetate condenses with oxalacetate to form citric or iso-citric acid. Subsequently, *cis*-aconitate, α-ketoglutarate, succinate, fumarate, malate, and finally oxalacetate are produced concurrently in a series of equilibrium reactions. This oxidative cycle, as shown in Fig. 5, is primarily a series of decarboxylation and dehydrogenation processes.

Relatively little energy is produced by this series of anaerobic reactions. The major part, 80 per cent or more, of the energy produced by the oxidation of carbohydrate is derived from the stepwise enzymatic passage to oxygen of the electrons liberated in the above catalyzed reactions. The bundles of energy liberated in this stepwise electron transfer are captured and utilized in the production of high-energy phosphate bonds, i.e., ADP + inorganic phosphate + energy → ATP. The energy thus stored is held in readiness for essential synthetic reactions required for cell repair and growth.

From the enzyme studies cited previously it is likely that in the

Lactic acid (3C)
⇅ (lactic dehydrogenase)
Pyruvic acid (3C)
↓
Acetic acid (2C) + CO_2
+
Oxaloacetic acid (4C)
⇅
Citric acid (6C)
⇅
cis-Aconitic acid (6C)
⇅ (aconitase)
iso-Citric acid (6C)
⇅ (iso-citric dehydrogenase)
Oxalosuccinic acid (6C)
⇅ (oxalosuccinic decarboxylase)
α-Ketoglutaric acid (5C) + CO_2
⇅ (oxidative decarboxylation)
Succinic acid (4C) + CO_2
⇅ (succinic dehydrogenase)
Fumaric acid (4C)
⇅ (fumarase)
Malic acid (4C)
⇅ (malic dehydrogenase)
Oxaloacetic acid (4C)

FIG. 5. Krebs tricarboxylic acid cycle.

malarial parasite the hydrogen is passed first to DPN, then to flavine adenine dinucleotide (flavoprotein) and finally, after a series of electron transfers via the cytochromes, to molecular oxygen. These transfers and the systems which may be involved are represented in Fig. 6. The protons are present in the environment of the cell and available for reaction when cytochrome oxidase passes its electron to oxygen.

Bovarnick, Lindsay, and Hellerman (1946a), approaching the problem from a somewhat different angle, arrived at the same conclusions regarding the glycolytic and Krebs cycle reactions of the parasite *P. lophurae*. They determined that this parasite, freed of the erythrocyte and previously depleted of substrate by incubation at 37° C. for 100 minutes, showed a slow resumption of oxygen uptake when glucose was added. The induction period was much less marked with lactate, pyruvate, succinate, or fumarate. The authors believe the shortening of the induction period may be due to an increased rate of ATP formation, previously described as being formed by the energy derived from the

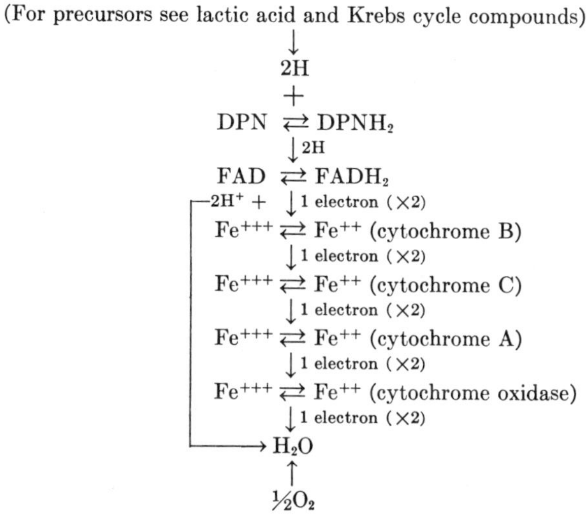

FIG. 6. Transfer of hydrogen to oxygen (oxidation of hydrogen).

cell oxidative processes. Analysis of the parasites for total, labile and inorganic phosphorus before and after incubation in the absence of glucose indicated a marked decrease in organic and labile phosphorus and a simultaneous increase in inorganic and total acid-soluble phosphorus. The situation was reversed by adding glucose. The logical conclusion of the workers was that, "the induction period in the oxidation of glucose by the substrate-depleted cells is attributed to the necessity for phosphorylation of glucose before this substance can be utilized. . . ."

It is apparent, from the above-mentioned studies, that glucose is the carbohydrate essential for long-term maintenance of plasmodia. Although many carbohydrates can be oxidized by the parasites, apparently the energy derived from the phosphorolative anaerobic glycolysis is required for growth and reproduction. Undoubtedly only a portion of this energy requirement is for the maintenance of the erythrocytes. Needless to say it is quite unlikely that the additional energy derived from the 25- to 100-fold increase in glucose consumption by the parasitic cell is utilized by the host red cell *per se.* Since growth and multiplication of plasmodia have not been attained in the absence of the red cell the metabolic activities of the parasite and of the erythrocyte cannot be divorced. The work of Evans and collaborators on enzyme preparations of *P. gallinaceum* certainly indicates the ability of the parasite itself to convert glucose to lactate.

Thermodynamically only about 10 per cent of the total energy obtainable from the complete oxidation of glucose is derived from the phosphorolative glycolytic processes. The malarial parasite is, however, a wasteful organism as indicated by the fact that it glycolyzes much more carbohydrate than it oxidized. Thus, a sizable portion of the energy which the plasmodia utilizes comes from the glycolytic processes.

D. Protein and Hematin Metabolism

Although studies on the protein metabolism of the plasmodia have not been as extensive as those for carbohydrate, this type of metabolism is no less important to the growth and reproduction of the organism. The plasmodia are parasitic and contain enzymes for the degradation of the proteins of the red cell and the synthesis of their own characteristic proteins. Since the parasites produce cytoplasm and nuclear material it is a foregone conclusion that these cells must have the ability to produce nucleoproteins as well as structural components. The appearance of additional enzyme activity by parasitic cells is additional evidence of protein synthesis.

It is evident from the work of Brown (1911), of Morrison and Jeskey (1947, 1948), of Moulder and Evans (1946), and of McKee, Geiman, and Cobbey (1947), and McKee and Geiman (1948) that the parasite utilizes the red cell hemoglobin as the main source of amino acids, the building blocks for parasite protein synthesis. In 1911 Brown showed that the hemoglobin of the host red cell was split to form hematin and globin; however, the fate of the protein moiety was not determined at that time. More recently the work of Moulder and Evans (1946) with *Plasmodium gallinaceum* and of Morrison and Jeskey (1947, 1948) on *P. knowlesi* has shown that the protein of the red cell is broken down by hydrolysis or phosphorolysis, and about half of the amino acids are reutilized by the parasite for synthesis of its own proteins, while the remainder of the amino acids diffuse out of the cell. In spite of this apparent excess of amino acids, methionine must be added to give a normal complement of essential building blocks for *P. knowlesi* (McKee and Geiman, 1948). This seems like a peculiar phenomenon, but the answer probably lies in the fact that hemoglobin contains only about 1 per cent methionine while most proteins contain 3 or 4 per cent of the amino acid. This would mean that there is present in the red cell inadequate amounts of methionine to produce parasite protein and the parasitic cell is forced to draw on the surrounding plasma and cells. This need for extra methionine has been shown for *P. knowlesi* both *in vitro* and *in vivo* (McKee and Geiman, 1948). As will be discussed in more detail later, monkeys infected with this parasite required methionine in their diet for growth and multiplication of the organisms. It is quite possible that the other species of malarial parasites require the addition of methionine and other amino acids as well. Thus the plasmodia are not parasites merely for their host red cell, but also for the environmental fluids and tissues.

That characteristic parasite protein is formed from the red cell protein has been shown by Morrison and Jeskey (1947). In addition to obtaining a decrease of 44 per cent in the nitrogen of the *P. knowlesi* infected red cell, they demonstrated in the parasite itself a protein mixture containing about 14 per cent nitrogen, which was separable into three components on electrophoresis at pH 8.6 in veronal buffer.

The formation of nucleoproteins has not been determined by direct chemical analysis, but Ball, McKee, Anfinsen, Cruz, and Geiman (1948) by phosphorus analysis of *P. knowlesi* obtained evidence for the synthesis of nucleic acids. The need for added purines and pyrimidines in cultures of parasites, as will be presented later, is further evidence in this direction. Calculations from the phosphorus data indicate the formation of some 5.46 g. of nucleic acids per 5×10^{12} parasitized cells. These calculations are based on the assumption that no phosphoprotein is present, and this may be incorrect. Be that as it may, nucleic acids do increase in quantity in the parasites *P. knowlesi* and *P. vivax* as shown by the Feulgen staining technique (Deane, 1945).

Indirect evidence to show the synthesis of enzymes (containing protein moiety) in malarial parasites may be garnered from changes in their metabolic rates. Christophers and Fulton (1938, 1939), Fulton (1939), Coggeshall (1940), Maier and Coggeshall (1941), Coggeshall and Maier (1941), Velick (1942), and McKee and Geiman (1948) have shown increases in the utilization of glucose and the consumption of oxygen by malarial parasites. Velick (1942) studying *P. cathemerium* obtained a slow increase in the oxygen uptake as the parasite grows and then a greatly accelerated respiration when nuclear division begins. They determined a simultaneous increase in the cytochrome oxidase activity. McKee *et al.* (1946) determined a two-and-one-half-fold increase in the lactic dehydrogenase content of the red cell infected with *P. knowlesi.* Ball *et al.* (1948) showed a 6- to 15-fold increase in the flavine adenine dinucleotide content of *P. knowlesi.* This compound is the coenzyme for several oxidative enzymes. The amount of increase for both of these enzyme constituents is determined by the degree of development of the parasites. There is then an accumulation of indirect evidence indicating the ability of

TABLE VI. Effect of Anaerobic Conditions upon Formation of Amino Nitrogen and Ammonia in Parasitized Erythrocytes

(Taken from Moulder and Evans, 1946, with the permission of the authors and the publisher)

0.01 M glucose	Fraction determined	Amount formed in micromoles per ml. parasitized erythrocytes		Inhibition in nitrogen
		Air	Nitrogen	per cent
+	Amino nitrogen	23.3	4.2	82
+	Amino nitrogen	1.4	3.3	−120
−	Amino nitrogen	15.8	3.7	80
−	Amino nitrogen	8.4	3.0	64

malarial parasites to produce enzyme proteins as well as nucleoproteins and other structural and functional proteins.

The next question is, how do the parasites metabolize amino acids? The utilization of the amino acids and peptides for synthesis of parasite protein depends apparently on oxidative reactions. Moulder and Evans (1946) found, as stated previously, that chicken erythrocytes infected with *P. gallinaceum* produced large amounts of amino nitrogen (amino acids and peptides) when incubated in air and in the presence of glucose. When glucose was absent, much of the amino nitrogen appeared as ammonia. However, when oxygen was absent, even in the presence of glucose, there was little splitting of the protein to produce amino nitrogen (Table VI). The indication from this work is that, although *P. gallinaceum* can deaminate amino acids, the normal course of events for growth and reproduction of parasites is the utilization of the amino acids for protein synthesis. Probably only very small amounts of amino acids are consumed in oxidative processes as indicated by the extremely small stimulation of the respiration of *P. knowlesi* with the addition of amino acids in Warburg respirometer experiments (McKee *et al.,* 1946). Moulder and Evans (1946) also prepared cell-free extracts of *P. gallinaceum* and determined that these extracts hydrolyze native hemoglobin at a very slow rate but that they split denatured globin much more rapidly. Unlike the anaerobic effects produced in infected cells, anaerobiosis in the cell-free extracts does not inhibit the production of amino acids from the denatured globin. This phenomenon is difficult to explain. The investigators suggest that in the intact parasite, protein breakdown is in some unknown way linked to carbohydrate oxidative processes. Undoubtedly the synthesis of parasite protein is also dependent on the energy derived from carbohydrate metabolism, perhaps in the form of high-energy phosphate bonds. As indicated under "Carbohydrate Metabolism," the classical oxidation of glucose involves the utilization of the energy evolved in the synthesis of high-energy phosphate bond compounds, i.e., ATP. Such compounds might then be the mediators of phosphorolative protein reactions (Lipmann, 1949). Extensive experimentation will be required, however, to establish such a hypothesis.

Let us now consider briefly the fate of the non-protein portion of hemoglobin, namely hematin. As the malarial parasite develops within an intact red blood cell, hemoglobin gradually disappears and a granular pigment is formed. This pigment is reddish brown by transmitted light and grayish black by reflected light. A non-protein moiety of hemoglobin, which was shown by Brown (1911) to be split from the hemoglobin of red cells infected with *P. knowlesi,* was later identified as hematin or ferrihemic acid by isolation and by spectroscopic examination (Sinton and Ghosh, 1934). This was confirmed by Morrison and Anderson (1942), and by Rimington, Fulton, and Sheinman (1947) also working with *P. knowlesi.*

It is believed by most investigators that this pigment (hemozoin), which is deposited as granules in the cytoplasm of malarial parasites, is hematin. Huff and Bloom (1935) have shown that *P. elongatum* when present in non-hemoglobin-containing erythroblasts do not produce pigment. Ghosh and Nath (1934) found that the iron, carbon, and hydrogen values for malarial pigment derived from *P. knowlesi* agreed with those for hematin. Thus there is good evidence that the parasite pigment and hematin are identical. In human plasmodia the color of the pigment varies with the species, it being much lighter in *P. vivax.* Nevertheless, even this pigment is believed to be hematin, although perhaps admixed with other materials.

The need for large quantities of protein (globin) by the parasite during its growth within the red blood cell necessitates the destruction of large quantities of hemoglobin. Ball, McKee, Anfinsen, Cruz, and Geiman (1948) presented data indicating a sizable amount of the hemoglobin being split to form hematin during the erythrocytic cycle of the parasites *P. knowlesi, P. vivax,* and *P. lophurae.* Morrison and Jeskey (1948) confirmed this finding with *P. knowlesi* and by calculation arrived at a value of 76 per cent as the amount of the red cell's hemoglobin destroyed during the life of a parasite.

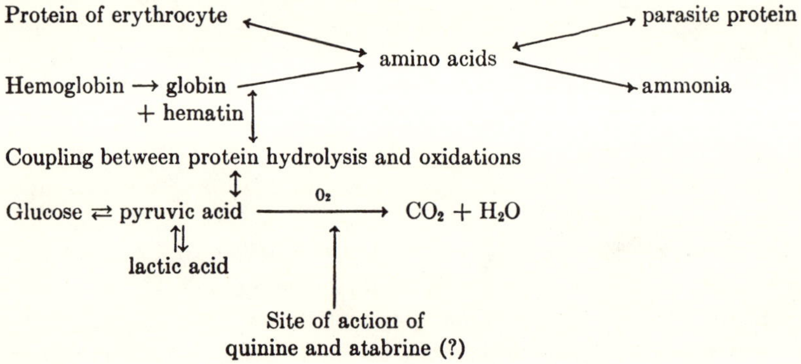

FIG. 7. Diagrammatic representation of protein and coupled oxidative metabolism in plasmodia. (Taken from Moulder and Evans, 1946, with the permission of the authors and the publisher.)

Upon rupture of the parasitic red cell the pigment is released into the circulating blood. Apparently most of the hematin is picked up by the reticulo-endothelial system, particularly the spleen and liver. In the infected animal as well as in cultures of parasites, the leucocytes become engorged with the pigment. Rigdon (1945) believes that this malarial pigment may be slowly oxidized within the endothelial cells to form hemosiderin which can be reused by the host animal.

Thus the protein metabolism of malarial parasites is an important part of the metabolic activity of these cells, being necessary for all the enzymatic processes, growth, and multiplication. Moulder and Evans (1946) have summarized their findings of the protein metabolism of *P. gallinaceum* in the schematic way shown in Fig. 7. The broken lines represent reactions which at that time had not been demonstrated experimentally. However, the work of Morrison and Jeskey (1947, 1948), as indicated above, furnish fair confirmatory evidence for those two reactions in question. Undoubtedly similar metabolic processes are active in other species of plasmodia.

E. Lipid Synthesis

Our knowledge regarding the lipid metabolism of *Plasmodium* is even scantier than that for protein metabolism, but data are available to show the synthesis of fatty acids and phospholipids. Morrison and Jeskey (1947, 1948) found for *P. knowlesi* infected red cells a 423 per cent increase in the total lipids, or about 28.8 per cent of the dry weight of the isolated parasites. About 25 per cent of this is non-saponifiable material, mainly cholesterol. The work of Ball *et al.* (1948) showed the fatty acid content of *P. knowlesi* infected cells to be 4 to 5 times and the phospholipid phosphorus 2 to 4 times that of normal monkey red blood cells (Table II). The ratio of fatty acids to phospholipid phosphorus for the parasitic cell was not appreciably altered from that of the normal cell.

The nature of the fatty acids present in *P. knowlesi* have been studied by Morrison and Jeskey (1947). They determined that less than 0.3 per cent was volatile acid, while 36 per cent consisted of stearic acid, and 41 per cent was a liquid, 18 carbon, unsaturated acid, probably oleic acid. Why the parasite produces this huge store of lipids is not known, although one might speculate that one function might be to serve as an intricate matrix for the many enzyme systems present in the cell.

Laser and Friedman (1945) and Laser (1948) have made the interesting observation that a lytic material is present in normal blood and in much larger quantities (25–75 times) in *P. knowlesi* infected monkey red cells. This substance they found to be soluble in ether and insoluble in water. The isolated compound contains no nitrogen, phosphorus, sulfur, or halogen and is apparently an unsaturated monocarboxylic fatty acid containing 18 carbons. These workers observed that the fatty acid is lytic to normal erythrocytes in a dilution of 1:100,000. Hematin, which has lytic properties at the low concentration of 1:50,000, potentiates the action of this fatty acid, even at a 1:200,000 dilution of the hematin. Laser (1946) has suggested that an accumulation of the C–18 fatty acid in the parasitic cell is the cause of cell rupture and release of merozoites and that the process is perhaps arrested by effective antimalarial drugs (Laser, 1946).

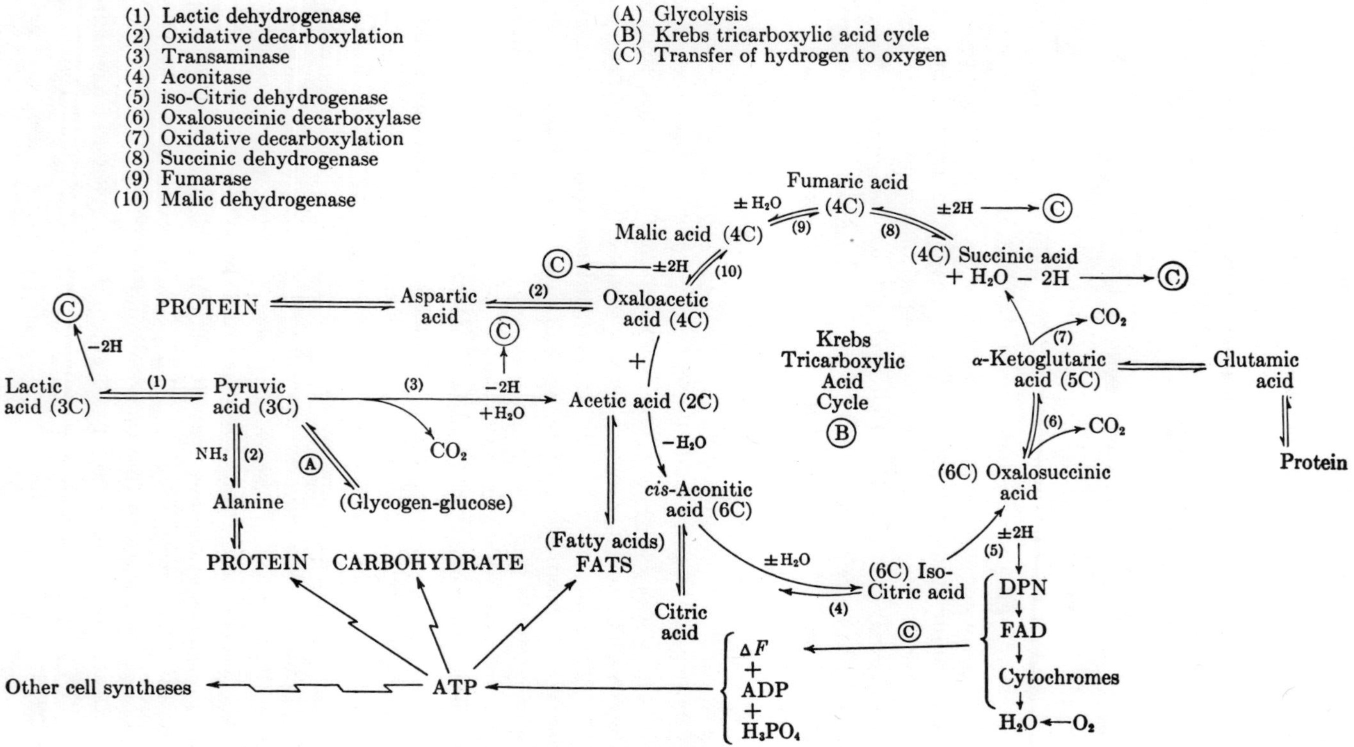

FIG. 8. Diagrammatic summary of metabolism.

From the foregoing considerations it is apparent that the plasmodia are able to glycolyze and to oxidize carbohydrate, to cleave hemoglobin, and to synthesize new proteins of a wide variety, and to produce large amounts of lipids. All these processes are highly integrated and of the utmost importance for the rapid growth and multiplication of the parasite. The metabolic processes which have been studied by numerous investigators are shown diagrammatically in Fig. 8. Such a figure should serve only as a working hypothesis for continued studies and not be considered as a final picture of the cell's activities. Since this representation might serve equally well as a pattern for the biochemical activity of many tissue cells it stresses the fact that our knowledge is not adequately definitive to allow us to show the characteristic differences between parasitic and normal tissue cells.

IV. Cultivation of Malarial Parasites

A. Early Cultivation Studies

Following the epochal discovery of malarial parasites by Laveran in 1880, little real progress was made in either metabolic or cultural studies until the work of Bass and Johns (1912), Bass (1913), and Thompson and Thompson (1913). They reported the *in vitro* survival of *Plasmodium falciparum* and *P. vivax* and the development of three successive generations of the parasites. Shortly after this Row (1917) reported similar results with *P. malariae*. There was then a considerable lull in experimental studies on malaria until 1932 when the simian parasite *P. knowlesi* was isolated by Sinton and Mulligan (1932). This discovery marked the beginning of an important era in malarial research because it yielded a pathogenic parasite which was highly infectious for an experimental animal, the Rhesus monkey (*Macaca mulatta*), and which produced large quantities of parasites in that host, a requirement essential for concentrated biochemical and cultural studies. The recent discovery of the rodent malarial parasite, *P. berghei* (Vincke and Lips, 1948; Van den Berghe, 1948), which is highly infectious for the white mouse, should further facilitate biological studies. The importance of chemical and biochemical knowledge for the advancement of cultural techniques is apparent from the realization that no progress was made toward *in vitro* cultivation until after the basic metabolic studies of Christophers and Fulton (1939), Maier and Coggeshall (1941), Velick (1942), and Wendel (1942).

Trager (1941, 1943b) revived the cultivation studies, taking cognizance of the nutritional requirements and intracellular environment of the parasite. Working with the avian parasite *P. lophurae* and employing a balanced salt solution containing a red cell extract, glutathione, glucose, calcium pantothenate, and serum embryo extract, he obtained survival up to 16 days. The culture material was placed in small Erlenmeyer flasks and gently agitated with different gas mixtures. Coulston (1941) achieved somewhat comparable results with *P. circumflexum.* Hawking (1944), working with *P. gallinaceum* and using the procedure of Trager, found little benefit from the culture medium constituents other than the red cell extract and glutathione. He obtained no increase in numbers of the parasites in cultures.

Working with the exoerythrocytic forms of plasmodia, Huff and Bloom (1935) reported experiments on tissue cultures of bone marrow in an attempt to observe the life cycle in living cells. They found that *P. elongatum* lived in such cultures for 48 hours, but *P. cathemerium* failed to survive.

The use of the developing chick embryo as a means of carrying malarial infections has been studied rather extensively; however, successful results have been obtained only with the avian species. Using duck embryos, Wolfson (1940) succeeded in obtaining infections with cultures of *P. cathemerium, P. elongatum,* and *P. lophurae,* and Stauber and Van Dyke (1945) obtained fatal infections with cultured *P. cathemerium.* Transmission of *P. gallinaceum* was obtained by the direct bite of the mosquito, *Aedes aegypti,* on developing chick embryos (Haas and Ewing, 1945). Demonstrable infection was obtained in the hatched chicks. Also, Haas, Feldman, and Ewing (1945) effected 14 serial passages of *P. gallinaceum* by the intravenous inoculation of 10- to 13-day-old chick embryos with parasitized blood.

B. Cultivation of Oocysts and Exoerythrocytic Stages

Making use of tissue culture techniques, G. H. Ball (1947, 1948) accomplished the Herculean task of obtaining and maintaining sterile oocysts from mosquitoes. The oocysts of the avian parasite *P. relictum,* attached to the stomach of the mosquito *Culex tarsalis,* were utilized for the study. Although the oocysts were maintained in cultures for 20 days, no visible development was obtained. Various gas mixtures and media, including completely synthetic media, were employed. These investigations clearly indicate the dual problem in the cultivation of plasmodia, namely the maintenance of the host cellular structure (stomach of mosquito) as well as the parasites themselves. The highly parasitic nature of the organism requires that the two processes go hand in hand. The only other reference to work of this nature which is available to the author is that of Ragab (1948).

The finding of Huff and Bloom (1935) that the avian parasite *P. elongatum* can live in all the blood and blood-forming organs of the canary and the discovery by James and Tate (1938) of the exoerythrocytic development of *P. gallinaceum* provided considerable stimulus for extended studies on the preerythrocytic stages of malarial parasites.

Hegner and Wolfson (1939) obtained survival of the exoerythrocytic stages of *P. cathemerium* in tissue cultures. Gavrilov, Bobkoff, and Laurencin (1938) also were successful in maintaining infective exoerythrocytic forms of *P. gallinaceum* in tissue cultures of bone marrow. However, Rodhain, Gavrilov, and Cowez (1940) and Parsense, Meyer, and Menezes (1942) failed to obtain development of the sporozoites of *P. gallinaceum* in chick embryonic tissue. The first really successful cultures of sporozoites were those reported by Hawking (1945) with *P. gallinaceum.* Eight- to ten-day-old infected chicks were used as the source of the infected tissues (spleen, liver, bone marrow, or brain) which were implanted into roller tubes or Carrel flasks. Survival, growth, and multiplication were obtained up to 89 days, with a maintenance of infectivity.

C. Cultivation of Erythrocytic Forms

World War II brought about an unprecedented urgency for the acquisition of further basic information about the biological processes of malarial parasites so that effective antimalarial drugs could be devised to replace the captured stores of quinine. This urgent need resulted in the previously described concentrated metabolic studies by a number of different laboratories both in America and Europe.

Simultaneous with this renewed interest in the metabolic activities of plasmodia new attempts were made to obtain *in vitro* growth and multiplication of malarial parasites. The major part of the cultivation studies have been carried out on the erythrocytic stages of plasmodia, and it is here that the best results have been achieved. Doctor Trager of the Rockefeller Institute and our group at Harvard University have been the principal investigators in this phase of malarial studies.

In carrying out cultivation studies it is important that the culture medium be composed of known ingredients and the physical conditions be well defined so that the procedures can be used for the effective evaluation of the nutrient requirements, metabolic processes, and the actions of antimalarial drugs. Consequently, experiments were first devised to investigate the chemical and physical requirements of the parasitic red blood cells. The inorganic and organic compositions of normal and parasitic cells were determined. The earlier studies of other investigators on respiration and glycolysis were extended so as to determine the amounts and kinds of nutrients to be supplied, the rates of waste material production, and the optimal gas phase. All these have been considered earlier in this chapter.

D. Composition of Media, Culture Vessels, Physical Conditions, and Methods of Parasite Evaluation

From the data which were obtained, principally on the blood of monkeys (*Macaca mulatta*) and from available literature data on simian and human bloods, a nutrient culture medium and physical conditions were formulated for *in vitro* cultivation studies (McKee *et al.*, 1946; Geiman *et al.*, 1946; and Anfinsen *et al.*, 1946). Table VII shows the composi-

Table VII. Composition of Culture Medium

(Taken from Anfinsen *et al.*, 1946, with the permission of the authors and the publisher)

		Inorganic			
$MgCl_2 \cdot 6H_2O$	0.203	g./liter	NaCl	5.825	g./liter
$CaCl_2 \cdot 2H_2O$	0.074		Na_2HPO_4	0.301	
KCl	0.410		$NaHCO_3$*	2.35	
		Organic			
Group 1			Group 3 (*Continued*)		
Glucose	2.50	g./liter	D-Ca pantothenate	500	γ/liter
Amino acids	0.056		Pyridoxine	500	
Glycerol	0.25		Ribose	500	
Sodium acetate·$3H_2O$	0.25		Riboflavine	500	
			Choline	500	
Group 2			Biotin	0.4	
Ascorbic acid	5000	γ/liter			
p-Amino benzoic acid	100		Group 4		
			Xanthine	250	γ/liter
Group 3			Guanine	250	
Thiamine	1000	γ/liter	Thymine	125	
Niacin	1000		Adenine	250	
Nicotinamide	1000		Uracil	250	
Cocarboxylase	400		Cytosine	125	

* Added as Na_2CO_3 (1.48 g./liter) and converted to bicarbonate by passing CO_2 through the solution.

tion of the medium which was devised and successfully used by our group, by Trager (1947c), and by Manwell and Brody (1949).

This bicarbonate medium when equilibrated with a gas phase containing 5 per cent carbon dioxide is a buffer system with a pH of 7.45 ± 0.10. It is also isotonic with blood, freezing point −0.60 ± 0.02° C. As will be discussed later, this medium was subsequently modified to make certain of its constituents more nearly optimal.

Two general types of culture vessels have been devised for studying the *in vitro* growth and metabolism of malarial parasites (Geiman *et al.*, 1946). The first consists of a dilution vessel (RD) which holds a mixture of the nutrient medium and cellular blood elements. The second is a perfusion type (PA) in which the cell mix is separated from the nutrient medium by means of a cellophane membrane which permits the passage of dialyzable nutrients and waste products through the membrane. In fact, two different varieties of the perfusion type vessel, giving widely different membrane surfaces (diffusion rates), have been devised and used. One is a self-contained apparatus (PA_1) and the other a more complex setup which maintains a constant flow of nutrient medium (PA_2). The RD and PA_1 vessels have been employed for most of the studies to be presented here, but the effects of excessive exchange of nutrients and waste products have been demonstrated with the PA_2 vessel. Each of the vessel types, RD and PA, have certain advantages and disadvantages and the choice of setup is governed by the kind of work that is to be performed. If one wishes to retain the nutrients, metabolic materials, or antimalarial drugs for an accurate determination of their rates of change, the RD vessel is the one of choice. However, if one is desirous of removing waste products at a maximal rate or of studying cellophane diffusibility of certain nutrients or test compounds, then the PA type of apparatus is preferred. Also for long-term cultures this latter type of setup has the advantage in that waste product and nutrient exchanges keep the culture in a more nearly physiological state.

With any of these culture vessels, particularly the RD type, it is essential that limited numbers of parasitic cells be placed in the vessel and for more than one generation of growth, and multiplication the numbers of parasites have to be periodically reduced. This is necessary because of the large quantities of nutrients utilized, particularly glucose, and the production of excessive amounts of waste products, for example, lactic acid. The accumulation of excess acid with a resulting increase in acidity of the culture has been shown to be detrimental to the red cell and the contained parasite. First there is a breakdown of 2,3-diphosphoglyceric acid to produce inorganic phosphorus, later a hydrolysis of adenosine triphosphate, and finally a stoppage of glycolysis with resulting leakage of erythrocyte potassium and hemoglobin and general disruption of the cell. For the successful cultivation of the erythrocytic stages of the malarial parasites these processes must be prevented. The other problem involved in parasite cultivation is the maintenance of proper conditions and the supplying of the essential nutrients for the parasite itself. As was stated earlier the parasite depends not just on the red cell but on the environmental plasma and neighboring cells.

The gas phase which was found the best of those tested was one of 5 per cent carbon dioxide—95 per cent air (20 per cent oxygen). While an oxygen content of 95 per cent was detrimental to parasite growth, an oxygen content of 0.4 per cent appeared as good as one of 20 per cent. These studies were made with a constant flow of gas across the culture fluid and thus a constant oxygen content of the culture material. When the oxygen was reduced farther, growth and multiplication of *Plasmodium knowlesi* were diminished (McKee and Geiman, 1950). This finding is consistent with the observations of several investigators that cyanide is inhibitory to respiration and that cytochrome enzymes are functional in parasite metabolism.

Various temperatures ranging from 30° to 40° C. have been tested and 37–39° C. appears to be nearly optimal for cultivation of the erythrocytic forms of human and simian plasmodia. A temperature of 30° C. slows metabolism, growth, and multiplication of *P. knowlesi* (Geiman and McKee, 1950).

A study of the total ionic strength of different media indicated that a relatively narrow range is conducive to maximal parasite growth and multiplication. Media with freezing points of −0.55° to −0.65° C. (±8 per cent of isotonic) are those which produce optimum results.

Agitation of the culture so that suspension of the erythrocytes is maintained has proved of considerable value for *in vitro* cultivation of the plasmodia (Trager, 1947b; Geiman *et al.*, 1946). Lack of agitation allows sedimentation of the red cells which causes multiple cellular invasion, resulting from a high localized concentration of merozoites. As many as five parasites have been found in a single red cell when the culture is not kept in motion.

Few studies of the effects of light waves on malarial parasites have been made. However, Ceithaml and Evans (1946a) found that exposure of chicken erythrocytes infected with *P. gallinaceum* to 100 r units of x rays had no effect, whereas 10,000 r units reduced the ability of the parasitized cells to increase in numbers when injected into chickens. Exposure of the parasitic red cells to 30,000 r units of x rays destroyed the ability of the parasite to infect normal chickens and increased the oxygen and glucose consumption of the treated cells.

E. Growth and Multiplication of Parasites

Employing the medium, culture vessels, and equipment (Fig. 9), and optimum conditions outlined above, *Plasmodium knowlesi* (Ball, Anfinsen, Geiman, McKee, and Ormsbee, 1945) and *P. lophurae* (Anfinsen *et al.*, 1946) were cultivated successfully. The average increase in parasite numbers in one generation was fourfold, although values up to elevenfold have been obtained. Using similar conditions, Trager (1947c) has

FIG. 9. Culture setup showing rocker, RD and PA_1 vessels and gas system used for cultivation studies. (Taken from Geiman, 1948b, and Geiman and McKee, 1948, with the permission of the authors and the publisher.)

obtained growth and multiplication of the avian parasite *P. lophurae*. In experiments performed prior to 1946 but as yet unpublished, the human malarial species *P. vivax* and *P. falciparum* have been cultivated with increases in parasite numbers (McKee, Geiman, Anfinsen, Ormsbee, and Ball, 1949). *Plasmodium relictum* has been cultured *in vitro* by Manwell and Brody (1949). Recently *P. cynomolgi* has been cultivated with multiplications up to fivefold (Geiman and McKee, 1950).

In many cases of human parasitic infections of *P. vivax* the numbers of parasites were too few to allow accurate counting. With such bloods the albumin flotation method of Ferrebee and Geiman (1946) was used to concentrate the parasites. The concentrate was subsequently diluted with normal human blood to give the proper numbers. This procedure accomplished three objectives, the first of obtaining adequate numbers of organisms, the second of removing the patient's red cells and plasma which might contain inhibitory materials and the third of allowing the addition of normal blood which contains adequate nutrients without inhibitory agents.

The techniques and procedures utilized for *in vitro* cultivation studies and the results obtained have been summarized and reviewed by Geiman (1949, 1951) and McKee (1951). Chemical and morphological controls were employed in all experiments so that both the increases in numbers of parasites and their physiological state could be assessed more accurately.

F. *In Vitro* Nutritional Studies

The nutrient culture medium as originally formulated (Ball *et al.*, 1945), although affording multiplication of *Plasmodium knowlesi,* did not produce maximum growth and multiplication. Also the nutrient medium contained Difco proteose-peptone which has many unknown components, an objectionable point when a precisely defined medium is desired. It was soon discovered that the proteose-peptone could be replaced by 1–10 γ per liter of *p*-aminobenzoic acid (Anfinsen *et al.*, 1946).

Growth and multiplication equal to that produced in the host animal still was not obtained. In attempting to evaluate further the relatively small effects produced by variation of the organic components present in the medium, it was necessary to remove the plasma which masked the changes. A dilution technique was devised utilizing normal parasitic cells freed of plasma by washing with an isotonic solution and suspending in the nutrient medium. With this mixture in RD culture vessels little, if any, multiplication of parasites was obtained until protein was added to the system. Purified human and bovine serum albumin and globulin have been used successfully, but purified bovine albumin, at a concentration in the medium of 1 per cent, has been used for the re-evaluation of the organic components of the medium.

Utilizing this plasma-free RD test system it was possible to establish an optimum level of amino acids for the *in vitro* growth of malarial parasites. Eight milligrams per cent of amino acids is the rather critical level required for *P. knowlesi* and about 16 mg. per cent for *P. vivax* (McKee, Geiman, and Cobbey, 1947). The amino acid mixture employed was

Stearn's "Parenamine," which is a 15 per cent solution of an acid hydrolyzate of casein. These more recent studies have shown that with cultures of *P. knowlesi* the hydrolyzate can be replaced completely by a mixture of synthetic L-amino acids, used in the proportion in which they occur in the casein hydrolyzate. It was also determined that this mixture of amino acids could be substituted by the single acid L-methionine in RD cultures of *P. knowlesi* (McKee and Geiman, 1948). The *in vitro* effect produced by methionine is enhanced with *p*-aminobenzoic acid (Table VIII). The interrelationship between these two compounds, apparent both *in vitro* and *in vivo*, will be discussed in more detail under "*In Vivo* Nutritional Aspects." In the perfusion apparatus, where there is outward diffusion and loss of the red cell components, undoubtedly all the

TABLE VIII. *In Vitro* Effects of Methionine (M) and *P*-Aminobenzoic Acid (PAB) on the Growth and Multiplication of *P. knowlesi*. 22-Hour Culture RD, M257 (McKee and Geiman, 1950)

Compound added to culture		Glucose utilized	Lactate utilized	Parasite differential (%)						Fold increase in parasites
Whole blood system	Washed cell system	mg. %	mg. %	Ring	Young troph.	Old troph.	Sch.	Seg.	Non-ident.	
PAB		155	22	2	28	53	14	1	2	5.0
	—	117	32	5	35	40	18		2	1.2
	PAB	135	37	8	43	35	10	1	3	2.0
	M	136	51	2	14	67	16		1	2.5
	PAB + M	139	40	3	27	53	15		2	4.4

amino acids have to be present in the reservoir medium so as to maintain an adequate cell concentration of all the building blocks for protein synthesis.

Ascorbic acid, as will be discussed later, has been studied extensively both *in vitro* and *in vivo*. The nutrient medium as originally prepared contained 0.5 mg. per cent of ascorbic acid. Although the vitamin has no appreciable effect on the multiplication of parasites it does play some *in vitro* role in maintaining the parasites in a good morphological state. The testing of a wide range of vitamin concentrations indicated that the 0.5 mg. per cent became depleted, principally by oxidative degradation, after only about 12 hours of cultivation. At the other extreme 10 mg. per cent was quite detrimental to the growth and multiplication. For *P. knowlesi* a 2 mg. per cent level appears optimum for 24-hour experiments.

Studies with biotin indicate it to be no more than an accessory factor for the *in vitro* growth and multiplication of parasites. Little effect was produced by the addition of biotin to cultures of *P. lophurae* (Trager, 1947c). In using the blood from biotin-deficient parasitic monkeys, no effect was observed on cultures of *P. knowlesi* (McKee and Geiman, 1950).

Although the blocks of vitamins and of purines and pyrimidines (Organic, groups 3 and 4 in Table VII) produce an overall enhancement of the growth of cultures of *P. knowlesi* (Anfinsen *et al.*, 1946), as shown in Table IX, and of *P. lophurae* (Trager, 1947c), it has been extremely

TABLE IX. Effects of Nutrients on Growth (PA_1 Apparatus)
Taken from Anfinsen *et al.*, 1946, with the permission of the authors and the publisher)

Tube	Glucose and PAB	Vitamins	Purines	Pyrimidines	Amino acids	Fold increase 24 hr.
1.	+					3.8
2.	+	+				3.5
3.	+	+	+			4.2
4.	+	+		+		4.2
5.	+	+	+	+	+	6.9

difficult to get detailed data on individual components of each of the two groups of nutrients. Competitive inhibition studies have yielded some interesting results which will be discussed under antimalarial drugs. In this regard Trager (1943b) has obtained an indication of the need of pantothenate for *P. lophurae.*

With this added information it has been possible to obtain growth and multiplication of *P. knowlesi* in excess of that obtained simultaneously in monkeys. Multiplication in the culture using the plasma-free system does not equal that obtained with the RD culture containing monkey plasma. Liver fractions as well as plasma and dialyzable components of plasma potentiate the culture system (McKee and Geiman, 1950). In other words, this nutrient medium is deficient in some of the components of normal plasma. It is the object of present research to determine the nature of these other plasma components which are essential for maximum *in vitro* growth and multiplication of malarial parasites.

V. *In Vivo* Nutritional Aspects

A. Influence of Animal's Diet on Parasite Growth and Multiplication

The nutrition of the host animal in relation to its ability to grow malarial parasites has been considered by a number of investigators. As stated previously, the vital importance of the amino acid methionine, for the growth and multiplication of *Plasmodium knowlesi* has been shown *in vivo* as well as *in vitro* (McKee and Geiman, 1948; Geiman and McKee, 1948). By fasting monkeys for 24 to 48 hours the parasite infection

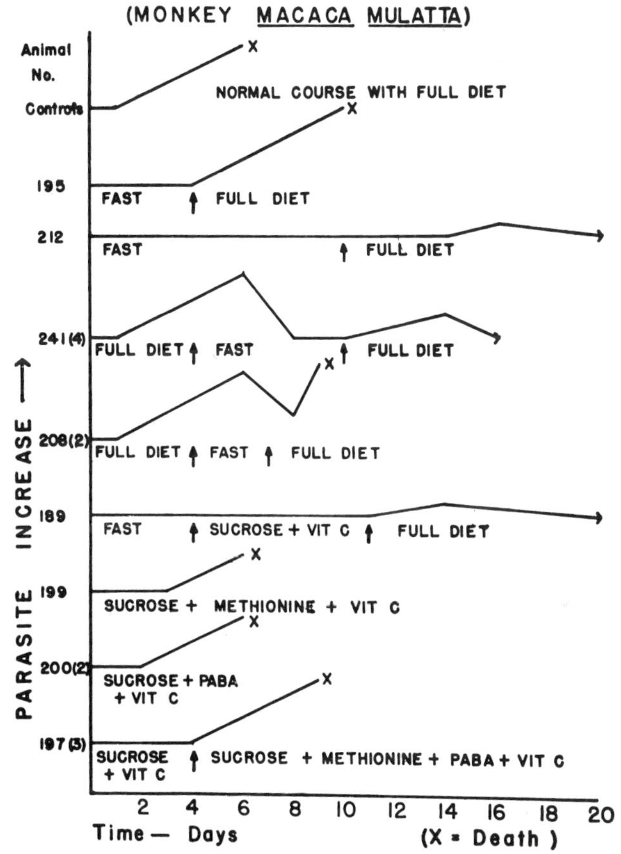

FIG. 10. Schematic representation of the effects of diets on the *in vivo* growth of the *P. knowlesi*. (Taken from Geiman and McKee, 1948, with the permission of the authors and the publisher.)

can be strikingly ameliorated. Likewise when the fasting is started simultaneously with the inoculation of the animal with parasites, only a few organisms appear in the blood during the period of continued fast. As soon as the animal is given a full diet, parasites again grow normally. Also, the oral or intramuscular administration of methionine or *p*-aminobenzoic acid (PAB), while continuing the fast, was found to reverse the parasite control in the fasted monkeys, and the animals died of overwhelming parasitic infections. Administered glucose and ascorbic acid had no effect on the animal's infection. It will be noted from the diagrammatic representation of the results in Fig. 10 that withholding food or methionine and PAB for more than 9 or 10 days produced a nonreversible situation. As will be described later the explanation appears to be that the additional time gives the animal a chance to produce antibody protein which controls the growth of parasites. This reversal is produced in spite of the normal levels of plasma methionine and abnormally high levels of PAB, even before treatment with these two compounds (McKee and Geiman, 1948). It is important to realize, however, that these analyses were for free or non-protein-bound methionine and PAB in the plasma. No analyses were made for protein-bound erythrocytic materials. Thus these results may not be too meaningful since it is the content of the parasitized cell *per se* and probably the bound forms present there which are of paramount importance. The presence of excessive quantities of free compounds in the plasma may mean merely an increased mobilization from the parasitic cell. The precise reasons for this acute need for methionine and PAB are not entirely clear; however, as suggested earlier, the low content of methionine in red cell protein may make it necessary for the parasite to draw on the environment (plasma and other tissues) for additional supplies of the amino acid. This may be equally true for PAB. It is the belief of the author that peptides containing methionine and PAB may be vital to the economy of the parasite. Similar requirements for the growth of different parasites in other animal species have not been determined. Trager (1947a, 1947b) had observed earlier a similar situation with plasma biotin in *P. lophurae* or *P. cathemerium* infected chicks or ducks. The infected birds showed an increase in the plasma level of biotin as the infections progressed.

Another dietary factor of considerable importance to the *in vivo* growth of parasites is ascorbic acid. A deficiency of this vitamin in Rhesus monkeys greatly reduces the numbers of *P. knowlesi* in the animals. Giving pure ascorbic acid to the animal reverses the situation and effects growth of erythrocytic parasites (McKee and Geiman, 1946). As indicated previously, the absence of ascorbic acid in cultures produces no appreciable effect on parasite multiplication. This means that the vitamin acts indirectly through the host animal. Attempts to answer the question of what influence vitamin C has on the growth of parasites *in vivo* have met with only partial success. Studies with scorbutic guinea pigs (McKee, Cobbey, and Geiman, 1947, 1949) indicate a sluggishness both in the production and in the mobilization of liver glycogen. Attempts to stimulate liver glycogen production by administering adrenal cortical hormones were unsuccessful. It is well known from the investigations of many workers that glucose is required for the growth and multiplication of malarial parasites and that liver glycogen drops to low levels during the normal course of a parasite infection (Christophers and Fulton, 1938; McKee and Geiman, 1946). Hegner and MacDougall (1926) showed with the avian parasite *P. gallinaceum* that insulin could markedly slow the growth of the parasites. Although the lowering of blood sugar may have some influence on *in vivo* multiplication of parasites other factors are believed to be important. There may be many other indirect effects, such as influencing the amounts of available methionine or peptides containing this amino acid, which we know to be of great importance in parasite growth and multiplication.

Josephson, Taylor, Greenberg, and Nadel (1949) studied the ascorbic acid content of the adrenal glands of chicks infected with *P. gallinaceum* but observed no detectable change, although adrenal hypertrophy was demonstrated (Nadel, Taylor, Greenberg, and Josephson, 1949). The explanation for the inapparent drop of adrenal ascorbic acid is undoubtedly that the chick produces its own ascorbic acid and the adrenal hypertrophy definitely indicated a state of stress due to the parasite infection.

A number of other vitamins appear to have significant influences on the *in vivo* growth of malarial parasites. Trager (1943b, 1947c) working with *P. cathemerium* infections in ducks and *P. lophurae* infection in both ducks and chickens, and Seeler and Ott (1944, 1946) studying *P. lophurae* infections in chickens have shown that multiplication in both chicks and ducks increases as the biotin deficiency becomes more and more severe. Similar effects have been observed with *P. lophurae* infections in chickens possessing a folic acid deficiency (Seeler and Ott, 1945a), with a protein deficiency (Seeler and Ott, 1945b), and with nicotinic acid deficiency (Roos, Hegsted, and Stare, 1946). These last workers also observed a slight increase in parasite growth in choline deficiency, the reverse effect with vitamin A deficiency, and no noticeable influence in thiamine deficiency. In ducks, however, none of the nutritional deficiencies influenced the course of infection to a marked degree. Definite controlling influences on *in vivo* growth of *P. lophurae* were observed in riboflavin deficiency (Seeler and Ott, 1944) and on *P. gallinaceum* growth in calcium pantothenate deficiency (Brackett, Waletzky, and Baker,

1946). Brooke (1945) observed that a diet of restricted quantity caused an increase in the numbers of *P. relictum* and *P. cathemerium* in canaries and of *P. relictum* in pigeons.

B. Relationship of Diet to *In Vitro* Cultivation

An interesting and instructive approach to the problem of determining whether or not certain nutrients are important for *in vitro* cultivation of malarial parasites is through animal nutrition studies. As stated previously (under "Cultivation"), plasma was removed from the blood used in some of the later cultivation studies in an attempt to remove the masking effects produced by the plasma constituents. This still leaves the components of the formed elements of the blood which may supply materials required for growth and multiplication. One way to obviate this difficulty is to make the animals deficient of the factor to be studied and then, if possible, to infect the animals with parasites and use the infected blood for *in vitro* studies with culture medium deficient in the dietary component being investigated.

Such studies have been made with *Plasmodium knowlesi* infections and ascorbic acid and biotin deficiencies in monkeys (McKee and Geiman, 1950). The results gave the strong indication that neither compound had any influence on the *in vitro* growth of the parasite. As stated previously, the *in vitro* effect of ascorbic acid was one of controlling the *P. knowlesi* infection in monkeys, whereas biotin produced no significant effect in the animals.

VI. Natural Control of Malarial Infections

A. Antibody Protein Production and Control of Malarial Infections

No attempt will be made to consider in detail the cellular defenses of the animal body against parasites, important as they are. The material presented will involve principally a consideration of plasma components which are effective in controlling the growth and multiplication, or influence the biochemistry, of malarial parasites.

Extensive studies with human, avian, and simian malarias have proved that immunities or controlling conditions are produced in the host animals during the course of active infections. The acquired immunity is effective only against homologous strains and not against heterologous strains. This defense depends primarily on the response of the lymphoid-macrophage system induced by the presence of the parasites. The stimulus brings about the production of protein or protein-containing substances released into the bloodstream which in some as yet unknown manner act on the parasitic cell. Concurrent with or subsequent to this action, a phagocytic response is elicited. This action is exemplified by a proliferation of macrophages and the appearance of more parasitic cells in the spleen, liver, and bone marrow.

In 1938 Eaton demonstrated the presence of agglutinins in sera from Rhesus monkeys possessing chronic *Plasmodium knowlesi* infections. The agglutination test was species specific, but no reaction was obtained with erythrocytes containing immature parasites, indicating a slow antigenic response. Complement-fixing antibodies were found to be present in the blood of monkeys with chronic *P. knowlesi* infections (Coggeshall and Eaton, 1938a). Recently a γ-globulin protein containing a high potency of antibody material has been isolated from the plasmas of chronically infected monkeys repeatedly inoculated with large quantities of *P. knowlesi* (Burnham, McKee, and Geiman, 1950). Monkey plasma with titers as high as 1:4,000 were obtained and the isolated protein accounted for essentially all the starting activity. This protein material, as will be described in detail later, is highly effective in controlling the *in vitro* growth of *P. knowlesi.*

Mulligan and Russell (1940) made the initial observation that there was specific agglutination of the sporozoites of *P. gallinaceum* with homologous antiserum in dilutions as high as 1:8,000. Working with sporozoites and immune sera from human and fowl malarias, they obtained strong evidence for the species specificity of the reactions.

Many antigen-antibody tests have been devised to indicate the presence of antibody substances in the blood of infected animals. A precipitin test was devised by Taliaferro and Taliaferro (1928) using an extract of human placenta from a case of malaria. Dulaney and House (1941) were able to obtain a precipitin reaction with an antigen from *P. knowlesi.* Macroscopic agglutination was obtained with the collodion methods of Cannon and Marshall (1940) and of Goodner (1941). Also, as mentioned above, there are the agglutination test (Eaton, 1938) and the complement-fixation reaction (Coggeshall and Eaton, 1938a). In this regard, however, the work of Mayer and Heidelberger (1946) showed that only positive tests indicated malaria while a negative test was inconclusive. This is due to the fact that the reaction diminishes 1–6 weeks after relapses and only increases during or immediately after a subsequent relapse. Thus the asexual parasites appear to be poor antigens. This has been indicated further by lack of protective action against *P. vivax* infections when concentrated antigenic material was administered (Heidelberger *et al.,* 1946). However, Freund, Thomson, Sommer, Walter, and Pisani (1948) reported that they were able to protect monkeys and ducks against malaria by using a vaccine containing parasites, an emulsifying agent, an adjuvant, and killed tubercle bacilli. The subject of immunization against malaria has been reviewed by Freund (1947).

A number of interesting chemical tests have been devised for testing plasmas from malarious animals. Non-specific ferro-flocculation and melano-flocculation tests for antigen-antibody reactions have been devised by Henry (1933, 1934), and subsequently a similar protein-tyrosine reaction was introduced by Proske and Watson (1939). An interesting colorimetric test for malaria has been devised by Carlson, Mueller, and Bissell (1946). The method makes use of solutions of Congo red dye, quinine hydrochloride, and the serum to be tested. Positive results were obtained with infections of *P. knowlesi, P. gallinaceum,* and *P. vivax.* This test is of considerable value for detecting malarial infections of long standing. Although the mechanism of these chemical reactions has not been worked out, they undoubtedly depend principally on an altered ratio of albumin to globulin and particularly in the formation of additional quantities and kinds of plasma globulin proteins.

B. Effects of Antibody Proteins on *In Vitro* Growth and Multiplication of Parasites

Interesting inhibitory effects on the *in vitro* growth of parasitic cells have been obtained by two groups of workers. Trager (1947c) working with cultures of *Plasmodium lophurae* found that the human plasma fractions of α- and β-globulins, which are relatively rich in the lipid-like material FSF to be described in the next section, have a pronounced influence on the parasites, while γ-globulin, fibrinogen, and albumin, which are quite poor in the FSF factor, produce little effect on the parasite cultures. When the protein fractions richer in FSF were added to the culture medium at a concentration of 2 per cent there was hemolysis of the red cells and degeneration of the parasites. At lower concentrations no hemolysis was produced, but there was inhibition of multiplication with the degeneration of many of the parasites. Similar results were obtained with plasma protein fractions from horse and pig bloods.

Making use of γ-globulin plasma protein fractions from chronic monkeys superinfected with *P. knowlesi,* Burnham, McKee, and Geiman (1950) obtained marked inhibition of the *in vitro* growth of *P. knowlesi*

The other protein fractions showed little activity. The isolated proteins were employed in amounts comparable to those present in the original

Table X. *In Vitro* Influence of γ-Globulin Antibody Protein on the Multiplication of *P. knowlesi*. 23-Hour RD Culture, M207 (Burnham, McKee, and Geiman, 1950)

Material added to washed cell system	Agglutination titer with *P. knowlesi*	Parasite differential %					Increase in parasites fold
		Ring	Young troph.	Old troph.	Sch.	Non-ident.	
Plasma #186	1:256	1	27	61	9	2	1.8
γ-Globulin Protein	1:256	3	38	49	9	1	2.1
Supernatant	1:2	1	30	61	7	1	4.0

serum and at levels in the cultures similar to the quantities of plasma present (0.8 ml. per RD culture). The results are shown in Table X.

C. Other Factors Influencing the *In Vivo* Control of Malarial Infections

We have already seen that the diet of the host animal is an important factor in the growth of malarial parasites. This may be due primarily to a direct nutritional effect on the parasites as discussed previously. However, a second factor may be the influence of the diet on the production of antibody protein. A great deal of difficulty is encountered in attempting to differentiate between these direct nutritional effects on the parasites and the influences on the production of specific humoral factors working to destroy the parasites. The striking influence that famines exert on the intensity of epidemic malaria in Asiatic countries, particularly India, has been noted by Gill (1928). However, Krishnan (1948) has indicated that the addition of protein components to the diets of Indians caused an upsurge of the dormant malarial infections. Thus the age-old question of whether famine or epidemic malaria comes first is still an unsolved riddle.

It is well known that the adequacy of the diet, particularly from the standpoint of sufficient quantities of dietary protein and its content of "essential" amino acids, is of paramount importance in the production of structural and functional cell proteins (Rose, 1949). Likewise the production of antibody proteins as well as the proteins of the malaria parasite are dependent upon the presence of essential building blocks which ultimately come into the body from the food eaten. The reservoir of tissue proteins and the ability of the animal body to produce the "nonessential'' amino acids may conceal for a period of time the need from outside sources. The fasting and methionine-PAB studies cited earlier are indicative of such a reservoir of material for the vital processes of the animal, but not for the parasites. Such a study indicated the ability of the host animal, even in the face of several days of fasting, to produce antibody protein which after a period of 8–10 days was adequate to control the infection even when nutrients required for parasite growth were again given to the animal. This simple type of experiment indicates the unsuccessful competition of the parasite with the host animal for the essential building materials, amino acids. In this instance then the restriction of the animal's diet was primarily one of starving the parasites, with apparently little effect on the ability of the host to produce antibody protein.

Likewise the ascorbic acid deficiency in monkeys, discussed earlier, is probably an indirect starvation of the parasites with little influence on the defense mechanisms of the animal body. There is no intention of implying here that the diet of the animal does not play a major role in the humoral control of diseases. Many investigators have conclusively shown important influences of the dietary on the humoral amelioration of bacterial and protozoal infections, but this is outside the scope of this chapter. The important point to stress here is that there are marked differences between the host and the parasite in their metabolic requirements, particularly in their competition for vital nutrients from the environmental body fluids and tissues. The successful parasite triumphs at the expense of the host, while the unsuccessful parasite is doomed to death, with many factors operating, including the phagocytic action of the lymphoid-macrophage system (Cannon and Taliaferro, 1931).

One of the most interesting studies on the interrelationship of blood factors to the *in vivo* growth of malarial parasites is that by Trager (1947a,b) on the biotin and biotin-active substances in the plasma of birds infected with *Plasmodium lophurae* and *P. cathemerium*. During the course of the infection of chickens with *P. lophurae* and of ducks with either *P. lophurae* or *P. cathemerium*, significant changes occurred both in the plasma concentrations of free biotin and of a material which on acid hydrolysis yields an ether-soluble substance (FSF) having the biological activity of biotin. In birds surviving *P. cathemerium* and *P. lophurae* infections the biotin concentration rose early in the infection, the concentration then fell slightly, but then increased again, reaching a peak concentration at about the time of the peak parasite infection, then returned to normal. In contrast to the biotin, the bound FSF material increased at first, then decreased slightly, then rose and remained at a high concentration throughout the period of decline of the parasitemia. The plasma levels of biotin, free FSF, and protein-bound FSF are shown in Table XI. In the animals which died of the parasite infection, the

TABLE XI. The Concentration of Biotin and of Ether-Soluble Factor (FSF) in the Plasma of Normal 40-Day-Old Ducks and of Ducks of the Same Age at the Peak of a Malarial Infection
(Taken from Trager, 1947c, with the permission of the author and the publisher)

Conditions of ducks	No.	Biotin activity in mγ per ml. Plasma		
		Free Biotin*	Free FSF†	Bound FSF‡
Not infected	1	1.8	1.0	11.3
	2	1.3	0.7	9.8
	3	1.8	0.2	9.3
	4	1.8	0.2	10.5
5th day of infection with *P. lophurae*	5	10.5	1.5	11.7
	6	3.3	0.2	8.5
	7	8.3	1.2	8.8
	8	5.0	0.5	7.8
3rd day of infection with *P. cathemerium*	9	12.3	2.4	12.9
	10	10.3	2.7	12.7
	11	9.5	1.5	16.2
	12	4.8	0.7	13.2

* Activity of unhydrolyzed plasma after ether extraction.
† Difference between activity of unhydrolyzed plasma before and after ether extraction.
‡ Difference between activity of hydrolyzed plasma before and after ether extraction minus the free FSF.

free biotin instead of returning to normal, rose to very high levels just before death, while simultaneously the FSF fell to very low levels terminally.

Trager also determined in cultures of *P. lophurae* that plasma protein fractions rich in FSF inhibited parasite multiplications. Thus his results indicate that the FSF substance in plasma is intimately concerned in resistance to avian malaria.

VII. Antimalarial Compounds and Their Influence on the Metabolism of Malarial Parasites

A. Chemotherapeutic Agents and Methods of Study

No attempt will be made to survey all the various chemical types of antimalarial compounds which have been synthesized and studied. This extensive subject, covering well over 15,000 different antimalarial drugs, has been reviewed by several investigators and authors, Wiselogle (1946), Blanchard (1947), and Lourie (1947). Also hundreds of natural plant products have been surveyed for antimalarial activity (Osborn, 1943; Lucas and Lewis, 1944; Spencer *et al.*, 1947). In this brief résumé

TABLE XII. Structural Configurations of Some Antimalarial Agents Employed in Metabolic Studies

Common name	Chemical name	Structural formula
Quinine, SN 359	6-Methoxy-α(5-vinyl-2-quinuclidyl)-4-quinoline-methanol	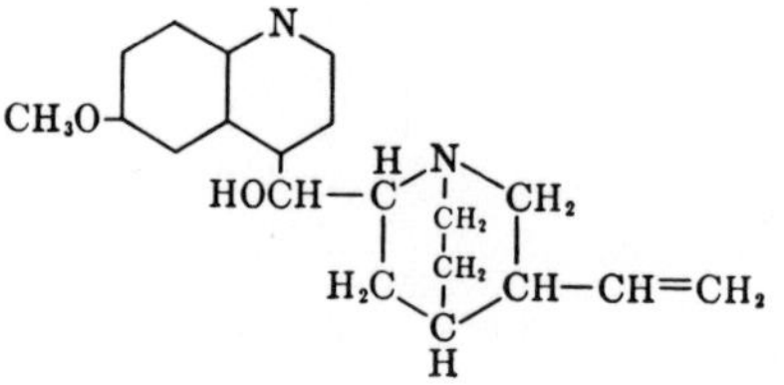
Quinacrine, Atabrine, SN 390	6-Chloro-9-(4-isoethylamino-1-methylbutylamino)-2-methoxyacridine	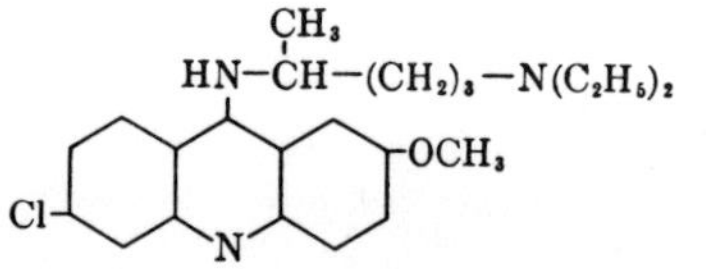
Paludrine, Chlorguanide, Sn 12,837, M 4888	1-(*p*-Chlorophenyl)-4-isopropylbiguanide	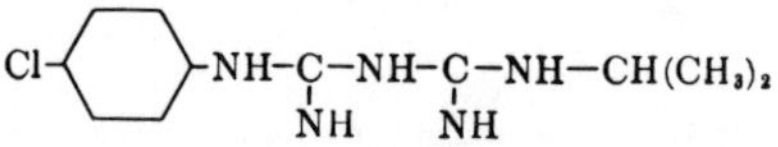
Sulfadiazine, SN 112	*p*-Aminobenzene sulfonamidopyrimidine	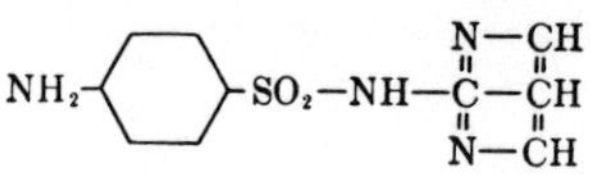
Chloroquine, SN 7618	7-Chloro-4-(4-diethyl-1-methyl-butylamino) quinoline	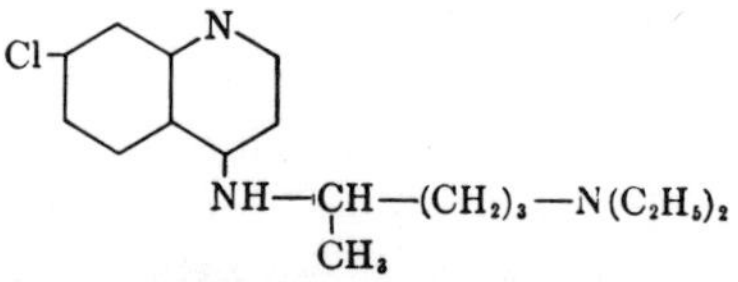
Pentaquine, SN 13,276	8-(5-Isopropylamino amylamino)-6-methoxyquinoline	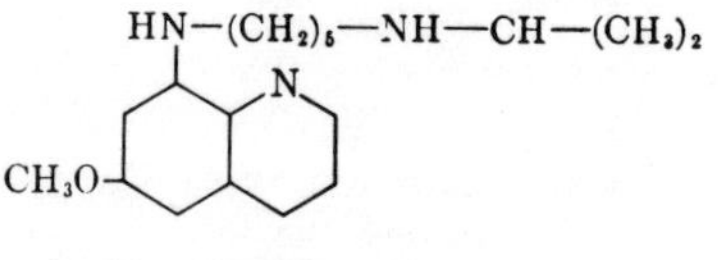
Isopentaquine, SN 13,274	8-(4-Isopropylamino-1-methylbutylamino)-6-methoxyquinoline	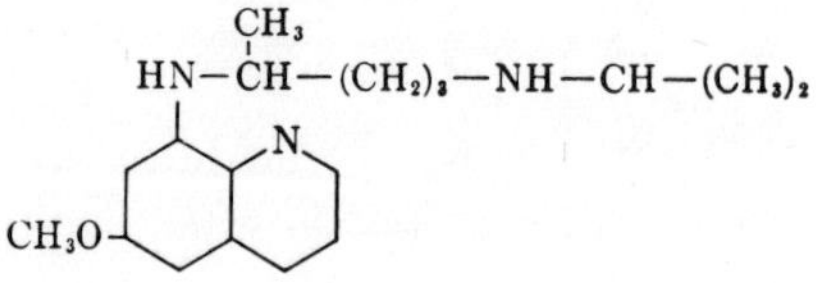

there will be considered only a few of the basic types of antimalarials and the known influences which they exert on the metabolic processes of the parasitic cell. Some of the biochemical effects produced by quinine, quinacrine, paludrine, the sulfonamides (sulfanilamide), the 4-aminoquinolines (chloroquine), the 8-aminoquinolines (pentaquine and isopentaquine), and the naphthoquinones (2-hydroxy-3-alkylnaphthoquinones) have been investigated adequately to warrant their discussion here. The names and structural configurations of these compounds, which are the major ones to be considered, are listed in Table XII.

In general, the methods employed for studying the metabolic and morphological influences of antimalarials may be classed as *in vitro* or *in vivo.* Combined *in vitro-in vivo* techniques have been utilized by some investigators. This latter type of combined experiment has been of considerable importance in evaluating the effectiveness of some of the compounds, of determining what metabolic systems are interfered with, and whether the drug, *per se,* or its metabolic and degradation products are the effective agents. Thus it can be seen that no single method can be used with maximum effectiveness in studying antimalarial agents. This fact is one which was not observed to its fullest extent during the recent wartime testing program for the screening of antimalarial drugs. Here the major emphasis was on the *in vivo* test system utilizing principally avian malarial species. The difficultness of the *in vitro* procedures and the lack of any culture method until late in the war program are the major reasons for the scarcity of data on *in vitro* antimalarial studies.

In addition to the uses of the Warburg respirometer and the usual chemical procedures for evaluating the chemical and metabolic changes over short periods of study, the longer-term culture technique already described has been employed for evaluating further the metabolic changes. Moreover, as indicated above, by comparing chemically and morphologically the culture with the simultaneously conducted animal experiment, more meaningful evaluations can be made.

The *in vitro* types of experiments to be presented here consist of the direct addition of the test compounds to the Warburg or culture vessels and then performing interval studies. The *in vivo* approach used was to administer the test antimalarial to the infected animal and then to draw blood samples at intervals for metabolic studies and chemical analyses. The combined *in vivo-in vitro* investigations involved either: (1) the withdrawing of blood from the infected animal and the simultaneous addition of the antimalarial compound to the animal and to the *in vitro* test system, or (2) the addition to the *in vitro* system of blood from a normal animal previously treated with the antimalarial. Metabolic, morphological, and chemical studies are then made on the animal and culture systems at intervals.

B. Distribution of Antimalarials between Blood Plasma and Erythrocytes

The presence of an effective concentration of the active chemotherapeutic agent at the site of action is of prime importance for effective inhibition of parasite growth. The work of Shannon, Earle, Brodie, Taggart, and Berliner (1944) has conclusively shown this by demonstrating the plasma level of quinacrine (atabrine) and other antimalarials to be an essential factor in the therapeutic action of the drugs. This means, of course, not just a high plasma level but also a high parasitic cell level. The former is, however, an essential prerequisite to the obtaining of the cell concentration of the agent. To this end a few workers have concerned themselves with plasma and red cell levels of the administered antimalarials, particularly quinine and atabrine.

Grenfell, Richardson, and Hewitt (1946) determined that in *Plasmodium lophurae* infected ducks treated with quinine the ratio of intracellular (I) to extracellular (E) quinine was about 4 compared to a ratio of 1.5 for uninfected birds. Likewise Oldham, Kelsey, Cantrell, and Geiling (1944) reported the I/E quinine ratios to be higher in *P. gallinaceum* infected chickens than for normal fowls. Geiman, McKee, Anfinsen, Ormsbee, and Ball (1949) found the I/E ratio for quinine in a perfusion culture of *P. knowlesi* infected monkey blood to be about 1.4 even when different blood quinine levels were used (13.3 and 34.3 mg. per liter). The parasite percentage was, however, only about 2 per cent, and the red cells, unlike the avian cells, are non-nucleated. Oldham *et al.* (1944) and Silverman, Ceithaml, Taliaferro, and Evans (1944) noted that although their ratios for quinine were fairly constant in normal animals, they were somewhat variable for infected bloods. Ceithaml and Evans (1946b) studied the factors governing the differences in the I/E quinine ratios between normal and parasitized blood samples. They found that many factors influenced the quinine distribution, and therefore, since these factors were more variable in blood containing parasitic cells, it explained the greater variabilities in the ratios obtained for infected bloods. Reticulocytes appeared to contain more quinine than mature red cells, and this may explain the marked differences, cited above, between nucleated bird red cells and non-nucleated monkey cells. These workers observed that higher pH values favor penetration of quinine and other basic drugs into the cells; and at a constant pH the concentration of pigments, particularly ferrihemic acid (hematin), influences the quinine distribution.

Similarly there is a concentration of atabrine in cellular elements of the blood (Shannon *et al.*, 1944). *In vitro* equilibration experiments by these workers showed I/E ratios in blood of 2.1 to 2.4 while blood from patients repeatedly treated with atabrine gave I/E valves of 1.3 to 1.7. It is interesting that the proteins of the blood bind as much as 90 per cent of the atabrine which was added so that for *in vitro* experiments using Ringer's solution in place of plasma the I/E values were 7.9–8.2.

Data on blood ratios are not available for the other antimalarial compounds to be discussed here, but quite likely I/E values, if known, would show values comparable to those for quinine and atabrine.

C. General Effects of Antimalarials, *In Vitro* and *In Vivo*

The early antimalarial studies consisted principally of observations on the morphological changes produced in the parasites by *in vivo* drug administration. The most important aspect of these early investigations was, however, the realization that the different drugs acted rather specifically on the different stages of the plasmodia. Wenyon (1926) described a reduction in the size of *Plasmodium vivax* schizonts and a decrease in the numbers of merozoites. Lourie (1934) working with *P. cathemerium* infections in canaries found that quinine caused a retardation of growth and the formation of a smaller number of merozoites. The production of pigmentary and cytoplasmic changes by atabrine on *P. vivax, P. malariae,* and *P. falciparum* were discussed by James (1934) and by Hühne (1942). Schwartz (1939) and Young, McKendon, and Smarr (1943) determined that thiobismol (sodium bismuth thioglycollate) has an inhibitory effect specifically on the half-grown trophozoites of *P. vivax.* Trager, Bang, and Hairston (1945) found that *P. vivax* lost its amoeboid activity an hour after an intramuscular injection of atabrine. Morphological changes and the stoppage of development immediately preceding nuclear division were observed by Fairley *et al.* (1946) in *P. vivax* patients treated with paludrine.

Numerous investigators have studied both *in vitro* and *in vivo* the effects of many of the antimalarial drugs on the metabolism of the parasitic red cell. It would be of value at this point to consider briefly the statement of Maegraith (1948) that, "When interest was first taken in the method of action of various specific drugs on the infecting organisms, it was found that these drugs caused a remarkably lethal action on the organism in the body but a like effect was not necessarily seen when a similar concentration of the drug was added to the organism *in vitro*." There are undoubtedly many reasons for discrepancies between the inhibitory effects in test tube and animal experiments, but they probably fall within the following two categories of reasoning. First, there have not been, until quite recently, *in vitro* methods for maintaining malarial parasites for extended periods of time in a condition approaching physiological. Thus the early *in vitro* experiments would allow at best only survival of the parasites, without significant growth and multiplication. It is well known that arrested metabolic processes are not as drastically influenced by metabolites as are those same processes in an active state. Secondly, there is the ability of the host animal's organs and tissues to metabolize and change the chemical structure as well as to influence the absorption of the active compounds into the parasitic cells. This is a remote possibility when the drug is added directly into the test tube. In addition, culture experiments essentially rule out humoral components (hormones, antibodies, etc.) and other indirect influences which may have a considerable effect in the animal. It is important to keep these two general factors in mind in a consideration of the *in vitro* versus *in vivo* action of any chemotherapeutic agent, particularly in the following discussions where the amounts of antimalarial compounds used are in most cases 10 to 1,000 times greater than a therapeutic concentration.

D. Effects of Antimalarials on Enzyme Systems

One of the earliest investigations of the effects produced by antimalarials on parasite metabolism was that of Fulton and Christophers (1938) in which they measured the inhibitory effects of quinine and atabrine on oxygen uptake by *Plasmodium knowlesi*. At added concentrations of 0.001 to 0.0001 M these two drugs inhibited oxygen consumption by 25 to 50 per cent. Similar results subsequently have been obtained for these two antimalarials on the avian parasites *P. lophurae* and *P. gallinaceum* by Coggeshall and Maier (1941) and by Silverman *et al.* (1944). Coggeshall and Maier concluded from their work that a respiratory method could not be used for evaluating the therapeutic effectiveness of antimalaria drugs, there being a poor correlation both among the various kinds of drugs and among the concentrations of compounds required to produce repiratory inhibition. The concentrations of quinine utilized were ten to one hundred times greater than the effective therapeutic blood levels (Kelsey, Oldham, and Geiling, 1943; Oldham *et al.*, 1944). However, Ceithaml and Evans (1946) found that at 0.00001 M quinine concentration, approximately the serum concentration in therapy, the oxygen consumption of *P. gallinaceum* infected chicken erythrocytes was inhibited.

At this low level (1×10^{-5} M) of quinine Silverman, Ceithaml, Taliaferro, and Evans (1944) found no inhibition to the formation of lactate from glucose. Thus they suggested that quinine interferes with some step in the oxidation of lactate or pyruvate to carbon dioxide and water. Moulder (1948, 1949) has recently determined this to be the case. In his studies he utilized the *in vivo* method, previously described, of pretreating the *P. gallinaceum* infected chickens with quinine. He has placed the site of metabolic inhibition at the first step after the removal of pyruvate, namely the formation of a reactive two-carbon particle.

The early *in vitro* experiments on quinacrine also were performed with about one thousand times the blood levels used in the treatment of malarial infections (Oldham and Kelsey, 1945). In these high concentrations, atabrine like quinine inhibits both the respiration and glycolysis of *P. gallinaceum,* while lower concentrations inhibit only respiration (Silverman, Ceithaml, Taliaferro, and Evans, 1944).

Considerable investigation has been carried out on the mechanism of enzyme inhibition by atabrine, particularly concerning the flavoprotein enzyme systems. Haas (1944) determined that quinacrine strongly inhibits the flavoprotein enzyme cytochrome reductase, and that the effect may be reversed by riboflavin phosphate. Wright and Sabine (1944) reported that quinacrine inhibits the flavoprotein enzyme D-amino acid oxidase and that the inhibition is antagonized by flavine adenine dinucleotide (FAD). From these results the suggestion was made that atabrine, because of its structural similarity to FAD, might specifically compete with the flavine coenzyme for the protein enzymes. Hellerman, Lindsay, and Bovarnick (1946) investigated this possibility and found that not only atabrine but also many structurally unrelated compounds, quinine, plasmochin, and certain quinoline derivatives, competitively inhibit D-amino acid oxidase. They concluded therefore that atabrine inhibits the action of flavoprotein enzymes because of its ability to combine with the enzyme proteins and not because of its structural resemblance to the coenzyme.

Speck and Evans (1945b) made a study of specific glycolytic enzymes of *P. gallinaceum* and of yeast and mammalian muscle which were inhibited by quinine and atabrine. Atabrine was found to inhibit the hexokinase and lactic dehydrogenase enzymes of the parasite preparations whereas quinine was less effective. Both atabrine and quinine inhibited yeast hexokinase. Neither drug effected any change in the activity of the enzyme, 3-phosphoglyceraldehyde dehydrogenase. Quinine inhibited the enzymes phosphorylase and phosphoglucomutase from rabbit muscle while the other glycolytic enzymes from muscle were not affected. The lactic dehydrogenase system from beef heart was strongly inhibited by atabrine but only slightly affected by quinine. Since the

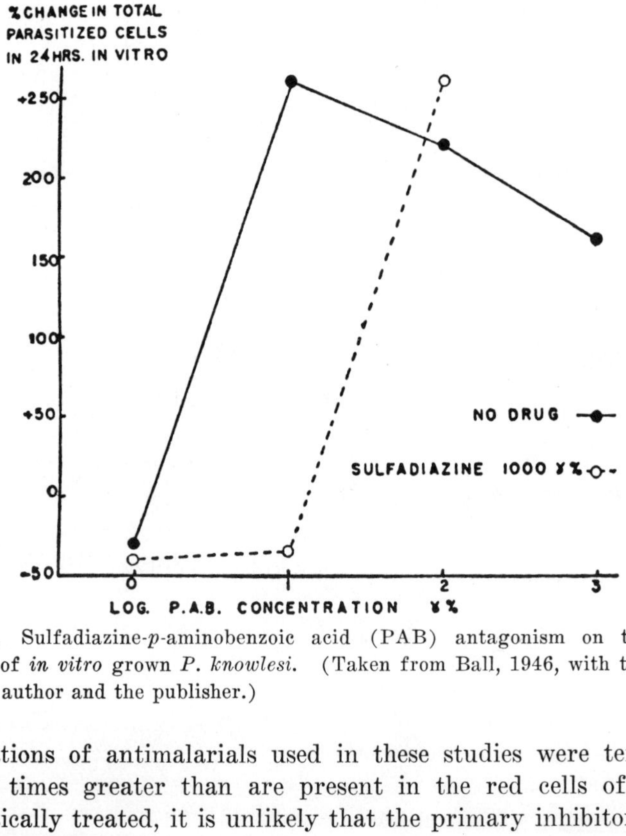

FIG. 11. Sulfadiazine-*p*-aminobenzoic acid (PAB) antagonism on the second generation of *in vitro* grown *P. knowlesi.* (Taken from Ball, 1946, with the permission of the author and the publisher.)

concentrations of antimalarials used in these studies were ten to one thousand times greater than are present in the red cells of animals therapeutically treated, it is unlikely that the primary inhibitory action is on the glycolytic systems. It seems probable from this work and the investigations of Moulder that quinine and atabrine inhibit principally oxidative processes in the malarial parasite. This is supported by the investigations of Bovarnick, Lindsay, and Hellerman (1946) mentioned earlier under "Carbohydrate Metabolism." They determined that quinacrine interferes with the phosphorylation of glucose in *P. lophurae.*

However, the phosphorolative processes appear to be oxidatively linked since substrate-depleted parasites oxidize glucose only after an induction period which is shortened by the addition of adenylic acid, adenosine triphosphate, succinate, or fumarate. This effect, however, may be as suggested by the investigators, only a competition of quinacrine with ATP for the enzyme hexokinase.

The metabolic studies with the sulfonamides have been numerous and of considerable interest. Coggeshall (1940) and Coggeshall and Maier (1941) found that the sodium salts of sulfathiazole and sulfapyridine (0.01 M) inhibit the respiration of *P. knowlesi,* but only by 40 per cent. This rather small amount of oxidative inhibition at the high drug level, together with its slow rate of effect on parasite control either *in vitro* or *in vivo,* suggests the primary role to be something other than a direct influence on the oxidative enzymes. In this regard Marshall, Litchfield, and White (1942) and Maier and Riley (1942) have demonstrated that the chemotherapeutic action of sulfonamides against malaria is antagonized by *p*-aminobenzoic acid (PAB). Ball (1946) has demonstrated, as shown in Fig. 11, a similar competitive inhibition of PAB and sulfadiazine for *in vitro* grown *P. knowlesi.* These observations substantiate the requirement of PAB for the cultivation of *P. knowlesi* (Anfinsen *et al.,* 1946). As indicated previously under "*In vitro* Nutritional Studies" there is a suggestion that there may be an interrelationship between PAB and the amino acids, particularly methionine. The relatively greater inhibitory effect of the sulfonamides in *P. knowlesi* infections in monkeys compared to human infections in man is also an intriguing problem.

The 2-hydroxy-3-alkylnaphthoquinones are an interesting group of antimalarial agents, a large number of which have been studied by Fieser (1946) and by Wendel (1946). These naphthoquinones inhibit the respiration of *P. lophurae* and *P. knowlesi* in concentrations as low as 0.000001 M. The oxidation of lactate appears to be the main inhibitory action of these compounds. The respiration of yeast cells and of beef heart succinoxidase likewise have been demonstrated to be inhibited by a number of different 2-hydroxy-3-alkylnaphthoquinones (Ball, Anfinsen, and Cooper, 1947). These investigators concluded that the naphthoquinones inhibit some step in the chain of electron transport below cytochrome C and above cytochrome B.

The comparative effects of atabrine, chloroquine, pentaquine, isopentaquine and paludrine, all at comparable blood levels, have been studied *in vitro* with the simian parasites, *P. cynomolgi* (Geiman, 1948) and *P. knowlesi* (Geiman and McKee, 1950). The results summarized for

P. cynomolgi in Table XIII indicate a rapid and pronounced inhibitory influence on the development and the glucose and lactate utilizations of the parasites. Isopentaquine has a slight inhibitory effect in 19 hours while pentaquine and paludrine have little, if any, influence during the

TABLE XIII. *In Vitro* Effects of Antimalarials on the Growth and Multiplication of *P. cynomolgi*. 19-Hour RD Culture, M235 (Adapted from Geiman, 1948b)

Antimalarials added to whole blood system	Glucose utilized mg. %	Lactate utilized mg. %	Parasite Differential, %						Increase in parasites fold
			Ring	Young troph.	Old troph.	Sch.	Seg.	Non-ident.	
—	156	81	5	88	1	3		3	2.9
Atabrine, 1000 γ/l.	112	34	15	61		2		22	0.2
Chloroquine, 1612 γ/l.	106	33	14	46	2	20	2	16	0.2
Pentaquine, 1250 γ/l.	162	72	9	78	1	3		9	2.3
Isopentaquine, 1250 γ/l.	160	67	7	87		4		2	1.5
Paludrine, 3000 γ/l.	167	37	7	73	2	12		6	2.0

same interval of time. It is interesting to note, in view of the similarities of structure between atabrine and chloroquine, that both drugs produce an early characteristic clumping of the pigment of *P. knowlesi* as well as of other parasites.

E. Combined *In Vitro–In Vivo* Antimalarial Studies

As stated previously, a comparison of the action of drugs both *in vitro* and *in vivo* is of value because of the metabolism of the compounds within the animal body. Often the question arises as to whether the drug *per se* or the degradation products elicit the predominant antimalarial response. With both atabrine and quinine it has been found (Ball, 1946; Geiman, McKee, Anfinsen, Ormsbee, and Ball, 1949) that therapeutic amounts of the agent will suppress the *in vitro* and *in vivo* growth of *Plasmodium knowlesi*. This suggests that it is the drug itself which exhibits the major action. Tables XIV and XV show data for a typical set of experiments with quinine. The experiment was carried out as follows. When about 2 per cent of the red cells of a monkey were parasitized with *P. knowlesi*, blood was removed by venapuncture and two RD culture vessels were set up. To one was added 10 mg. of quinine per liter while the other served as the control. The parasitized monkey was given quinine dihydrochloride and 1 hour later blood was withdrawn and an additional culture vessel was set up with the quinine-containing parasitic blood. Blood samples were then withdrawn at intervals from all three cultures and from the animal to determine the blood levels of quinine and the changes in parasite counts and morphology. The differential counts, not

TABLE XIV. *In Vitro* Action of Quinine on *P. knowlesi*
(see Table XV for Action of Quinine *in Vivo*)
(Taken from Ball, 1948, with the permission of the author and the publisher)

Time	Control—No Quinine		Quinine added *in Vitro* to Give 10 mg./l.		Quinine containing blood drawn from animal*	
	Total Parasites	Abnormal Forms	Total Parasites	Abnormal Forms	Total Parasites	Abnormal Forms
Hours	%	%	%	%	%	%
0	1.3	1	1.3	1	1.9	1
5	2.1	5	2.3	8	1.9	7
24	3.4	13	0.7	24	0.4	36

* Quinine concentration by determination 5.6 mg./liter.

TABLE XV. *In Vivo* Action of Quinine on *P. knowlesi*
(See Table XIV for Action of Quinine *in Vitro*)
(Taken from Ball, 1946, with the permission of the author and the publisher)

Time after initial injection Hours	Blood quinine mg./l.	Total parasites %	Abnormal forms %
0	0	2.4	3
1	22.4	4.5	2
4	22.8	6.8	2
12	11.0	5.5	8
22	5.4	1.6	18
24*		1.1	5
28*	17.2	0.7	43
48	3.8	0.1	60

* At 24 hours, there was injected intramuscularly 300 mg. and at 28 hours 100 mg. of quinine dihydrochloride.

shown in Tables XIV and XV, indicated a definite slowing of the parasite's life cycle. Similar data have been obtained for atabrine. However, paludrine (Geiman, 1948b; Geiman and McKee, 1950) exhibits an entirely different picture. With this drug it was found that during the *in vitro* growth of two generations of *P. knowlesi* there was no effect, except a slight enhancement by the compound itself, whereas whole blood or plasma taken from a normal animal previously treated with paludrine elicited a rapid and marked *in vitro* inhibitory influence (Table XVI). The inhibitory effects consist of retardation of growth and death of the asexual forms, a differentiation of many sexual parasites (gametocytes) and decreases in the glucose and lactate utilizations by the cultured parasites. Similar inhibition of growth and the formation of gametocytes have been reported by Hawking and Perry for *P. cynomolgi*

TABLE XVI. *In Vitro* and *in Vivo* Effects of Paludrine and Its Metabolic Products on *P. knowlesi*. 46-Hour RD Culture and Subculture, M286 (Geiman and McKee, 1950)

Material added to whole blood system	Glucose utilized mg. %	Lactate utilized mg. %	Parasite differential, %						Increase in parasites fold
			Ring	Troph.	Sch.	Seg.	Gam.	Non-ident.	
—	182	51	—	96	4	—	—	—	3.7
—(subculture)	173	—	5	67	19	1	1	7	1.3
Paludrine, 2250 γ/l.	254	113	3	90	3	1	2	1	4.4
Paludrine (subculture)	130	56	5	59	24	3	3	6	1.7
Paludrine blood containing 92 γ/l.	118	24	1	17	28	—	38	16	1.0
Paludrine (subculture)	91	—	—	20	—	—	28	52	0.3
In vivo Paludrine, 600 γ/l.			1	24	13	—	57	5	0.2 (6.5 to 1.5%)
Same (2nd 24 hr.)			—	2	6	—	76	16	0.1 (1.5 to 0.02%)

(1948). These investigators have determined that rat liver slices can convert paludrine to active products. Likewise Wendel (1946) has shown that drugs present in the serum after administration to the host depressed the respiration of parasites *in vitro*. Schmidt, Hughes, and Smith (1947) were able to account for only a small portion of the administered drug as paludrine *per se* in the urine of monkeys. The nature of these degradation materials has not yet been determined, but it emphasizes the importance of this approach in the study of antimalarial compounds.

F. Influence of Metabolic Antagonists on *Plasmodium*

The theory of the interference with the action of essential metabolites in metabolic processes by the competitive inhibition with analogs of metabolites has been the basis of a number of research studies in chemotherapy. These investigations have resulted in the devising of some unusual types of antimalarials.

The report by Trager (1943b) that calcium pantothenate is a growth factor for *Plasmodium lophurae* led to the investigation of analogs of pantothenate as possible antimalarials. Mead, Rapport, Senear, Maynard, and Koepfli (1946) determined that D-pantoyltaurine showed definite activity in avian malaria. Marshall (1946) reported that pantothenone was as active as quinine against *P. gallinaceum* in the chick and that activity was antagonized by pantothenic acid. This subject has been reviewed by Lourie (1947).

It was a similar consideration which led the English workers (Curd, Davey, and Rose, 1945) to the rational development of the antimalarial agent paludrine. The structure of this compound was patterned after that of the pyrimidine compounds with the thought that there might be competitive inhibition in the synthesis of nuclear material of which the pyrimidines are essential components.

The requirement of *P. knowlesi* for methionine has resulted in the observation that methoxinine and ethionine, analogs of methionine, inhibit the growth of the simian parasite *in vitro*. The inhibition is overcome by the addition of extra methionine (McKee and Geiman, 1948).

Seeler (1945) determined that the administration of pyridoxine inhibited the actions of quinine and atabrine on *P. lophurae* and *P. cathemerium*.

The inhibition of the action of sulfonamides on PAB was reported by Marshall, Litchfield, and White (1942), Maier and Riley (1942), and Seeler, Graessle, and Dusenbury (1943) for *P. gallinaceum* and *P. lophurae* infections in birds and by Richardson, Hewitt, Seager, Brooke, Martin, and Maddux (1946) for *P. knowlesi* infection in monkeys. PAB has no effect on the actions of quinine and atabrine with *P. gallinaceum*. Tonkin (1946) determined that the inhibitory action of sulfathiozole in cultures of the exoerythrocytic forms of *P. gallinaceum* was antagonized by PAB. Ball (1946), as already stated, has reported the antagonism between sulfadiazine and PAB in cultures of *P. knowlesi*, with a ratio of Drug/PAB of about 1/100 to 1/1000.

G. Discussion of Antimalarial Action

The results obtained from the investigations conducted on antimalarial compounds suggests a specific combination of the drug with essential components of the parasitic cell. This most likely is a combination with protein enzymes and genes or other vital nucleoproteins, which in turn affect the metabolism of the parasite, i.e., consumption of nutrients, glucose, and lactate. The reactions in most instances are of the competitive type and can be antagonized by the addition of coenzymes or metabolites, which are essential components of the metabolic system. Also a specific combination is suggested by the intensification of antimalarial action due to slight modifications in the structures of compounds (Findlay, 1939). This is demonstrated in the greater inhibitory action of isopentaquine over pentaquine.

Although much has been learned about the mode of action of antimalarials there is as yet no detailed explanation for their selective action upon the plasmodia. It is generally believed that for an antimalarial agent to be selective and have a maximum effect one must choose an agent which attacks a unique metabolic process—a parasitic process not having a parallel in the host. To the author this does not seem essential since many factors determine the mode of action of a drug. Most of these considerations have been discussed previously, and they will be listed only in review. The "exposed" position of the parasitized red cell in the circulating blood, the differential concentration of the drug in the parasitic cell and the metabolic conversion of the non-active administered drug to active compounds are important considerations. Above all, however, is the fact that the *Plasmodium* is a parasitic agent and not an integral part of the animal organism which makes it particularly dependent not only on the environmental tissue but also upon the circulating body fluids which contain the effective agent. Thus the parasite is vulnerable to attack by administered chemical agents—much more vulnerable than integral organ tissue components. This may be a partial explanation of the lack of attack of such agents on the tissue stages of the plasmodia.

The production of metabolic degradation products from the administered drugs and the acquisition of resistance to certain drugs are two of the most important practical problems remaining to be solved in malariology. The resistance of *P. gallinaceum* to paludrine has been described by Bishop and Birkett (1947, 1948) and by Williamson, Bertram, and

Lourie (1947). Likewise Fairley *et al.* (1946) have reported atabrine-resistant strains of *P. falciparum.*

Perhaps an even more important observation than the resistance of parasites to drugs is the differentiation of the sexual forms of malarial parasites in paludrine treatment of *P. knowlesi* (Geiman and McKee, 1950). The production of such a reservoir of sexual forms in the human malarias may make for a hazard of inestimable size. Such factors must be of paramount consideration in an effective attack on malarial parasites.

VIII. Discussion—Unsolved Biochemical Problems

A detailed comparison of the metabolic aspects of plasmodia with other parasitic protozoa will be presented in the other chapters of this book and in reviews on the subject (Bueding, 1950; Brand, 1950; Moulder, 1950; and Geiman and McKee, 1950). However, a few brief remarks are in order, particularly regarding the blood-dwelling protozoa, so as to indicate biochemical problems requiring further investigation.

Blood as a transport agent indicates a priori that it is a rich source of nutrients and metabolites—a composite of the building blocks of the animal body. The variations in the metabolic processes of parasitic organisms will depend to a great extent on whether they are obligate intracellular parasites or extracellular bodies, and finally upon the inherent ability of the organisms to utilize the erythrocyte and plasma components. In retrospect it seems that the classification of any organisms as parasitic is a purely arbitrary one and is based on its inability to live free of a host and not on its inability to metabolize (degrade and synthesize) carbohydrate, protein, and fat, and to synthesize enzyme systems and other metabolic components. The malarial parasites, apparently unlike many of the other blood protozoa, have most of the metabolic processes of their vertebrate hosts.

Although the malarial parasite is an obligate intracellular organism its metabolic processes, as already described, are quite extensive. On the other hand the trypanosomes, although not directly dependent on the body cells for their growth and multiplication, are lacking many of the enzyme processes which the malarial parasites possess. The trypanosomes tested thus far do not hydrolyze proteins (von Brand, 1933; Krijgsman, 1936), and their endogenous oxidative metabolism is negligible, oxidizing only in the presence of added substrates, with an incomplete oxidation of glucose. The pathogenic species of trypanosomes, however, produce almost no carbon dioxide from the metabolism of glucose (Christophers and Fulton, 1938; Reiner, Smythe, and Pedlow, 1936) and do not have cyanide-sensitive respiratory enzymes (von Brand and Johnson, 1947). Unlike plasmodia the recurrent spirochetes (*Borrelia novyi*) and the treponemes (*Treponema pallidum*) require little, if any, oxygen for their metabolic processes (McKee and Geiman, 1950). Their chief source of energy, as with the trypanosomes, is glucose; however, their oxidative metabolism is insignificant (von Fenyvessy and Scheff, 1930). Although little is known about the ability of the spirochetes to produce or utilize protein, it is inconceivable that any organism that divides does not have at least elementary nucleoprotein synthetic processes. Leishmania, however, utilize both carbohydrate and protein, and in the absence of glucose the proteolytic action of the organisms is increased, indicated by an increased formation of protein degradation products. Adding glucose to cultures of *Leishmania tropica* and *L. donovani* causes a marked sparing of the protein, indicating a preference for the carbohydrate (Salle and Schmidt, 1928; Salle, 1931). Cultivation of Leishmania has been obtained in the presence of blood, and the need for hemin and ascorbic acid has been shown (M. Lwoff, 1940). However, no concentrated attempt has been made to evaluate the protein and amino acid requirements.

Thus a large amount of further study is required for all the blood-dwelling protozoa. Precise nutrient requirements for the growth and multiplication of each species of the *Plasmodium,* more detailed knowledge regarding the lipid and protein composition of each species, including enzymes and nucleoproteins, and further information regarding the mode of action and antimalarial drugs are examples of areas which require more detailed investigation. The complete solution of these problems very likely will require a characterization of the presently unknown components of plasma and the cellular elements of blood. Knowledge regarding chemical differences between the bloods of various animal species likewise will be of inestimable value for evaluating host-parasite specificity. It should be emphasized, however, that it is both likenesses and dissimilarities among the various parasitic protozoa and between the protozoa and their hosts with which one must concern himself for a more complete understanding of the life processes of the parasites and a more rational approach to their control through chemotherapy.

ACKNOWLEDGMENTS

The author is grateful to all those investigators who have so generously permitted the use of their experimental data, to Drs. Gordon H. Ball and Clay G. Huff for their suggested references, and particularly to Drs. Quentin M. Geiman, Eric G. Ball, and A. Baird Hastings for their valuable suggestions and criticisms of the manuscript.

The investigative work of the author and collaborators reviewed in this chapter was done under contract with the Office of Scientific Research and Development, Committee on Medical Research and under grants-in-aid from the United States Public Health Service, Division of Research Grants and Fellowships.

REFERENCES

Anfinsen, C. B., Geiman, Q. M., McKee, R. W., Ormsbee, R. A., and Ball, E. G. (1946). *J. Exptl. Med.* **84**, 607.

Ball, E. G. (1946). *Federation Proc.* **5**, 397.

Ball, E. G., Anfinsen, C. B., and Cooper, O. (1947). *J. Biol. Chem.* **168**, 257.

Ball, E. G., Anfinsen, C. B., Geiman, Q. M., McKee, R. W., and Ormsbee, R. A. (1945). *Science* **101**, 542.

Ball, E. G., McKee, R. W., Anfinsen, C. B., Cruz, W. O., and Geiman, Q. M. (1948). *J. Biol. Chem.* **175**, 547.

Ball, G. H. (1947). *Am. J. Trop. Med.* **27**, 301.

Ball, G. H. (1948). *Am. J. Trop. Med.* **28**, 533.

Bass, C. C. (1913). *Am. J. Trop. Diseases Preventive Med.* **1**, 546.

Bass, C. C., and Johns, F. M. (1912). *J. Exptl. Med.* **16**, 567.

Bishop, A., and Birkett, B. (1947). *Nature* **159**, 884.

Bishop, A., and Birkett, B. (1948). *Parasitol.* **39**, 125.

Blanchard, K. C. (1947). *Ann. Rev. Biochem.* **16**, 587.

Bovarnick, M. R., Lindsay, A., and Hellerman, L. (1946a). *J. Biol. Chem.* **163**, 523.

Bovarnick, M. R., Lindsay, A., and Hellerman, L. (1946b). *J. Biol. Chem.* **163**, 535.

Boyd, M. F. (1949). Malariology, Vols. 1 and 2. W. B. Saunders Co., Philadelphia.

Brackett, S., Waletzky, E., and Baker, M. (1946). *J. Parasitol.* **32**, 453.

von Brand, T. (1933). *Z. vergleich. Physiol.* **19**, 587.

von Brand, T. (1950). *J. Parasitol.* **36**, 178.

von Brand, T., and Johnson, E. M. (1947). *J. Cellular Comp. Physiol* **29**, 33.

Brooke, M. M. (1945). *Am. J. Hyg.* **41**, 81.

Brown, W. H. (1911). *J. Exptl. Med.* **13**, 290.

Bueding, E. (1950). *J. Parasitol.* **36**, 201.

Burnham, E., McKee, R. W., and Geiman, Q. M. (1950). Unpublished.

Cannon, P. R., and Marshall, C. E. (1940). *J. Immunol.* **38**, 365.

Cannon, P. R., and Taliaferro, W. H. (1931). *J. Preventive Med.* **5**, 37.
Carlson, H. J., Mueller, M. G., and Bissell, H. D. (1946). *J. Lab. Clin. Med.* **31**, 677.
Ceithaml, J., and Evans, E. A., Jr. (1946a). *J. Infectious Diseases* **78**, 190.
Ceithaml, J., and Evans, E. A., Jr. (1946b). *Arch. Biochem.* **10**, 397.
Chopra, R. N., Mukherjee, S. N., and Sen, B. (1935). *Indian J. Med. Research* **22**, 571.
Christophers, S. R., and Fulton, J. D. (1938a). *Ann. Trop. Med. Parasitol.* **32**, 43.
Christophers, S. R., and Fulton, J. D. (1938b). *Ann. Trop. Med. Parasitol.* **32**, 257.
Christophers, S. R., and Fulton, J. D. (1939). *Ann. Trop. Med. Parasitol.* **33**, 161.
Coggeshall, L. T. (1940). *J. Exptl. Med.* **71**, 13.
Coggeshall, L. T., and Eaton, M. D. (1938a). *J. Exptl. Med.* **67**, 871.
Coggeshall, L. T., and Eaton, M. D. (1938b). *J. Exptl. Med.* **68**, 29.
Coggeshall, L. T., and Maier, J. (1941). *J. Infectious Diseases* **69**, 108.
Coulston, F. (1941). *J. Parasitol.* **27**, 38.
Coulston, F., and Huff, C. G. (1947). *J. Infectious Diseases* **80**, 209.
Curd, F. H. S., Davey, D. G., and Rose, F. L. (1945). *Ann. Trop. Med. Parasitol.* **39**, 208.
Deane, H. W. (1945). *J. Cellular Comp. Physiol.* **26**, 139.
Dulaney, A. D., and House, V. (1941). *Proc. Soc. Exptl. Biol. Med.* **48**, 620.
Eaton, M. D. (1938). *J. Exptl. Med.* **67**, 857.
Fairley, N. H., and Bromfield, R. J. (1934). *Trans. Roy. Soc. Trop. Med. Hyg.* **28**, 307.
Fairley, N. H. (1946). *Med. J. Australia*, 234.
von Fenyvessy, B., and Scheff, G. (1930). *Biochem. Z.* **221**, 206.
Ferrebee, J. W., and Geiman, Q. M. (1946). *J. Infectious Diseases* **78**, 173.
Fieser, L. F. (1946). *Record Chem. Progress (Kresge-Hooker Sci. Lib.)* **7**, 26.
Findlay, G. M. (1939). Recent Advances in Chemotherapy, 2nd ed. J. and A. Churchill Ltd., London.
Flosi, A. Z. (1944). Contibuicao par o estudo da insuficiencia super-renal paliduca. Renascenca, Sao Paulo, S. A.
Freund, J. (1947). *Ann. Rev. Microbiol.* **1**, 291.
Freund, J. Thomson, K. J., Sommer, H. E., Walter, A. W., and Pisani, T. (1948). *Am. J. Trop. Med.* **28**, 1.
Fulton, J. D. (1939). *Ann. Trop. Med. Parasitol.* **33**, 217.
Fulton, J. D., and Christophers, S. R. (1938). *Ann. Trop. Med. Parasitol.* **32**, 77.
Gavrilov, W., Bobkoff, G., and Laurencin, S. (1938). *Ann. soc. belge méd. trop.* **18**, 429.
Geiman, Q. M. (1943). *New Engl. J. Med.* **229**, 283.
Geiman, Q. M. (1944). In Simmons, J. S., and Gentzkow, C. J., Laboratory Methods of the U. S. Army, pp. 627–664. Lea and Febiger, Philadelphia.
Geiman, Q. M. (1948a). *New Engl. J. Med.* **239**, 18, 58.
Geiman, Q. M. (1948b). *Proc. Fourth Internatl. Cong. Trop. Med. Malaria* **1**, 618.
Geiman, Q. M. (1949). Boyd, M. F., Malariology, Vol. 1. W. B. Saunders Co., Philadelphia.
Geiman, Q. M. (1951). Most, H., Parasitic Infections in Man, pp. 130–149. Columbia University Press, New York.
Geiman, Q. M., Anfinsen, C. B., McKee, R. W., Ormsbee, R. A., and Ball, E. G. (1946). *J. Exptl. Med.* **84**, 583.

Geiman, Q. M., and McKee, R. W. (1948). *Sci. Monthly* **67**, 217.
Geiman, Q. M., and McKee, R. W. (1950a). *J. Parasitol.* **36**, 211.
Geiman, Q. M., and McKee, R. W. (1950b). Unpublished.
Geiman, Q. M., McKee, R. W., Anfinsen, C. B., Ormsbee, R. A., and Ball, E. G. (1949). In preparation.
Ghosh, B. N., and Nath, M. C. (1934). *Records Malaria Survey India* **4**, 321.
Gill, C. A. (1928). Genesis of Epidemics and the Natural History of Disease. William Wood and Co., New York.
Goodner, K. (1941). *Science* **94**, 241.
Grenfell, N. P., Richardson, A. P., and Hewitt, R. I. (1946). Unpublished. Cited by Ceithaml, J., and Evans, E. A., Jr. (1946). *Arch. Biochem.* **10**, 397.
Haas, E. (1944). *J. Biol. Chem.* **155**, 321.
Haas, V. H., and Ewing, F. M. (1945). *U. S. Pub. Health Service, Pub. Health Repts.* **60**, 185.
Haas, V. H., Feldman, H. A., and Ewing, F. M. (1945). *U. S. Pub. Health Service, Pub. Health Repts.* **60**, 577.
Hawking, F. (1944). *Lancet* **246**, 693.
Hawking, F. (1945). *Trans. Roy. Soc. Trop. Med. Hyg.* **39**, 245.
Hawking, F., and Perry, W. L. M. (1948). *Brit. J. Pharmacol.* **3**, 320.
Hegner, R. W., and MacDougall, M. S. (1926). *Am. J. Hyg.* **6**, 602.
Hegner, R. W., and Wolfson, F. (1939). *Am. J. Hyg.* **29**, 83.
Heidelberger, M., Mayer, M., Alving, A. S., Craige, B. Jr., Jones, R. Jr., Pullman, T. N., and Whorton, M. (1946). *J. Immunol.* **53**, 113.
Hellerman, L., Lindsay, A., and Bovarnick, M. R. (1946). *J. Biol. Chem.* **163**, 553.
Henry, A. F. X. (1933). *Compt. rend. soc. biol.* **112**, 765.
Henry, A. F. X. (1934). *Arch. Schiffs-u. Tropen-Hyg.* **38**, 93.
Huff, C. G. (1947). *Ann. Rev. Microbiol.* **1**, 43.
Huff, C. G., and Bloom, W. (1935). *J. Infectious Diseases* **57**, 315.
Huff, C. G., and Coulston, F. (1944). *J. Infectious Diseases* **75**, 231.
Huff, C. G., and Coulston, F. (1948). *J. Parasitol.* **34**, 264.
Huff, C. G., Coulston, F., Laird, R. L., and Porter, R. J. (1947). *J. Infectious Diseases* **81**, 7.
Hühne, W. (1942). *Deut. Tropenmed. Z.* **46**, 385.
James, S. P. (1934). *Trans. Roy. Soc. Trop. Med.* **28**, 3.
James, S. P., and Tate, P. (1938). *Parasitology* **30**, 128.
Johns, F. M. (1930). *Proc. Soc. Exptl. Biol. Med.* **28**, 743.
Josephson, E. S., Taylor, D. J., Greenberg, J., and Nadel, E. M. (1949). *J. Natl. Malaria Soc.* **8**, 132.
Kehar, N. D. (1936). *Records Malaria Survey India* **6**, 499.
Kelsey, F. E., Oldham, F. K., and Geiling, E. M. K. (1943). *J. Pharmacol.* **78**, 314.
Krijgsman, B. J. (1936). *Z. vergleich. Physiol.* **23**, 663.
Krishnan, K. V. (1948). Personal communication.
Krishnan, K. V., Ghosh, B. M., and Bose, P. N. (1936). *Records Malaria Survey India* **6**, 1.
Laser, H. (1946). *Nature* **157**, 301.
Laser, H. (1948). *Nature* **161**, 560.
Laser, H. and Friedmann, E. (1945). *Nature* **156**, 507.
Laveran, A. (1880). *Bull. acad. med.* (Paris) **9**, 1235, 1268, 1346.

Lipmann, F. (1949). *Federation Proc.* **8**, 597.
Lourie, E. M. (1934). *Ann. Trop. Med. Parasitol.* **28**, 255.
Lourie, E. M. (1947). *Ann. Rev. Microbiol.* **1**, 237.
Lucas, E. H., and Lewis, R. W. (1944). *Scicnce* **100**, 597.
Lwoff, M. (1940). *Mono. l'institut Pasteur* 1.
McKee, R. W. (1951). Most, H., Parasitic Infections in Man, pp. 114–129. Columbia University Press, New York.
McKee, R. W., Cobbey, T. S., Jr., and Geiman, Q. M. (1947). *Federation Proc.* **6**, 276.
McKee, R. W., Cobbey, T. S., Jr., and Geiman, Q. M. (1949). *Endocrinology* **45**, 21.
McKee, R. W., and Geiman, Q. M. (1946). *Proc. Soc. Exptl. Biol. Med.* **63**, 313.
McKee, R. W., and Geiman, Q. M. (1948). *Federation Proc.* **7**, 172.
McKee, R. W., and Geiman, Q. M. (1950). Unpublished.
McKee, R. W., Geiman, Q. M., Anfinsen, C. B., Ormsbee, R. A., and Ball, E. G. (1949). Unpublished.
McKee, R. W., Geiman, Q. M., and Cobbey, T. S., Jr., (1947). *Federation Proc.* **6**, 276.
McKee, R. W., Ormsbee, R. A., Anfinsen, C. B., Geiman, Q. M., and Ball, E. G. (1946). *J. Exptl. Med.* **84**, 569.
Maegraith, B. (1948). Pathological Processes in Malaria and Blackwater Fever. Charles C Thomas, Springfield, Illinois.
Maier, J., and Coggeshall, L. T. (1941). *J. Infectious Diseases* **69**, 87.
Maier, J., and Riley, E. (1942). *Proc. Soc. Exptl. Biol., Med.* **50**, 152.
Manwell, R. D., and Brody, G. (1949). Unpublished.
Manwell, R. D., and Feigelson, P. (1949). *Proc. Soc. Exptl. Biol. Med.* **70**, 578.
Marshall, E. K. (1946). *Federation Proc.* **5**, 298.
Marshall, E. K., Jr., Litchfield, J. T., Jr., and White, H. J. (1942). *J. Pharmacol. Exptl. Therap.* **75**, 89.
Mayer, M. M., and Heidelberger, M. (1946). *J. Immunol.* **54**, 89.
Mead, J. F., Rapport, M. M., Senear, A. E., Maynard, J. T., and Koepfli, J. B. (1946). *J. Biol. Chem.* **163**, 465.
Miyahara, H. (1936). *J. Med. Assoc. Formosa* **35**, 1092.
Morrison, D. B., and Anderson, W. A. D. (1942). *U. S. Pub. Health Service, Pub. Health Repts.* **57**, 161.
Morrison, D. B., and Jeskey, H. A. (1947). *Federation Proc.* **6**, 279.
Morrison, D. B., and Jeskey, H. A. (1948). *J. Natl. Malaria Soc.* **7**, 259.
Moulder, J. W. (1948a). *Ann. Rev. Microbiol.* **2**, 101.
Moulder, J. W. (1948b). *J. Infectious Diseases* **83**, 262.
Moulder, J. W. (1949). *J. Infectious Diseases* **85**, 195.
Moulder, J. W. (1950). *J. Parasitol* **36**, 193.
Moulder, J. W., and Evans, E. A., Jr. (1946). *J. Biol. Chem.* **164**, 145.
Mulligan, H. W., and Russell, P. F. (1940). *J. Malaria Inst. India* **3**, 199.
Nadel, E. M., Taylor, D. J., Greenberg, J., and Josephson, E. S. (1949). *J. Natl. Malaria Soc.* **8**, 70.
Oldham, F. K., and Kelsey, F. E. (1945). *J. Pharmacol. Exptl. Therap.* **83**, 288.
Oldham, F. K., Kelsey, F. E., Cantrell, W., and Geiling, E. M. K. (1944). *J. Pharmacol. Exptl. Therap.* **82**, 349.
Osborn, E. M. (1943). *Brit. J. Exptl. Path.* **24**, 227.

Overman, R. R. (1946). *J. Lab. Clin. Med.* **31**, 1170.
Overman, R. R., Bass, A. C., Davis, A. K., and Golden, A. (1949). *Am. J. Clin. Path.* **19**, 907.
Overman, R. R., and Feldman, H. A. (1947). *J. Clin. Invest.* **26**, 1049.
Overman, R. R., Hill, T. S., and Wong, Y. T. (1949). *J. Natl. Malaria Soc.* **8**, 14.
Parsense, L., Meyer, H., and Menezes, V. (1942). *Rev. brasil. biol.* **2**, 89.
Pinelli, L. (1929). *Rev. Malariol.* **8**, 310.
Proske, H. O., and Watson, R. B. (1939). *U. S. Pub. Health Service, Pub. Health Repts.* **54**, 158.
Ragab, H. A. (1948). *Trans. Roy. Soc. Trop. Med. Hyg.* **41**, 434.
Reiner, L., Smythe, C. V., and Pedlow, J. T. (1936). *J. Biol. Chem.* **113**, 75.
Richardson, A. P., Hewitt, R. I., Seager, L. D., Brooke, M. M., Martin, F., and Maddux, H. (1946). *J. Pharmacol. Exptl. Therap.* **87**, 203.
Rigdon, R. H. (1945). *Am. J. Clin. Path.* **15**, 489.
Rimington, C., Fulton, J. D., and Sheinman, H. (1947). *Biochem. J.* **41**, 619.
Rodhain, J., Gavrilov, W., and Cowez, S. (1940). *Compt. rend. soc. biol.* **134**, 261.
Roos, A., Hegsted, D. M., and Stare, F. J. (1946). *J. Nutrition* **32**, 473.
Rose, W. C. (1949). *Federation Proc.* **8**, 546.
Ross, R. (1897). *Brit. Med. J.* **2**, 1786.
Row, R. (1917). *Indian J. Med. Research* **4**, 388.
Russell, P. F., West, L. S., and Manwell, R. D. (1946). Practical Malariology. W. B. Saunders Co., Philadelphia.
Salle, A. J. (1931). *J. Infectious Diseases* **49**, 481.
Salle, A. J., and Schmidt, C. L. A. (1928). *J. Infectious Diseases* **43**, 378.
Schmidt, L. H., Hughes, H. B., and Smith, C. C. (1947). *J. Pharmacol. Exptl. Therap.* **90**, 233.
Schwartz, W. F. (1939). *J. Pharmacol. Exptl. Therap.* **65**, 175.
Seeler, A. O. (1945). *J. Natl. Malaria Soc. Tallahassee* **4**, 13.
Seeler, A. O., Graessle, O., and Dusenbury, E. D. (1943). *J. Bact.* **45**, 205.
Seeler, A. O., and Ott, W. H. (1944). *J. Infectious Diseases* **75**, 175.
Seeler, A. O., and Ott, W. H. (1945a). *J. Infectious Diseases* **77**, 82.
Seeler, A. O., and Ott, W. H. (1945b). *J. Infectious Diseases* **77**, 181.
Seeler, A. O., and Ott, W. H. (1946). *J. Natl. Malaria Soc. Tallahassee* **5**, 123.
Shannon, J. A., Earle, D. P., Jr., Brodie, B. B., Taggart, J. V., and Berliner, R. W. (1944). *J. Pharmacol. Exptl. Therap.* **81**, 307.
Shortt, H. E., and Garnham, P. C. C. (1948). *Nature* **161**, 126.
Shortt, H. E., Garnham, P. C. C., and Malamos, B. (1948). *Brit. Med. J.* **1**, 192.
Silverman, M., Ceithaml, J., Taliaferro, L. G., Evans, E. A., Jr. (1944). *J. Infectious Diseases* **75**, 212.
Sinton, J. A., and Ghosh, B. N. (1934). *Records Malaria Survey India* **4**, 15, 43, 205.
Sinton, J. A., and Mulligan, H. W. (1932). *Records Malaria Survey India* **3**, 357, 381.
Speck, J. F., and Evans, E. A., Jr. (1945a). *J. Biol. Chem.* **159**, 71.
Speck, J. F., and Evans, E. A., Jr. (1945b). *J. Biol. Chem.* **159**, 83.
Speck, J. F., Moulder, J. W., and Evans, E. A., Jr. (1946). *J. Biol. Chem.* **164**, 119.
Spencer, C. F., Koniuszy, F. R., Rogers, E. F., Shavel, J., Easton, N. R., Kaczka, E. A., Kuehl, F. A. Phillips, R. F., Walti, A., and Folkers, K., Malanga, C., and Seeler, A. O. (1947). *Lloydia* **10**, 145.

Stauber, L. A., and van Dyke, H. B. (1945). *Proc. Soc. Exptl. Biol. Med.* **58**, 125.
Taliaferro, W. M., and Taliaferro, L. G. (1928). *J. Preventive Med.* **2**, 147.
Thomson, J. G., and Thomson, D. (1913). *Ann. Trop. Med. Parasitol.* **7**, 509.
Tonkin, I. M. (1946). *Brit. J. Pharmacol.* **1**, 163.
Trager, W. (1941). *J. Exptl. Med.* **74**, 441.
Trager, W. (1943a). *Science* **97**, 206.
Trager, W. (1943b). *J. Exptl. Med.* **77**, 411.
Trager, W. (1947a). *Proc. Soc. Exptl. Biol. Med.* **64**, 129.
Trager, W. (1947b). *J. Exptl. Med.* **85**, 663.
Trager, W. (1947c). *J. Parasitol.* **33**, 345.
Trager, W., Bang, F. B., and Hairston, N. G. (1945). *Proc. Soc. Exptl. Biol. Med.* **60**, 257.
Van den Berghe, L. (1948). *Proc. Fourth Internatl. Cong. Trop. Med. Malaria* **2**, 1436.
Velick, S. F. (1942). *Am. J. Hyg.* **35**, 152.
Velick, S. F., and Scudder, J. (1940). *Am. J. Hyg.* **31c**, 92.
Vincke, I. H., and Lips, M. (1948). *Ann. soc. belge méd. trop.* **28**, 97.
Wakeman, A. M., and Morrell, C. A. (1929). *West African Med. J.* **3**, 6.
Wats, R. C., and Das Gupta, B. M. (1934). *Indian J. Med. Research* **21**, 475.
Wendel, W. B. (1943). *J. Biol. Chem.* **148**, 21.
Wendel, W. B. (1946). *Federation Proc.* **5**, 406.
Wenyon, C. M. (1926). Protozoology. Bailliere, Tindall, and Cox, London.
Williamson, J., Bertram, D. S., and Lourie, E. M. (1947). *Nature* **159**, 885.
Wiselogle, F. Y. (1946). A Survey of Antimalarial Drugs 1941–1945. Edwards Bros., Inc., Ann Arbor, Mich.
Wolfson, F. (1940). *Am. J. Hyg.* **32**, 60.
Wright, C. I., and Sabine, J. C. (1944). *J. Biol. Chem.* **155**, 315.
Young, M. D., McKendon, S. B., and Smarr, R. G. (1943). *J. Am. Med. Assoc.* **122**, 492.
Zwemer, R. L., Sims, E. A. H., and Coggeshall, L. T. (1940). *Am. J. Trop. Med.* **20**, 687.

The Biochemistry of Ciliates in Pure Culture

GEORGE W. KIDDER AND VIRGINIA C. DEWEY

Amherst College, Amherst, Massachusetts

CONTENTS

I. Introduction

No review which includes the biochemistry of ciliated protozoa appears to have been written since that of Hall in 1943. What little was then known of these animal microorganisms was obscured by concepts which are no longer tenable. At the present time, it must be stated immediately that our knowledge of the biochemistry of the ciliates has not progressed much beyond that of 1943, except for that relating to the single genus, *Tetrahymena*. Because this is so, by far the greater portion of this review will be confined to that genus.

As was pointed out many years ago by Lwoff (1923) and restated later by Kidder (1941), certain rigid conditions must be met if the investigator is going to solve any of the biochemical problems connected with a microorganism of this type. The first and most important obstacle to be surmounted is the establishment of the organism in pure (bacteria-free) culture. This condition has long been recognized as primary by bacteriologists and mycologists, for an environment containing more than one type of microorganism is hopeless from an analytical standpoint. Under these conditions no growth response or any other type of measurement of metabolic activity can be related to less than the entire microcosm.

It is also obvious that, even though the microorganism is freed from all associated forms, conditions must be provided which will allow growth and multiplication. When these conditions have been attained, no matter how crude or complex the culture medium, then the investigator has material with which to start his biochemical researches.

The next step is to determine the synthetic capacities of the organism under investigation. This may be started by attempts to substitute known chemical compounds for various portions of the initial crude medium. During the early phases of the work on ciliates this method did not always meet with success (Lwoff, 1924; Dewey, 1941, 1944) owing largely to our incomplete knowledge of many of the chemical factors now recognized as being important in animal metabolism. However, this mode of attack has led eventually to the recognition of new growth factors.

Many chemical substances of unknown nature which are important in animal metabolism remain to be identified, a fact which explains in large measure the many failures to establish all but a few ciliates in pure culture. Of these unknown factors the most difficult with which to cope are the labile complexes which either break down and become useless during attempts at chemical separation or which are heat labile and therefore cannot be sterilized by autoclaving. Much information can be obtained about such labile complexes when they are not absolute requirements, but the task of their separation and identification becomes great indeed if the organisms fail entirely to carry out metabolic activities in their absence.

The failure to establish satisfactory cultures for biochemical work with such ciliates as *Paramecium* (Johnson and Baker, 1942, Johnson and Tatum, 1945) and *Colpoda* (van Wagtendonk and Taylor, 1942; Taylor and van Wagtendonk, 1941; Tatum *et al.*, 1942; Garnjobst *et al.*, 1943) apparently stems from their requirement for at least one labile compound or complex, although van Wagtendonk and Hackett (1949) have reported the growth of *Paramecium aurelia* in autoclaved (though complex) media.

Of all the ciliates investigated, seemingly only three genera have so far been established in successful pure culture using heat sterilized media, and of these three only one has been investigated thoroughly enough to warrant lengthy discussion.

The first ciliate (Lwoff, 1923) to be grown in heat-sterilized bacteria-free culture was what we, in America, now call *Tetrahymena geleii,* but which was referred to by Lwoff as *Glaucoma piriformis.** Much later, successful cultures of *G. scintillans* were established (Kidder, 1941) and these cultures are still maintained. *Colpidium campylum* was established in 1941 (Kidder, 1941) and preliminary studies (Peterson, 1942) indicated that its nutritional requirements were complex. These cultures have not been maintained.

The genus *Tetrahymena* (Furgason, 1940) has proved to be the most

* Some protozoologists are of the opinion that the correct name of this ciliate should be *Leucophrys piriformis* Ehrenberg 1830.—Editor.

useful for biochemical studies, because its synthetic capacities allow it to multiply in a medium devoid of heat labile compounds or complexes.

The justification of intensive biochemical investigations of *Tetrahymena* will be realized from the following points: (1) it is the only animal which has been grown in the absence of all other living organisms, on a defined medium, the components of which (with one exception) are known chemicals; (2) its nutritional requirements, while differing in some details, correspond surprisingly to what we know of the nutritional requirements of higher animals (birds, mammals); (3) its growth rate is the highest of any animal known, and it therefore possesses technical advantages shared by the bacteria, yeasts, and certain fungi; (4) being a microorganism it is possible to investigate biochemical problems with relatively little outlay for expensive and rare chemicals, when compared to higher animals, and experimental series may be expanded, therefore, to a degree impossible where large quantities of chemicals are required per animal; (5) quantitation of results is very precise and experiments may be duplicated at will; (6) *Tetrahymena,* when grown in pure culture (the cultures being originally started from a single organism) reproduce by binary fission. So far there has been no evidence of sexuality in this genus. This is of advantage in biochemical studies when the end point is protoplasm production, as it is known that reproductive rates vary, in sexual ciliates, with the complex cellular reorganizations which periodically take place.

Considerable confusion has resulted from the appearance of reports in the literature dealing with pure cultures of ciliates under a variety of names. This problem was investigated by Furgason in 1940. He made an extensive taxonomic survey of a number of ciliates and established the genus *Tetrahymena* with the type species *T. geleii.* Subsequent papers have dealt with this problem pointing out that these ciliates cannot be included in either of the well-established genera *Glaucoma* or *Colpidium* (Kidder, 1941; Kidder and Dewey, 1945a; Kidder *et al.,* 1945). Although it is not the purpose of this review to enter into the taxonomy of ciliates, it must be pointed out that many reports of a somewhat biochemical nature claim to be dealing with a species of *Glaucoma* or of *Colpidium,* when in reality the organism used was obviously *Tetrahymena* and most frequently *T. geleii.* It should also be stated that our stand on this matter is simply against the use of either *Glaucoma* or *Colpidium* as generic designations for ciliates of this type. Lwoff (personal communication) has expressed the opinion that while use of the generic names *Glaucoma* and *Colpidium* is certainly wrong for these ciliates, he believes they should be included in the older genus *Leucophrys*. The name *Tetrahymena* has become so firmly established in the biochemical literature, however, that we feel it would only add to the confusion if we were to change our designation. For a review of the various names which have been used for *Tetrahymena* until 1945 see our previous reports (Kidder, 1941; Kidder and Dewey, 1945a).

The present review, which deals largely with *Tetrahymena*, will follow the plan of taking up the various classes of chemical compounds which have been investigated. We wish to beg the reader's indulgence for omitting many references to investigations with other types of organisms, but space does not allow more than the briefest comparisons.

II. Inorganic Requirements

A large body of information has been accumulated on the importance of certain inorganic ions in the metabolism of organisms. This information has been gained (1) by noting deficiencies on certain diets (e.g., vasodilation and growth failure in rats on a magnesium deficient diet) and by counteracting these deficiencies by the addition of specific ions, usually in the form of salts; (2) by studying isolated enzymes or enzyme systems where activation can be demonstrated by specific ions (e.g., the activation of cocarboxylase by Mg^{++} or Mn^{++}); (3) by detection of certain inorganic constituents of biologically important compounds (e.g., calcium in bone; $PO_4^{\equiv}$ in phosphoproteins, nucleotides, etc.). It has become axiomatic that "physiological salts" are necessary in foods and culture media if one is to avoid inorganic ion limitations to growth. And yet, by cultivation in synthetic media, such as are used in microbiology, it is often extremely difficult to demonstrate certain inorganic requirements. This can almost always be traced to contaminations, often minute, of the chemicals used to compound the medium. Inorganic chemicals, even of the highest purity obtainable on the market, may contain traces of a number of unsuspected compounds. Amino acids often contain enough copper and iron to satisfy the requirements of many microorganisms.

For these reasons, very few critical experiments have been conducted on the inorganic requirements of ciliates. Some definite requirements are known, some are inferred, many are suspected without definite evidence. What little knowledge we have deals with *Tetrahymena*. It has been known for some time that various strains require salts for growth in purified proteins such as casein (Dewey, 1944; Hall and Cosgrove, 1945) and gelatin (Kidder and Dewey, 1944) and more recently in synthetic media (Kidder *et al.*, 1951). Just which ions are necessary and which can be dispensed with in the salt mixtures is still only incompletely known.

A. Calcium

Hall and Cosgrove (1945) made an attempt to determine the requirement of *Tetrahymena geleii* GP for certain minerals by using a "vitamin-free" casein medium. They found transplantable growth was possible in this casein only if salts were added. By adding salts singly and in various combinations, they found that casein supplemented with $FeCl_3$ or with $CaCl_2$, or $CaSO_4$ supported low but transplantable growth. A combination of both the iron and the calcium salts raised the growth to the concentration obtained with all the salts originally used. They conclude that calcium and iron represent requirements, these being the only ones demonstrable in their medium.

The above result with calcium is surprising when one considers the high calcium concentration of milk and the difficulty of removing this element during casein preparation. Moreover, Kidder, Dewey, and Parks (1951a) were unable to demonstrate a calcium requirement for *T. geleii* W in their synthetic medium. It was found that the addition of oxalate produced an inhibition which could be reversed by Mg^{++}, but not completely by Ca^{++}. Thus, at high levels of both oxalate and $MgSO_4$, normal growth resulted and under these conditions essentially all the calcium would be removed as calcium oxalate. It appears likely from these results that calcium is not an indispensable element for the growth of this ciliate.

B. Potassium

When all potassium salts are omitted from the synthetic medium of Kidder, Dewey, and Parks (1951a), the growth of *Tetrahymena geleii* W is reduced to less than one-fourth of normal (Fig. 1). Addition of K^+ in the form of KCl restores the growth to normal. Under these conditions 5 γ/ml. of K^+ raises the growth to half maximum and 20 γ/ml. of K^+ results in maximum yields. These quantities should not be considered absolute even in this medium, however, for two reasons. First, it appears likely that some potassium was present as a contaminant of other constituents of the medium which accounts for the residual growth without K^+ additions. Second, the medium contained an appreciable quantity of

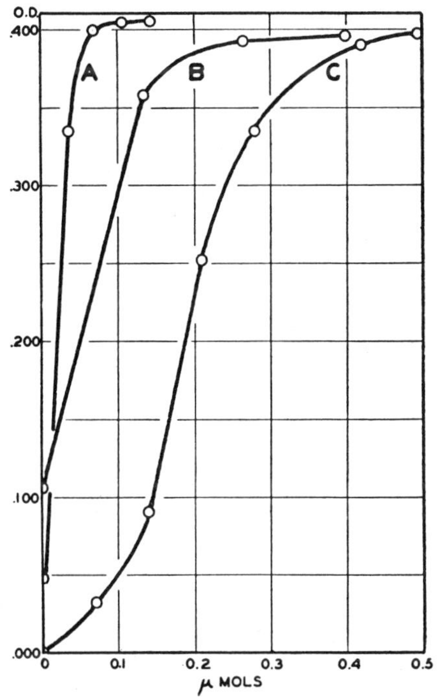

FIG. 1. Dose response of *Tetrahymena geleii* W to increasing amounts of (A) Mg^{++}, (B) K^+ and (C) $PO_4^{\equiv}$ plotted on a molar basis. In this figure and in all subsequent figures growth is represented as optical density (O.D.) and, unless otherwise stated, the incubation time was 96 hours at 25° C. Due to the fact that maximum yields vary with the different basal media used, the basal medium designation, corresponding to those given in Table IV will be stated below each figure. The basal medium (minus the appropriate ions) used for the experiments shown here was A.

Na^+, brought in during neutralization, and it is well known that the potassium requirement may vary with the amount of sodium present.

C. MAGNESIUM

A definite response to Mg^{++} is demonstrated for *Tetrahymena geleii* in the synthetic medium (Kidder *et al.,* 1951). Essentially no growth results when magnesium is omitted from the medium (Fig. 1). The addition of 2 γ/ml. of Mg^{++} (as $MgSO_4 \cdot 7H_2O$) restores growth to half maximal.

Demonstration of a magnesium requirement was also shown by the

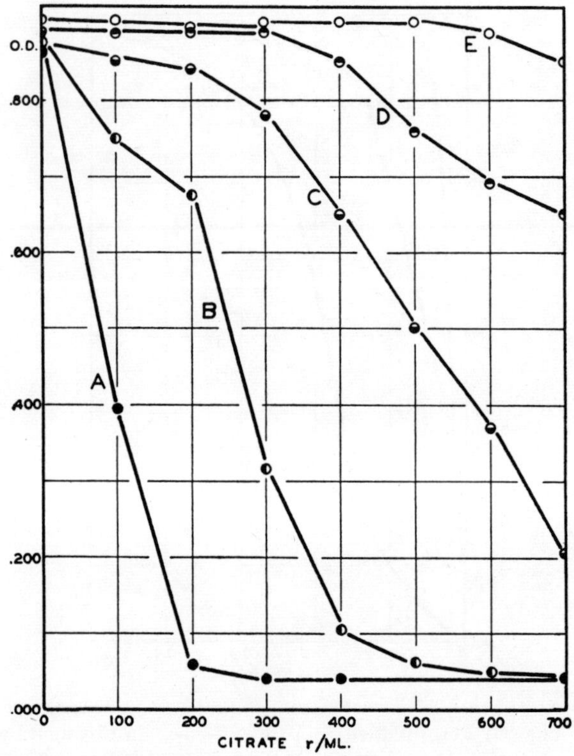

FIG. 2. Response of *Tetrahymena geleii* W to Na citrate in basal medium A minus $MgSO_4 \cdot 7H_2O$ and containing 10 mg./ml. Tween 80 instead of Tween 85. The amounts (γ/ml.) of magnesium sulfate are: A = 25; B = 50; C = 100; D = 150; E = 200.

addition of either oxalate (mentioned earlier) or citrate. Magnesium is known to form ion complexes with both of these compounds, and it was found that the inhibitory effects of both were completely removed by the addition of higher levels of Mg^{++} (Fig. 2). Both oxalate and citrate were competing with the ciliate enzymes for Mg^{++}.

D. Iron and Copper

Iron was found to be a possible requirement for *Tetrahymena geleii* GP by Hall and Cosgrove (1945). It was also found to stimulate growth of *T. geleii* W when added to a synthetic medium lacking what was earlier called Factor IIB (Kidder and Dewey, 1949a). In spite of its stimulatory nature, it has been found impossible to demonstrate its indispensability (Kidder *et al.,* 1951). This only reflects the contaminations with iron of the various constituents of the medium.

Likewise, copper was shown to be limiting when certain preparations of Factor IIB were used and rather large amounts were needed for optimum growth (Kidder and Dewey, 1949a). Changes in the brand of amino acids used in the earlier synthetic medium has made it impossible to show a requirement of *T. geleii* W for copper (Kidder *et al.,* 1951). It must be concluded, therefore, that the organism requires copper, but that one is fortunate to be able to prove such a requirement and can only do so with media containing amino acids of certain brands.

E. Trace Elements

No requirement for Mn^{++}, Zn^{++}, Co^{++}, Fl^-, $BO_3^=$, or MoO_3 could be demonstrated for *Tetrahymena geleii* W in the synthetic medium (Kidder *et al.,* 1951). The probable explanation for this is the contamination of salts or amino acids used in the medium, for all of these elements are known to be important in many organisms. Critical results on the question of trace elements in the metabolism of *Tetrahymena* must, therefore, await more highly purified chemicals.

F. Sulfates and Chlorides

No attempt has been made to test the requirement for Cl^- or Na^+. Certain amino acids (especially lysine) are only obtainable commercially as the hydrochlorides, and as these must be added, a Cl^--free medium would not be possible at the present time. Likewise, constructing a Na^+-free medium has not been attempted.

It has been possible, however, to demonstrate that *Tetrahymena* does not require the $SO_4^=$ ion. It is possible that $SO_4^=$ requirements are met by oxidation of the S of sulfur-containing amino acids.

G. Phosphate

It is not surprising that Hall and Cosgrove (1945) failed to find a $PO_4^≡$ requirement for *Tetrahymena.* It is known that casein is a phosphoprotein. In the synthetic medium used by Kidder and Dewey (1949a), however, no growth of *T. geleii* W takes place (even in the first transplant) unless $PO_4^≡$ is present (Kidder *et al.,* 1951). The requirement for maximum growth is high (50 γ/ml.) (Fig. 1). This high requirement is not strange considering the many phosphorylated compounds found in living organisms. In addition to the compounds acting in catalytic amounts (ATP, ADP, cocarboxylase, etc.), the phosphoproteins, phospholipids, and nucleotides of the nucleoproteins can account for relatively large amounts of $PO_{4\equiv}$

In constructing the medium for *Tetrahymena,* of course, high levels of KH_2PO_4 and K_2HPO_4 are added mainly for buffering (Dewey *et al.,* 1950).

III. Carbon Metabolism and Respiration

Tetrahymena geleii, unlike certain bacteria and most of the colorless flagellates, has no requirement for a source of organic carbon other than the carbon present in the amino acids which must be added in order that growth may occur. Nevertheless, it has been of interest to determine the effects of organic carbon compounds upon the growth and respiration of the ciliate and also the effects produced by the ciliate upon such substrates. The data upon this subject are somewhat scattered and, therefore, do not present a perfectly coherent picture. More work must still be done, particularly in the fields of respiration and intermediary metabolism before the pathways of carbon metabolism in the ciliate can be determined.

A. Carbohydrate Fermentation

Although it has long been recognized (Lwoff, 1932; Jahn, 1936; Phelps, 1936) that *Tetrahymena* is aerobic, Pace and Ireland (1945), demonstrated in a quantitative manner the complete dependence of the ciliate upon oxygen. At an oxygen pressure of 0.5 mm. of mercury the ciliates all died. With increasing oxygen pressures up to one atmosphere they found that growth increased progressively and was still increasing at 739 mm. of mercury. The minimum at which growth occurred was 10 mm. of mercury. These same authors found that carbon dioxide depressed growth of the ciliate and at relatively high concentrations (over 120 mm. of mercury) the organism is unable to metabolize.

In spite of its strict aerobic tendencies, *Tetrahymena* is able to live for short periods under anaerobiosis and to carry out anaerobic glycolysis. In the presence of adequate amounts of oxygen, carbohydrates are oxidized to carbon dioxide and water. Under anaerobic conditions dextrose is fermented to L-lactic, acetic, and succinic acids (Van Niel *et al.,* 1942; Thomas, 1943). From quantitative determinations of the acids produced,

Thomas (1943) concluded that lactic, on the one hand, and acetic and succinic on the other, are produced by independent processes.

$$\underset{\text{Pyruvic acid}}{CH_3COCOOH} + H_2O \longrightarrow 2H + CO_2 + \underset{\text{Acetic acid}}{CH_3COOH}$$

$$\downarrow$$

$$\underset{\text{Pyruvic acid}}{CH_3COCOOH} \longrightarrow \underset{\text{Lactic acid}}{CH_3CHOHCOOH}$$

It is suggested that lactic acid is produced from pyruvic acid in the same manner as in bacterial fermentations and vertebrate muscle glycolysis. The acetic acid appears to be formed from pyruvic acid by oxidative decarboxylation. Thomas also showed that *Tetrahymena* readily decarboxylated pyruvic acid. Succinic acid is formed by condensation of CO_2 with pyruvate to form oxaloacetate which is then reduced to succinic acid. The utilization of CO_2 for this process was conclusively demonstrated by the incorporation by the ciliate of isotopic carbon dioxide into the carboxyl group of succinic acid (Van Niel *et al.*, 1942).

In view of these facts it is not surprising that there is not always agreement in the literature regarding the abilities of different strains of *T. geleii* to ferment various carbohydrates when this is tested by the production of acid. It was pointed out by Kidder and Dewey (1945a) that tests of fermentation capacities must be carried out under strictly controlled conditions. The conditions which have the largest influence on the result are these: (1) diameter of tube; (2) volume of medium; (3) concentration of carbohydrate; (4) concentration of peptone (amino acids); (5) growth period; (6) initial pH; (7) amount of buffering. The first two conditions are obviously concerned with surface-volume relationships. Unless this ratio is maintained below a certain maximum value, the amount of fermentation occurring is insufficient to keep pace with oxidation and acid does not accumulate. In other words, conditions must be sufficiently anaerobic for some fermentation to occur.

It is also obvious that small concentrations of carbohydrate would be more readily oxidized than the higher concentration and detection of acid production would be, therefore, more difficult. The effect of the concentration of peptone or of amino acids is not so apparent. It has been demonstrated that ammonia accumulates as cultures of *Tetrahymena* grow (Lwoff and Roukhelman, 1926). This ammonia tends to obscure acid produced, particularly in high concentrations of amino acids or other metabolites which could serve as a source of ammonia. This is particularly true when the carbohydrate is fermented slowly. When the carbohydrate is not fermented, cultures become alkaline in reaction. Therefore, it is significant when the pH remains near 7.0 for periods longer than the controls, since it may indicate slow attack on the carbohydrate in question. This appears to be the case with galactose with certain strains of *Tetrahymena* (Kidder and Dewey, 1945a).

The importance of the initial pH of the medium scarcely needs discussion. It is obvious that either a high or low initial pH would have the effect of obscuring acid production as would large amounts of buffer. A nearly neutral reaction would be most sensitive. Since not all of these factors have been sufficiently controlled by the various authors who have reported upon carbohydrate fermentation in this species, it is not surprising that there is some lack of uniformity. Data on this subject have been presented by Colas-Belcour and Lwoff (1925), Elliott (1935b), Johnson (1935), Glaser and Coria (1935), Kidder and Dewey (1945a), Hall (1933), and Loefer (1938). There is no advantage to be gained from a detailed discussion of their results. It is generally agreed that all strains ferment dextrose, maltose, levulose, mannose, and glycogen. There is disagreement regarding dextrin, starch, and galactose. No other carbohydrate or polyhydric alcohols were fermented, although large numbers of them have been tested, including pentoses, disaccharides, trisaccharides, polysaccharides, and polyhydric acids. Glaser and Coria (1935) report fermentation of cellulose. *Tetrahymena vorax* (Kidder and Dewey, 1945a) ferments lactose and galactose.

A more adequate knowledge of the capacities of *Tetrahymena* could be gained by analysis of the medium before and after growth of the organism. Disappearance of added substrate, e.g., carbohydrate, would be a clear indication of its utilization. Such a study has been made by Loefer (1938) using dextrose as the substrate. He has calculated that the ciliates utilize $2.34–3.92 \times 10^{-7}$ mg. of dextrose per organism per hour. He also states that carbohydrates such as sucrose "accelerate" the growth, that is, increase the yield, of the ciliates. Loefer attributes this to a favorable effect of the carbohydrate on the oxidation-reduction potential of the medium. This observation has been made by others of the authors cited above, but we failed to confirm this stimulation by non-utilizable carbohydrate. Using Loefer's (1938) formula, it is possible to calculate that in a medium used by Dewey, Parks, and Kidder (1950), which contains 2.5 mg. of dextrose per milliliter, about half of the dextrose is consumed in a 4-day period. This should be subjected to experimental test, since it has been found that the larger amount of dextrose gives better growth than the amount which it is calculated is utilized. It is even possible that in this medium, less dextrose than the amount calculated is used, since the ciliate is grown in tubes slanted to give a larger aeration surface. Under these conditions, there should be relatively less fermentation and greater oxidation of the dextrose. Since more energy is obtained from complete oxidation than from fermentation of a given amount of dextrose, less dextrose should be required. That fermentation is at a minimum under these conditions is evident from the alkaline reaction of the medium after growth of the ciliates.

B. Organic Acids

With regard to carbon sources other than carbohydrates, very little work has been done with ciliates. The effects of acetate and other lower fatty acids on the growth of both colorless and green flagellates is well known. Kidder and Dewey (1949a) found that acetate has a stimulatory effect upon *Tetrahymena*. Although its effect on growth rate was not studied, it increased the yield of cells. The optimum concentration was about 1 mg./ml. Later (Dewey *et al.*, 1950) acetate was found to have no effect on growth in a medium considerably richer in all nutrients than had previously been used. It is possible that the presence of increased amounts of protogen (Kidder and Dewey, 1949a; Stokstad *et al.*, 1949) makes the addition of acetate unnecessary, since it has been shown (Snell and Broquist, 1949) that protogen can replace acetate in the growth of *Lactobacillus casei*. Protogen is the same factor which has been described by O'Kane and Gunsalus (1948) as a requirement for oxidation of pyruvate by *Streptococcus faecalis* (see below). Kidder, Dewey, and Parks (1950a) have shown that, while acetate cannot replace protogen as in the case of *L. casei*, its presence in the medium reduces the protogen requirement (Fig. 3).

Of a number of other organic acids tested (Kidder and Dewey, unpublished), none was stimulatory with the exception of pyruvate. Here again the effect may have been due to acetate which is produced by decarboxylation and oxidation of the pyruvate. Oxalate, glycerate, glutarate, α-ketoglutarate, succinate, fumarate, citrate, *cis*-aconitate, tartrate, lactate, propionate, butyrate, glycolate, formate, glycerol, ethanol, malate, malonate, and laurate were also tested. The only effect noted was inhibition of growth at high concentrations of certain of these compounds. The effect obtained depends to some extent upon whether or not dextrose and/or acetate is present in the medium. Much more work is required in order to demonstrate the roles which these compounds may play in the metabolism of *T. geleii*. Certain acids such as oxalic and citric are quite inhibitory to growth. Calcium salts reduce their toxicity, but manganese salts are without effect. This was a surprising result in view of the report of MacLeod and Snell (1947) with regard to *Lactobacillus*. On the other hand (Kidder *et al.*, 1951), complete reversal can be obtained with magnesium ions (Fig. 2). The relative effectiveness of Mg^{++} as opposed to Mn^{++} is therefore just the opposite for *Tetrahymena* to what it is for *Lactobacillus* (see above).

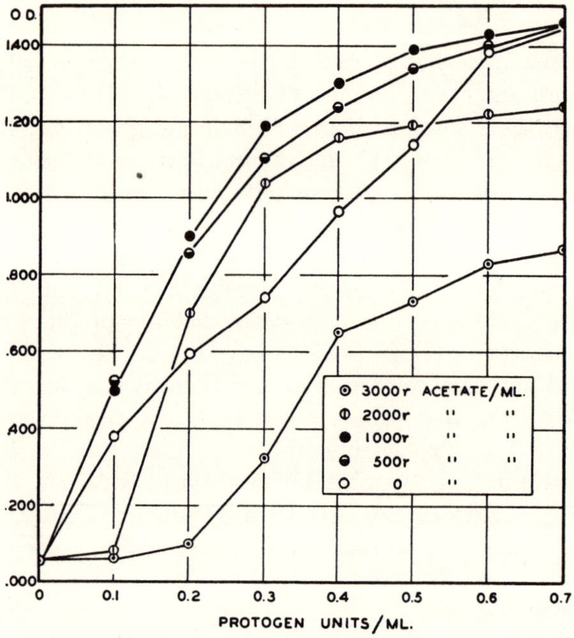

FIG. 3. Sparing action of acetate on protogen. Concentrations of 3 mg./ml. or more of acetate result in inhibition of maximum yield which is not overcome by increased protogen. Medium 2C, minus protogen.

The fact that succinic, fumaric, and other acids of the Krebs cycle do not function as growth stimulants is not surprising. They do not generally serve as carbon sources even in organisms in which the Krebs cycle is known to operate. Ryan, Tatum, and Giese (1944) were able to show that when respiration of the mold *Neurospora* was inhibited by iodoacetic acid, release of the inhibition could be obtained with succinic acid, or to some extent with pyruvic acid. Since it has been found (M. Lwoff, 1934) that respiration of the ciliate is also sensitive to iodoacetate as well as to arsenious acid and urethane, and the ciliate is known to accumulate succinic acid (Thomas, 1943), it may be tentatively assumed that the Krebs cycle operates in *Tetrahymena.*

In view of the requirements for fatty acids both in mammalian nutrition and in bacterial nutrition, a study was made of their effect upon the

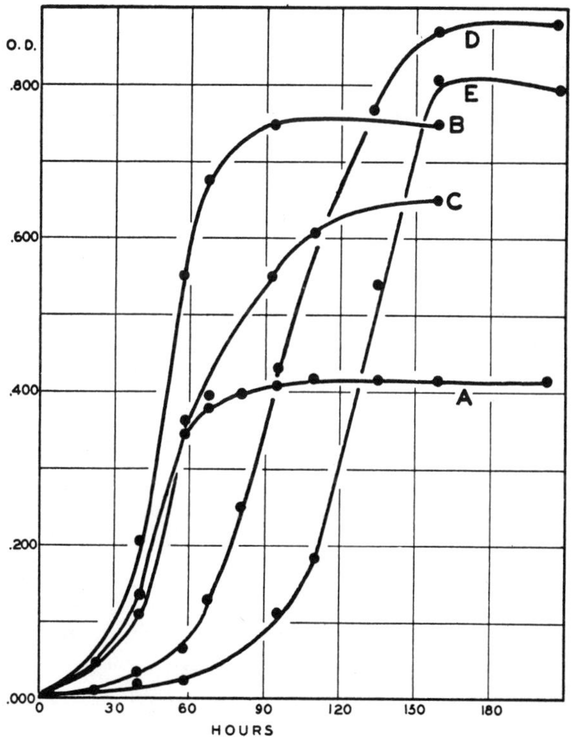

FIG. 4. Growth curves, plotted in hours, showing the effect of Tween 80 (B = 8 mg./ml., C = 14 mg./ml.) and Myrj G2144 (D = 10 mg./ml., E = 14 mg./ml.) as compared with control sets having no Tween or Myrj (curve A). Medium A minus Tween.

growth of *Tetrahymena.* No requirement for fatty acids has been found in the ciliate, although the stimulatory effect of acetate is discussed above. Propionate shows a slight stimulatory effect in low concentration, but becomes inhibitory in higher concentrations, while butyrate shows no stimulation at any concentration and soon becomes inhibitory. Lauric acid is also quite inhibitory as are all the higher unsaturated fatty acids tested (Kidder and Dewey, 1949a). Concentrations as low as 30–70 γ/ml. are sufficient to suppress growth completely. Such effects were obtained with oleic, linoleic, linolenic, arachidonic, vaccenic, and a C_{20}-unsaturated fatty acid.

On the other hand, it has been found that certain Tweens (polyoxy-alkylene derivatives of the sorbitan esters of fatty acids) show a marked stimulatory effect on growth (Kidder and Dewey, 1949a). Tween 85 (trioleate) had the most pronounced effect, but the turbidity of its solutions prevented studies of high concentrations of this material. Tween 80 (monooleate) gives clear solutions, and it has been found to increase growth tremendously at very high concentrations (10 mg./ml. or more) (Dewey *et al.,* 1950). A similar effect is given by Myrj G2144 (polyoxy-alkylene derivative of oleic acid). Although growth is delayed for three to four days, it is sufficiently rapid in the next two days to surpass that given by Tween 80 (Fig. 4). Preliminary tests appear to indicate that Tween esters of other fatty acids, e.g., stearic, are also stimulatory in high concentration, though somewhat less so than the oleates. It seems evident that the effectiveness of the Tweens is due either to a physical change in the medium or to the fact that in this form the fatty acids they contain are a good source of energy. This conclusion is reinforced by the fact that free fatty acids are inhibitory even in low concentration. It appears, however, that the effect of the Tweens cannot be attributed solely to their well-known surface activity, since such surface-active compounds as taurocholic and glycocholic acids are quite inhibitory.

C. Respiration

An investigation of the effect of various organic substrates upon the respiration of *Tetrahymena geleii* was made by Chaix, Chauvet, and Fromageot (1947). Of all the substances tested, butyric acid had the greatest stimulatory effect, although it was followed closely by acetate. Both these compounds gave a stimulation of about 50% of the respiration. Glucose, ethanol, propionic acid, isobutyric acid, caprylic acid, and palmitic acid, homocysteine and methionine gave moderate stimulation, while formate, succinate, citrate, stearate, α-amino butyrate, acetoacetate, α-hydroxy butyrate, and crotonic acid had little or no effect. Butyl alcohol, caproic acid, and malic acid had a slight depressing effect. Pyrocatechol was very inhibitory, and linoleic acid caused inhibition and lysis.

These results point to the importance of the lower fatty acids in the carbon metabolism of this organism. They also show that compounds having an effect on growth are not necessarily those which are important in respiration and vice versa.

A number of other studies have been made upon the respiration of *Tetrahymena*. As has been shown by Chaix *et al.* (1947), these are essentially in agreement when various factors are taken into account. M. Lwoff (1934) gives a value of 35 mm.^{3} O_2 per hour per milligram dry weight for the Q_{O_2} of *Tetrahymena*. These data agree less well than the others. According to Chaix *et al.* (1947), this is subject to an error of 3.5-fold in the determination of the dry weight from which the Q_{O_2} is calculated. Making this correction a Q_{O_2} of 10 is obtained. Since M. Lwoff was using organisms from a culture 150 hours old, a low value for the Q_{O_2} is to be expected. With organisms 48 to 70 hours old, values from 13.9 to 19.6 have been obtained (Baker and Baumberger, 1941; Chaix *et al.*, 1947; Hall, 1938, 1941; Ormsbee, 1942). The remarkable fact about these results is their uniformity since the data were obtained at different temperatures, pH, kind and concentration of buffer and some in the presence of substrate. Some of the data had to be recalculated from that given by the original authors since there was no uniformity of expression. Another factor which does not seem to have been taken into account is the concentration of the organisms present in the respirometer vessel. Pace and Lyman (1947) have shown that as the number of organisms per milliliter in the vessel increases there is a relative decrease in oxygen consumption.

From studies of the respiration, a few more facts regarding the metabolism of *Tetrahymena* have emerged. Hall (1941) showed that the respiration of *T. geleii* is cyanide sensitive. This fact was confirmed by Baker and Baumberger (1941) who also showed that the organism contains cytochromes. Since it is also somewhat sensitive to CO, it must utilize the cytochrome-cytochrome oxidase system. Ormsbee and Fisher (1944) have also studied the urethane-sensitive respiration.

IV. Nitrogen Metabolism

The first attempts to elucidate the nitrogen requirements of *Tetrahymena geleii* were made by Lwoff (1924, 1932). Since he was unable to obtain growth of the ciliate in peptones produced by the successive treatment of muscle with pepsin, trypsin, and erepsin, whereas good growth occurred in peptones prepared by only peptic digestion of muscle, the tentative conclusion was reached that *T. geleii* requires complex peptides for growth. It has now become apparent that the "ereptone" of Lwoff failed to support growth because it lacked some material essential for growth, since amino acids as such can be easily utilized. It is possible that the erepsin contained enzymes capable of digesting pteroylglutamic acid, thus destroying a growth factor required by the ciliate. This is merely one of several possibilities, such as destruction of vitamins, pyrimidines, or purines by prolonged maintenance of the material at the alkaline pH required for digestion by trypsin and erepsin.

Little profit is found in prolonged discussion of attempts to obtain growth of the ciliate in amino acid mixtures or "incomplete" proteins which were made before the identification or even recognition of some of the vitamins and growth factors required by *T. geleii* (Elliott, 1935a; Hall, 1939, 1942; Hall and Elliott, 1935). It is sufficient to say that what growth occurred in the media then used was due to the use of impure protein products. Far greater stimulation of growth occurred upon the addition of minute amounts of yeast extract to "incomplete" proteins than upon the addition of large amounts of the missing amino acids. Needless to say, the amino acid mixtures alone completely failed to support growth.

A. Essential Amino Acids

Upon the recognition of the fact that *Tetrahymena* requires materials not included among the then known vitamins (Dewey, 1941, 1944) and the additional fact that the ciliate would grow in an acid hydrolyzate of casein supplemented with tryptophane and a source of these factors (Kidder and Dewey, 1944, 1945b), it became apparent that peptides are not essential for growth. This made far simpler the investigation of the amino acid requirements of *Tetrahymena* than had at first been anticipated.

Since the use of "incomplete" proteins as a means of determining amino acid requirements is fraught with danger and the removal of single amino acids from protein hydrolyzates is tedious and not always completely successful, the technique of single omissions from a mixture of amino acids imitating a "complete" protein hydrolyzate was used. This is a time-honored procedure and one which was used by Rose (1938) and by him and his associates (1948) in studies on the amino acid requirements of mammals. It is not, however, completely satisfactory. The information it gives concerns only the requirement for a given amino acid in the presence of certain concentrations of a given number of other amino acids. Thus, for example, Kidder and Dewey (1945a,b) found that glycine appeared to be a requirement for *Tetrahymena* in the presence of all the other amino acids of casein at the levels at which they are found in casein. The amino acids required by *Tetrahymena* were otherwise the same as those needed by the growing rat (Rose, 1938; Rose *et al.*, 1948): arginine, histidine, isoleucine, leucine, lysine, methionine, phenylalanine, threonine, tryptophan, and valine. Growth in a medium containing only these amino acids was possible but poor. Therefore glycine cannot be considered an essential amino acid. The addition of rather large amounts of glycine was somewhat stimulatory, but very small amounts of serine gave a much larger effect. This effect appears to be a property of the hydroxyamino acids, since it can also be obtained by increasing the amount of threonine present in the mixture (Kidder and Dewey, 1947b). All the strains of *Tetrahymena* which have so far been tested require the same group of amino acids (Kidder and Dewey, 1945a). Strains E and T-P appear to have an additional requirement for serine, although cystine can replace serine for strain T-P, and for certain of the other strains in which serine is only stimulatory. The serine requirement for the growth of strain E has recently been confirmed by Elliott (1949b).

A paper by Kline (1943) purporting to be a determination of the amino acid requirements of *T. geleii* (*Colpidium striatum*) has also appeared. This author reports that his strain requires fifteen amino acids, but grows in the absence of histidine, leucine, lysine, and threonine. As is pointed out below, the medium used by Kline also lacked four other growth essentials. Calculations from the data presented by the author indicate that "growth" in his amino acid media was dependent upon the size of the inoculum introduced and was steadily declining from one transfer to the next. On the other hand, growth in the tryptone controls increased steadily from one transfer to the next until it reached the unprecedented height of 1.2 million organisms per milliliter. This medium is capable of supporting a population of about 60,000 to 120,000 ciliates per milliliter at best, as has been reported by a number of authors. In Kline's medium containing only the ten essential (for the rat) amino acids, dextrose was increased to 1%, which alone would have been sufficient to inhibit growth. For these reasons this paper should be dismissed from consideration. The report by Seaman (1949) that *Tetrahymena* (*Colpidium campylum*) can fix atmospheric nitrogen probably belongs in the same category.

That the story of amino acid requirements is not so simple as would appear at first glance becomes apparent when the interrelationships of the amino acids are investigated. Some indication of this complexity was given by Kidder and Dewey (1945a). The requirement for valine was shown to be dependent upon the absence of glycine, cystine, and serine. In the presence of certain combinations of these amino acids in addition to the other essentials, valine was not required. The key position of serine in the amino acid metabolism of *Tetrahymena* was demonstrated even more definitely in a later paper (Kidder and Dewey, 1947b). In the following section the amino acids will be considered separately. Optimal concentrations for a medium low in amino acids and containing only the essentials plus serine are given in parentheses. These values differ in media of varying total amino acid concentration.

B. Arginine

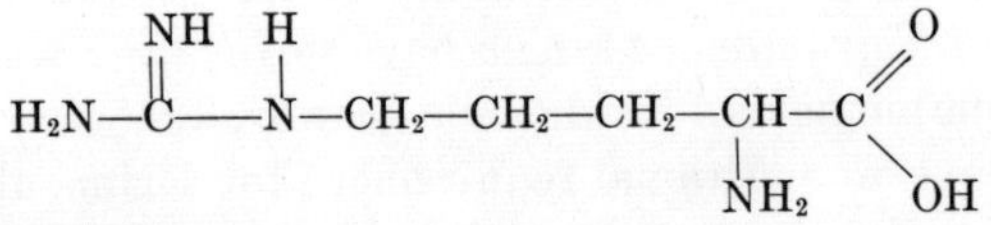

Arginine (20 γ/ml.) is synthesized slowly by *Tetrahymena* either in the simulated casein hydrolyzate medium or in the medium containing

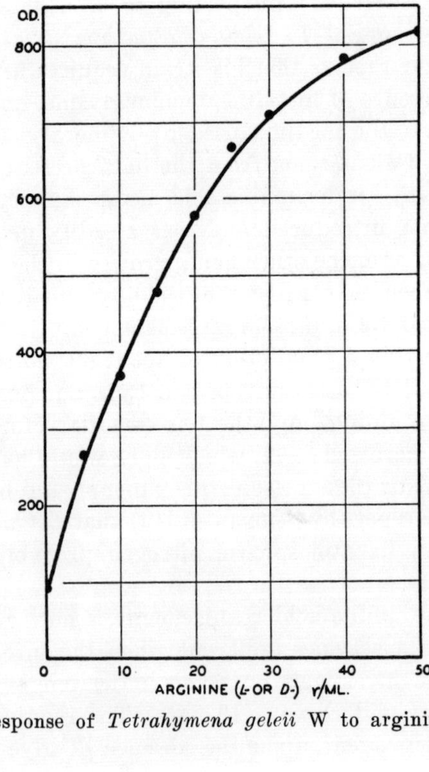

Fig. 5. Dose response of *Tetrahymena geleii* W to arginine. Medium S minus arginine.

only the essential amino acids plus serine. However, the rate of synthesis of arginine must be greater than the rate of synthesis of serine since arginine fails to stimulate in the absence of serine (Kidder and Dewey, 1947b). A larger requirement (Fig. 5) is indicated in a medium containing higher concentrations of amino acids.

The D-isomer of arginine is now available commercially. It has been found to have an activity equal to that of the L-isomer. Arginine cannot be replaced by either ornithine or citrulline, although both of these amino acids are capable of sparing it. This seems to indicate clearly that the Krebs-Henseleit urea cycle does not operate in the ciliate—a fact further borne out by their failure to produce urea (Lwoff and Roukhelman, 1926). It is of interest to note that the organism also fails to produce a urease (Lwoff and Roukhelman, 1926). There appears to be, however, a necessity for the production of ornithine and/or citrulline for some metabolic purpose.

C. Histidine

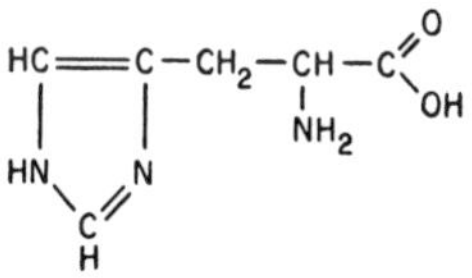

Histidine is an absolute requirement for growth of *Tetrahymena*. The amount required for optimum growth (7.5 γ/ml.) is small and is the same whether or not serine is present (Kidder and Dewey, 1947b), but is increased in more concentrated media. The D-isomer of histidine appears to be completely inert, since the optimum amount of DL-histidine is double that of L-histidine (Fig. 6). At concentrations above 50% of the total histidine, the D-isomer is inhibitory. *Tetrahymena* is unable to utilize imidazole in place of histidine, nor does imidazole spare the amino acid (Kidder and Dewey, 1947b).

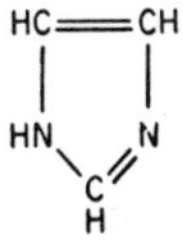

D. Isoleucine

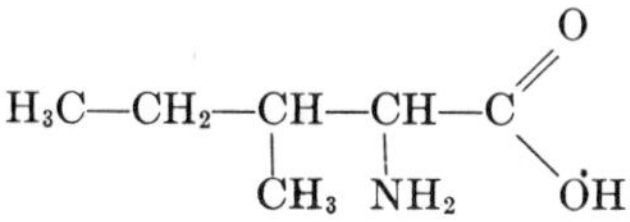

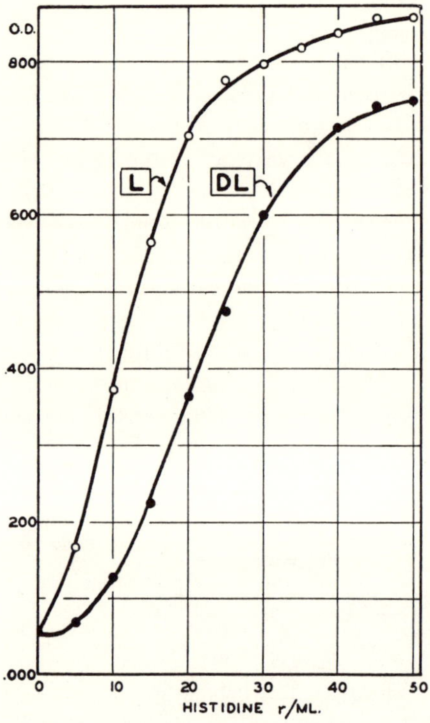

FIG. 6. Dose response of *Tetrahymena geleii* W to histidine. Medium S minus histidine.

The situation with regard to isoleucine is, at present, rather difficult to interpret. It was earlier stated (Kidder and Dewey, 1947b) that in the presence of serine, D-isoleucine is partially utilized (17.5 γ/ml. for L-isoleucine; 25 γ/ml. for DL-isoleucine), but no growth whatsoever occurred using DL-isoleucine in the absence of serine, while 200 γ/ml. of L-isoleucine were required.

In an attempt to confirm these results a number of samples of both L- and DL-isoleucine as well as one of D-isoleucine were tested. It was found that D-isoleucine had no activity whatever, with the exception of a very slight inhibition. A sample of L-isoleucine gave a half-maximal growth response at a concentration of 17.5 γ/ml. In all, four samples of DL-isoleucine were examined. These were all obtained from different sources. Two of the samples gave a half-maximal response at 26.5 γ/ml. and two at 37.5 γ/ml. (see Fig. 7).

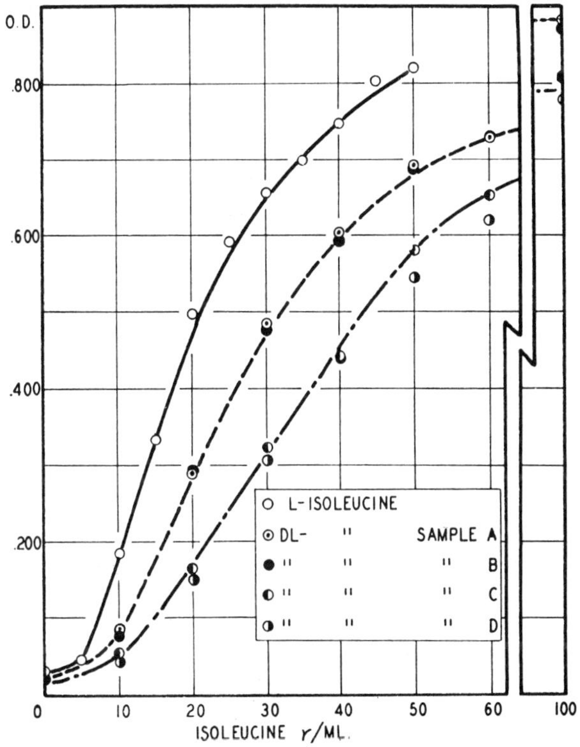

FIG. 7. Dose response of *Tetrahymena geleii* W to isoleucine. Medium S minus isoleucine.

It is not probable that the samples of DL-isoleucine which had the least activity were reduced in activity because of contamination with other essential amino acids, since one preparation at least is used routinely in making up basal media minus each of the other amino acids. In none of the media mentioned above was an unaccountably high blank reading obtained (Kidder and Dewey, 1945b). There remains the possibility that certain synthetic samples of isoleucine contain allo-isoleucine. On the basis of results with allo-threonine (see below), allo-isoleucine would be expected to be inactive and would therefore reduce the activity of the samples. There is also a possibility that the sample of L-isoleucine contained impurities which reduced its activity.

The hydroxy analog of isoleucine has proved to be completely unavailable to the ciliate.

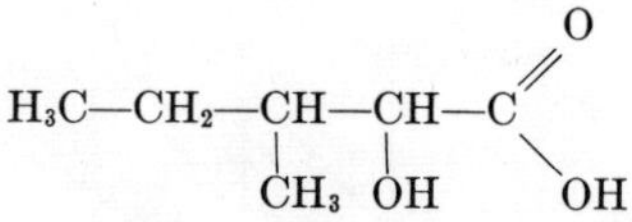

β-Methyl-α-hydroxyvaleric acid

This indicates some deficiency in the transaminating enzymes.

It is generally accepted that in organisms which can utilize D-amino acids, the amino acid is first oxidized to the keto acid and then reaminated to form the L-amino acid. However, the action of D-amino acid oxidase appears to occur in two steps. First a dehydrogenation to form the α-imino acid, then hydrolysis to the α-keto acid.

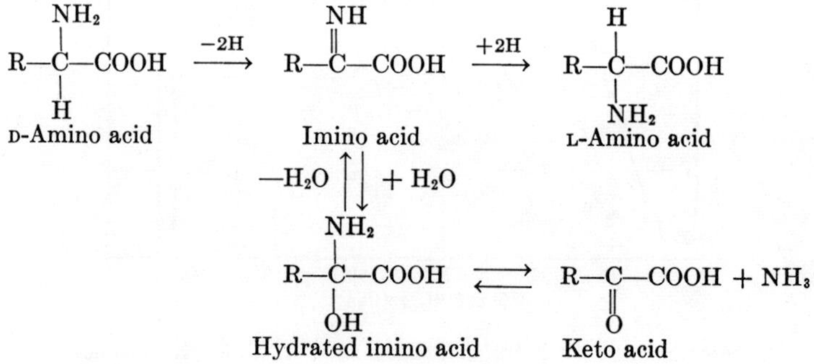

In this case it would be possible to obtain an α-amino acid by rehydrogenation of the α-imino acid. Since production of an imino acid destroys the optical activity of the α-carbon atom, rehydrogenation could theoretically lead to the production of either a D- or an L-amino acid. Possibly the enzymes responsible for reduction of the imino acid are such that only L-amino acids result. Such an explanation implies that *Tetrahymena* has lost the ability to reaminate an α-hydroxy or an α-keto acid, while retaining the ability to reduce an imino acid. This still remains in the realm of hypothesis, however, since α-imino acids have not been available for testing.

E. Leucine

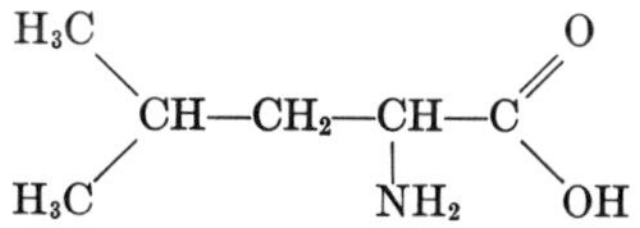

Like isoleucine, L-leucine is required at the same level whether or not serine is present (25 γ/ml.; 80 γ/ml. in a more concentrated medium). The ciliate differs in the greatly increased requirement for DL-leucine (140 γ/ml.) (Fig. 8). In this case the difference appears to be due to the inhibitory effect of D-leucine. This was postulated in 1947 (Kidder and

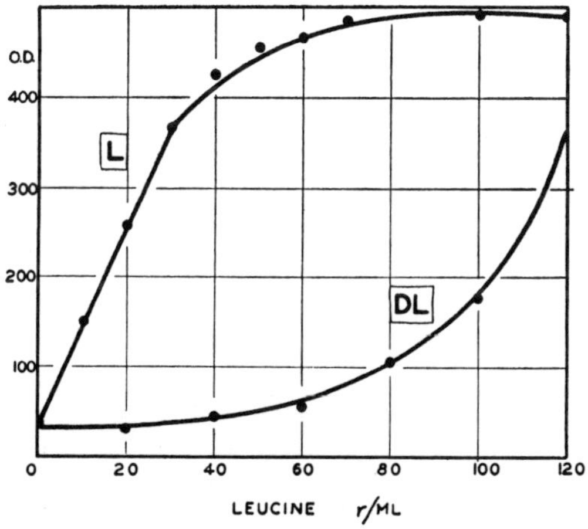

Fig. 8. Dose response of *Tetrahymena geleii* W to leucine. Medium A minus leucine.

Dewey, 1947b) to account for the high requirement for DL-leucine. The fact that at sufficiently high levels of DL-leucine growth did occur in spite of the presence of what should be inhibitory levels of D-leucine was explained on the basis of the disappearance of the natural form at low concentrations of the DL-mixture, by utilization by the organisms. Thus, in spite of a reversible enzyme-substrate combination, the available enzyme comes to be more and more saturated at all times with the D-isomer, and growth is inhibited. As soon as the absolute concentration of L-isomer approaches the optimal, the inhibitory effect of the D-isomer decreases and, at supraoptimal concentrations, disappears. This was demonstrated conclusively by the use of D-leucine. Inhibitory indices of about 0.4 were obtained at low levels of L-leucine, and no significant inhibition at ratios as high as 1:20 (L:D) at a supraoptimal level of L-leucine (Fig. 9).

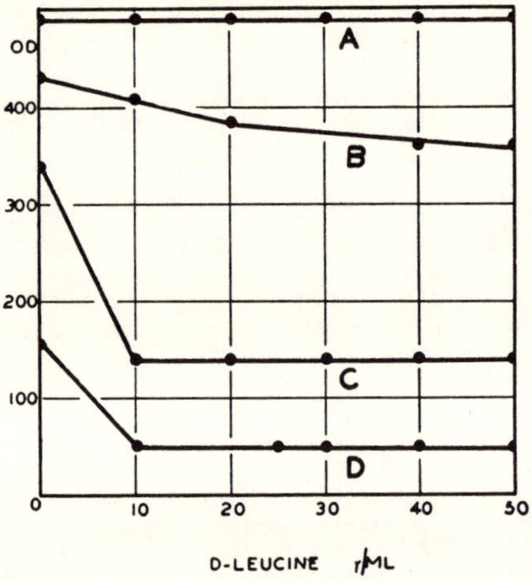

FIG. 9. Inhibitory action of D-leucine to *Tetrahymena geleii* W and the release of inhibition by L-leucine. The experimental sets contain the following amounts of L-leucine: in γ/ml. A = 200; B = 50; C = 25; D = 10. Medium A minus leucine.

The hydroxy and keto analogs of leucine were ineffective in replacing leucine for growth, nor did they spare leucine, again indicating a deficiency in transamination.

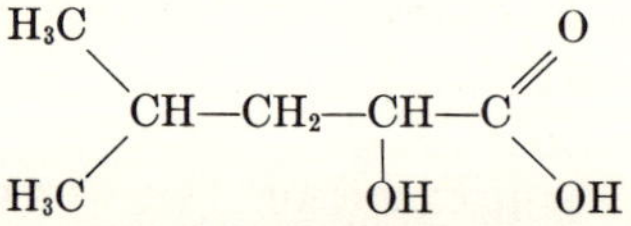

α-Hydroxyisocaproic acid

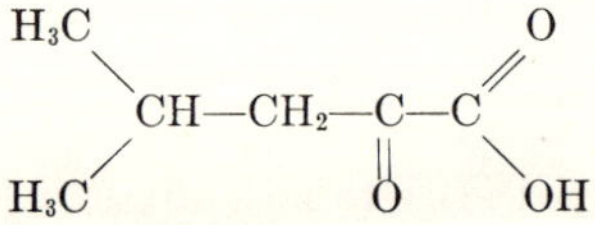

α-Ketoisocaproic acid

A number of N-substituted analogs of L-leucine were also tested. Both N-acetyl-L-leucine and N-propionyl-L-leucine would replace leucine, but their activity was of a very low order of magnitude. At low concentrations they were completely inert, their activity gradually increasing with the concentration from about 1.5% to 3.5% at 1 mg./ml. At the lower concentrations N-propionyl-L-leucine was somewhat more active than the acetyl derivative. N-butyryl-L-leucine did not replace leucine, but, like the other analogs mentioned, was active in sparing it, all three compounds being of about equal activity on a molar basis.

On the other hand, N-lauryl-L-leucine was extremely inhibitory to growth. This inhibition was found to be entirely non-competitive as far as leucine was concerned. At any concentration of leucine tested, complete inhibition occurred at 30 γ/ml. of N-lauryl-L-leucine. That this inhibition was largely due to lauric acid itself is demonstrated by the fact that this compound almost completely suppressed growth at a concentration of 25 γ/ml. Therefore, on a molar basis, N-lauryl-L-leucine is nearly twice as inhibitory as free lauric acid. There is at present no explanation for this effect.

The inhibition caused by N-salicyl-L-leucine appears to be due to still another cause. Salicylic acid itself is not inhibitory to the growth of *Tetrahymena*. The peptide, however, suppresses growth at concentrations of from 700 to 800 γ/ml. This inhibition is not released by leucine to any appreciable extent. It seems improbable, then, that the peptide is competing for enzymes which deal with natural peptides of leucine, although this possibility has not been thoroughly tested (see below under "Glycine"). So far it has been impossible to release this inhibition.

F. Lysine

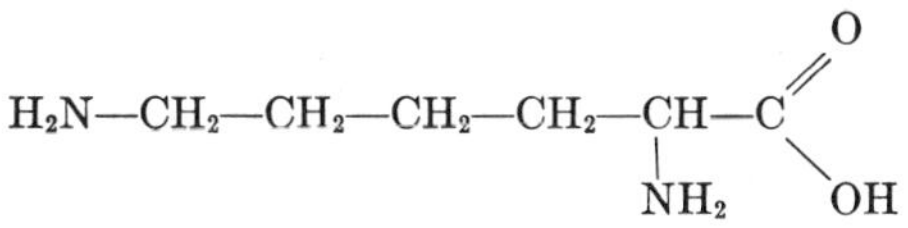

Lysine is required in the same concentration (15 γ/ml.) whether serine is present or absent and only the L-isomer is active. No analogs of any type have been available for testing (Fig. 10).

G. Methionine

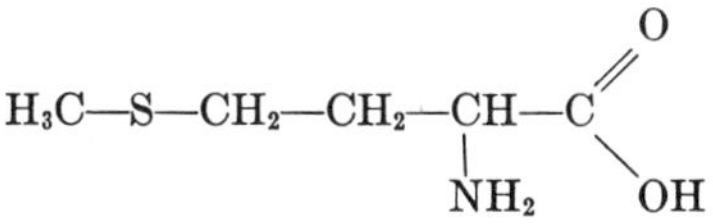

Although it has been reported (Kidder and Dewey, 1947b) that both D- and L-isomers of methionine are active (20 γ/ml. L-methionine, 30 γ/ml. DL-methionine), the former somewhat less than the latter, later results indicate that D-methionine alone is unavailable (Fig. 11). This discrepancy is apparently due to the fact that the "L-methionine" used in the earlier work was not pure and contained an appreciable amount of the D-isomer. This would tend to vitiate results by giving false high activities for racemic methionine.* The presence or absence of serine makes no difference in the requirement (Kidder and Dewey, 1947b). Cystine, homocystine and D-methionine spare L-methionine, the first being somewhat more active on a molar basis. Camien and Dunn (1950) have recently shown that the ability of *Lactobacillus arabinosus* to utilize D-methionine is dependent upon the presence of relatively large amounts

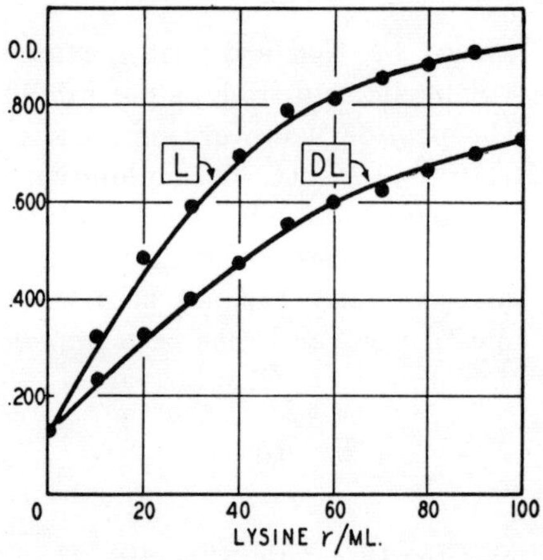

FIG. 10. Dose response of *Tetrahymena geleii* W to lysine. Medium S minus lysine.

of pyridoxamine in the medium. Pyridoxal was less effective. This probably does not hold true of *Tetrahymena,* however, since the medium used contains a large excess of both of these factors.

It was reported (Kidder and Dewey, 1948a) that homocystine would replace methionine provided a factor present in a liver extract were added. To explain this effect, a "co-transmethylase" was postulated. It was later shown that the effect of the liver extract was to introduce sufficient methionine into the medium to permit growth (Genghof, 1949). The error was the result of false methionine values obtained on assaying

* The sample of "L-methionine" used was obtained from Bios Laboratories and was kindly analyzed for us by Dr. Gerrit Toennies. He found it to contain approximately 85% L-methionine. This would account for the activities observed.

the liver extract with *Leuconostoc mesenteroides.* It now appears that *Tetrahymena* cannot utilize homocystine to satisfy its methionine requirement even in the presence of choline, dimethylthetin, betaine and/or vitamin B_{12}. The sole effect of homocystine is to spare methionine (at suboptimal levels of the latter) probably for the formation of cystine.

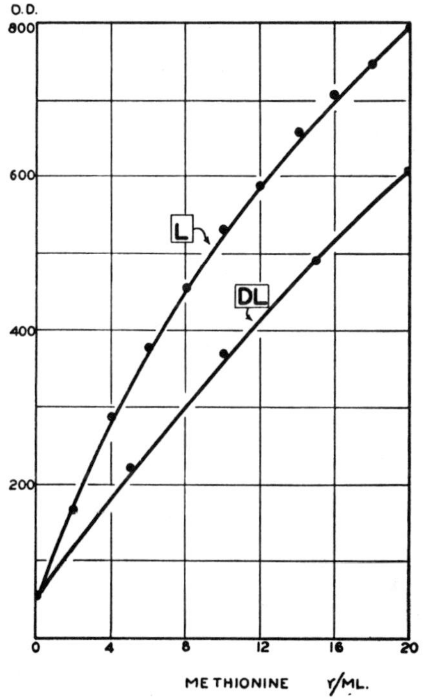

FIG. 11. Dose response of *Tetrahymena geleii* W to methionine. Medium S minus methionine.

Although there is no experimental proof, it is possible that this change is effected via cystathionine (Binkley, 1944) formation by a reversal of the process which Horowitz (1947) has demonstrated to occur in methionineless *Neurospora.*

It is of interest in this connection to note that neither choline nor dimethylthetin spares methionine, although the organism synthesizes choline as shown by choline-less-*Neurospora*-assay of the medium before and after growth of the ciliate (Kidder and Dewey, unpublished).

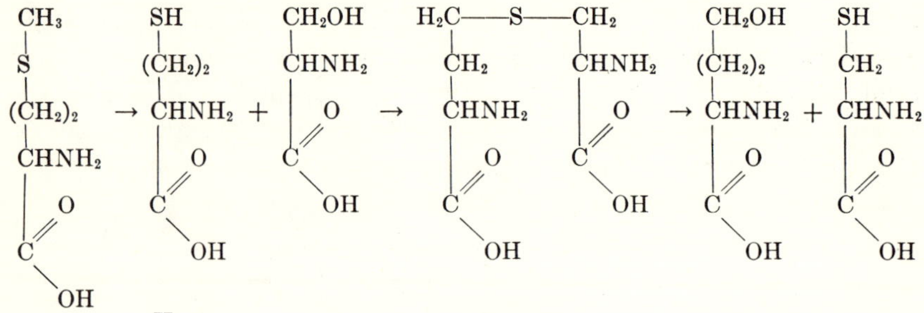

$$\begin{array}{ccccccc} CH_3 & SH & CH_2OH & H_2C\!\!-\!\!S\!\!-\!\!CH_2 & CH_2OH & SH \\ S & (CH_2)_2 & CHNH_2 & CH_2 \quad\; CHNH_2 & (CH_2)_2 & CH_2 \\ (CH_2)_2 \to & CHNH_2 + & \mid \quad O \to & CHNH_2 \quad\; \mid \quad O \to & CHNH_2 + & CHNH_2 \\ CHNH_2 & \mid \quad O & C\!\!=\!\! & \mid \quad O \quad\; C\!\!=\!\! & \mid \quad O & \mid \quad O \\ \mid \quad O & C\!\!=\!\! & OH & C\!\!=\!\! \quad\;\; OH & C\!\!=\!\! & C\!\!=\!\! \\ C\!\!=\!\! & OH & & OH & OH & OH \\ OH & & & & & \end{array}$$

Methionine | Homo-cysteine | Serine | Cystathionine | Homoserine | Cysteine

As was the case with the leucine derivatives, N-acetyl-DL-methionine, N-propionyl-DL-methionine and N-butyryl-DL-methionine were all weakly active in replacing methionine. On a molar basis all had an equal activity (about 4.5%) (Fig. 12). N-lauryl-DL-methionine was completely and

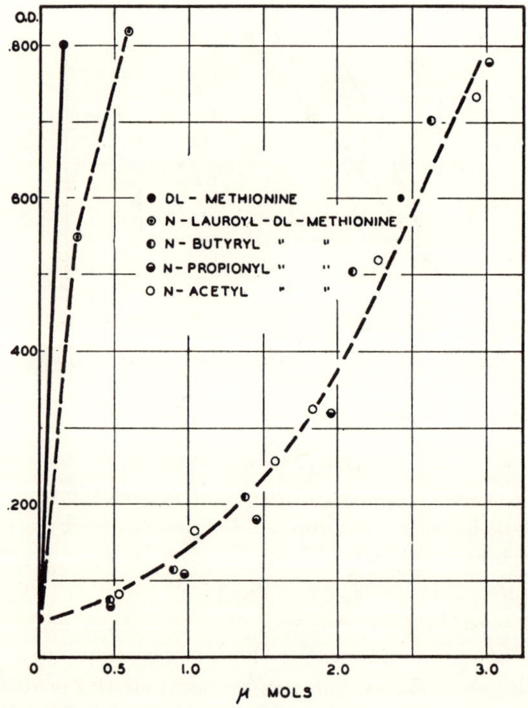

FIG. 12. Comparative activities of N-substituted methionines plotted on a molar basis. Medium S minus methionine.

irreversibly inhibitory at a concentration of 40 γ/ml. However, in a medium containing large amounts of Tween 80, N-lauryl-DL-methionine had a replacement activity of about 37% (Fig. 12), and complete inhibition occurred only at a concentration of 200 γ/ml. N-salicyl-DL-methionine had little if any effect in sparing, replacement or inhibition.

The methionine analog, DL-ethionine, has also been tested for its effect upon methionine metabolism.

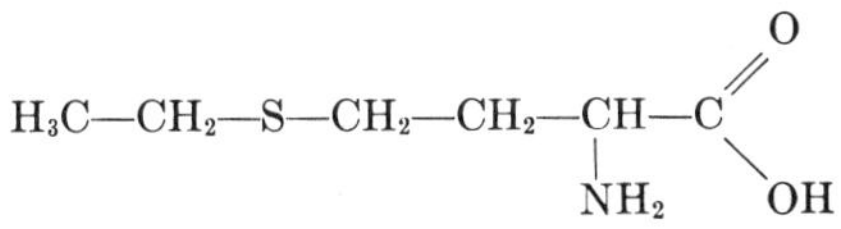

Ethionine

It was found that this compound is a potent and competitive methionine antagonist. The inhibition indices at suboptimal levels of methionine were in all cases less than 1.0 and progressively increased with increasing methionine concentration. At supraoptimal methionine levels the index approaches 3.0 (Fig. 13). This phenomenon is again attributed to the using up of methionine from suboptimal concentrations, so that the effective concentration is less than that used in calculating the inhibition index. This would give values lower than are the actual case.

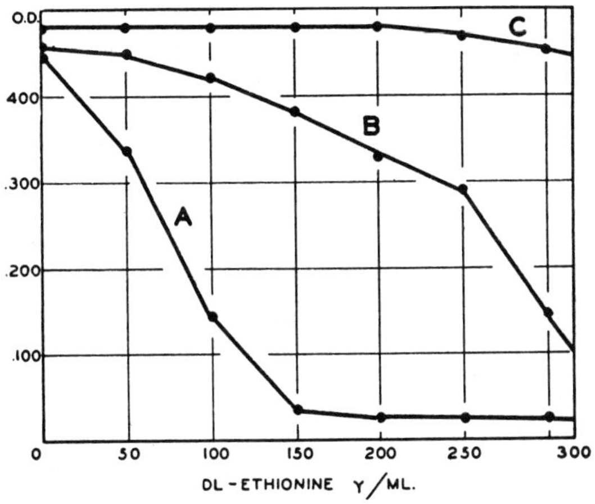

FIG. 13. Inhibition of the growth of *Tetrahymena geleii* W by ethionine and its release by methionine. A contains 50 γ/ml. of DL-methionine, B contains 100 γ/ml. DL-methionine and C contains 200 γ/ml. DL-methionine. Medium A minus methionine.

H. Phenylalanine

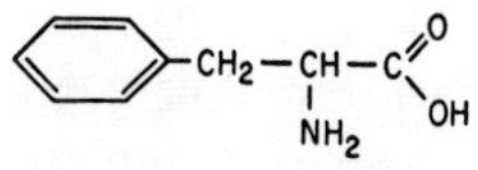

Phenylalanine resembles lysine in the behavior of the D-isomer. Although the omission of serine from the medium makes no change in the level required, in all cases double the amount of DL-phenylalanine is necessary as compared with the L-isomer (Fig. 14). The D-isomer has

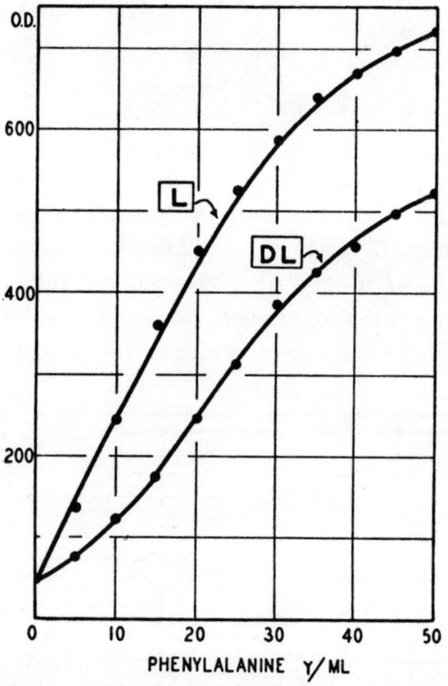

FIG. 14. Dose response of *Tetrahymena geleii* W to phenylalanine. Only the L-isomer is active. Medium S minus phenylalanine.

been tested and found not to be inhibitory. This is in contradiction to earlier results (Kidder and Dewey, 1947b) where larger than double the amounts of racemic mixture were required (40 γ/ml. DL-phenylalanine, 10 γ/ml. L-phenylalanine).

As might be expected, tyrosine is effective in sparing phenylalanine, indicating that the latter is the precursor of tyrosine in the metabolism of the ciliate. Approximately 50% of the phenylalanine can be spared by tyrosine.

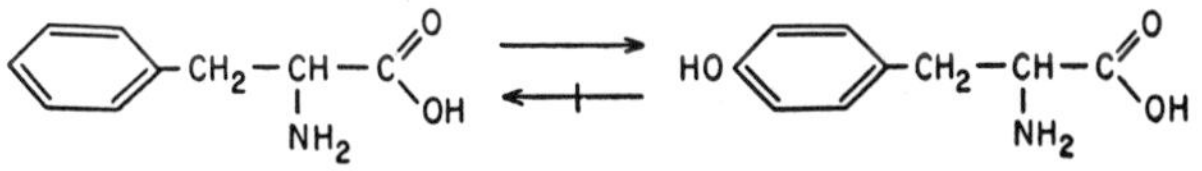

Preliminary tests with β-thienylalanine indicate that it is a potent and competitively reversible inhibitor of phenylalanine.

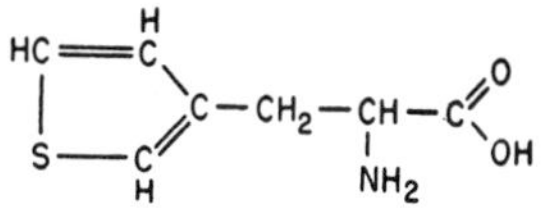

The same phenomenon of increase in inhibition index with increasing phenylalanine concentration was again observed, although the index did not rise above 1.0 (Fig. 15) except at supraoptimal levels of phenylalanine. Tyrosine proved to be somewhat more effective in releasing the inhibitory effects of β-thienylalanine than is phenylalanine.

I. Threonine

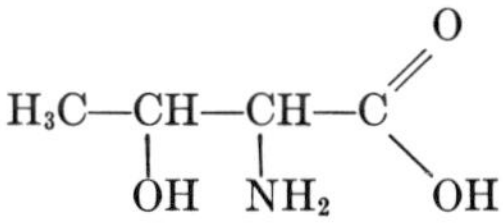

In an early report (Kidder and Dewey, 1947b) it was stated that the D-isomer of threonine appears to be about 50% active in the presence of serine (15 γ/ml. of DL-threonine required as compared to 10 γ/ml. of L-threonine). In the absence of serine the requirement for either L- or DL-threonine is tremendously increased. This is related to the fact that threonine can replace the serine effect at least as far as maximum yield is concerned. It cannot reproduce the stimulating effect of serine on the growth rate. This again indicates the dual nature of the function of serine: (1) to antagonize inhibitions produced by other required amino acids and (2) to relieve the need for its synthesis, which appears to occur at a very low rate. That other amino acids are concerned in one or another of these functions is indicated by the fact that omission of serine from the complete mixture does not affect growth.

Recent experiments (Kidder and Dewey, unpublished) have demonstrated conclusively that the D-isomer of threonine is without activity (Fig. 16). The amount of DL-threonine required is just double the

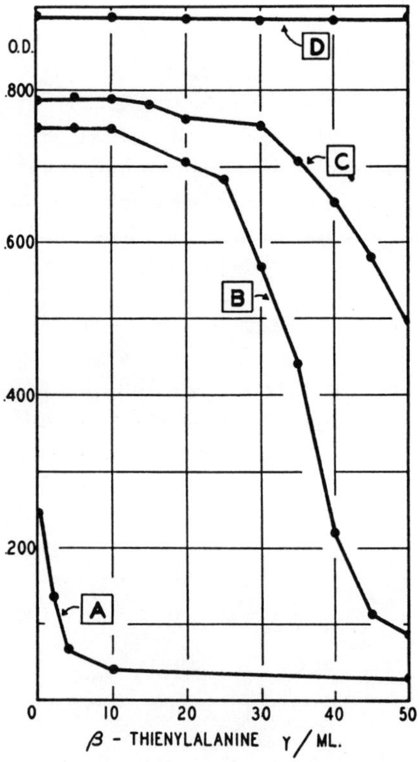

FIG. 15. Inhibition of the growth of *Tetrahymena geleii* W to β-2-thienylalanine and its release by phenylalanine. The amount of L-phenylalanine in γ/ml. in the different sets is as follows: A = 10; B = 50; C = 60; D = 200. Medium S minus phenylalanine.

amount of L-threonine required. Whether the discrepancies observed here as well as elsewhere in the work on amino acid requirements can be attributed to the use of impure compounds in the early experiments, or to the use of crude extracts as a source of Factor II, or to the fact that at present a more accurate means of measuring growth is used, it is not now possible to say. In any case the more recent data are the more reliable because of the factors noted above.

Under certain conditions threonine is not active in replacing the "serine effect" (Kidder and Dewey, unpublished). These effects are largely dependent upon both total and relative concentrations of amino

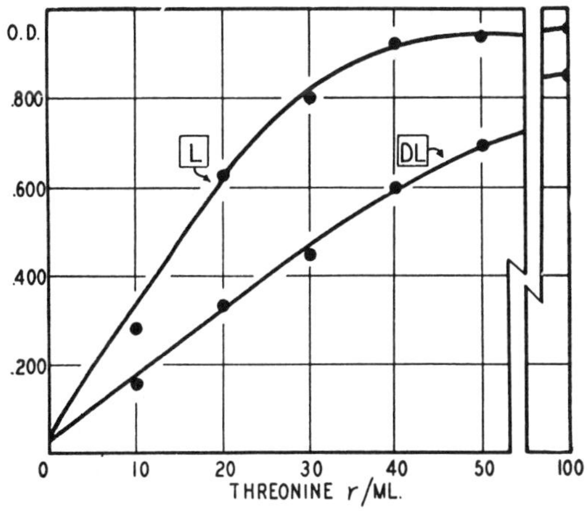

FIG. 16. Dose response of *Tetrahymena geleii* W to threonine. Medium S minus threonine.

acids as well as the presence or absence of certain dispensable amino acids and earlier conditions have not been exactly duplicated.

It has also been found (Kidder and Dewey, unpublished) that DL-allothreonine cannot replace threonine, nor does it cause inhibition of growth. The same is true of α-amino-*n*-butyric acid.

J. TRYPTOPHAN

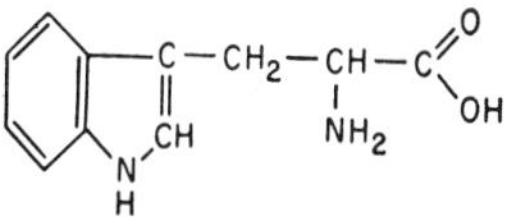

The D-isomer of tryptophan has been reported to be about 80% effective as compared to L-tryptophan (12 γ/ml. DL- and 10 γ/ml. L-tryptophan required) (Kidder and Dewey, 1947b). The requirement is unaffected by the presence or absence of serine.

Probably for the same reasons discussed above under threonine, it appears that the earlier results (Kidder and Dewey, 1947b) regarding the availability of D-tryptophan are in error (Kidder and Dewey unpublished). DL-tryptophan has one-half (or possibly slightly less) of the activity of L-tryptophan (Fig. 17). This is in accord with the observation that the keto analog of tryptophan, indole-3-pyruvic acid, cannot replace the amino acid for the growth of *Tetrahymena*. Tryptophan is

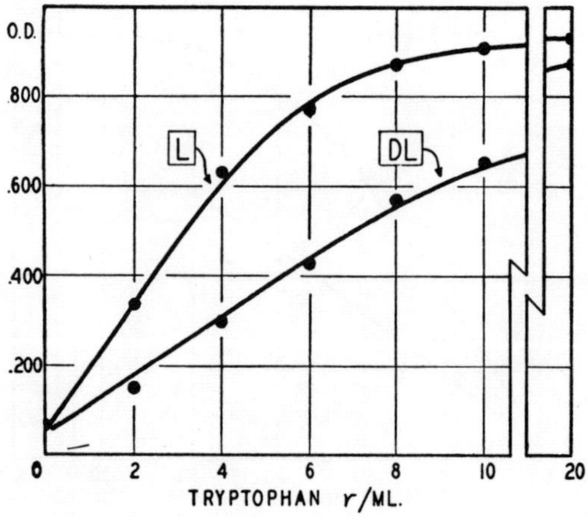

FIG. 17. Dose response of *Tetrahymena geleii* W to tryptophan. Medium S minus tryptophan.

an absolute requirement for *Tetrahymena.* The organism cannot utilize indole plus serine, as do certain mutant strains of *Neurospora,* nor is it able to transaminate indole-3-propionic acid. It has been shown, in addition, that tryptophan does not serve as a precursor of nicotinic acid (Kidder *et al.,* 1949) as it does in mammals and *Neurospora* (see below). Also in contrast to the condition found in mammals is the inability of the ciliate to utilize N-acetyltryptophan. The availability of the N-acetyl derivatives of leucine and methionine is also to be contrasted with this result.

K. Valine

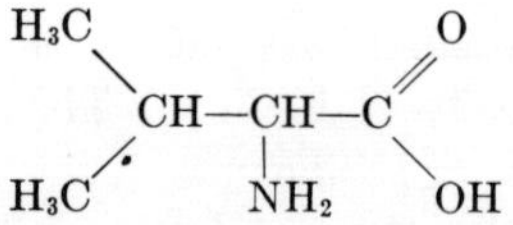

A number of interesting problems are still unsolved in connection with valine. In a mixture of only the essential amino acids plus serine it appears to be synthesized slowly (Kidder and Dewey, 1945a; 1947b). Only 2.5 γ/ml. of L-valine is required for complete stimulation. The D-isomer, however, again appears to be inhibitory, since 7.5 γ/ml. of the DL-mixture is required for maximal response. Valine is not stimulatory in the absence of serine (Kidder and Dewey, 1947b). Retesting of valine requirements recently appears to indicate a complete lack of activity for D-valine (Fig. 18). No inhibitory effect of D-valine was obtained and no growth-stimulating activity.

Since an absolute requirement for valine can be demonstrated in a

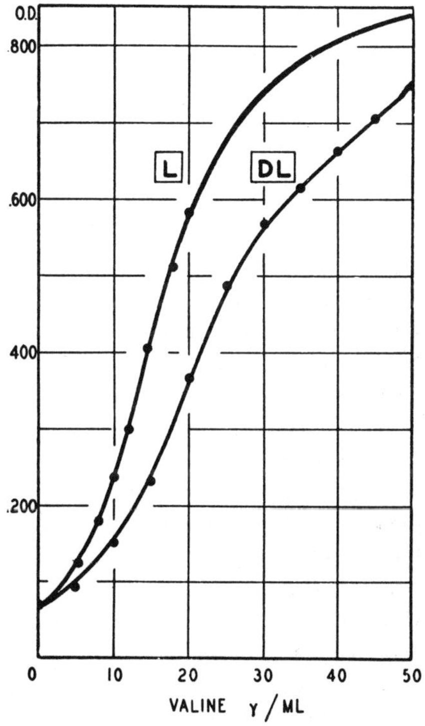

FIG. 18. Dose response of *Tetrahymena geleii* W to valine. Medium S minus valine.

complete mixture of amino acids (including serine) (Kidder and Dewey, 1945b), it follows that the synthesis of valine is inhibited by one or more of the amino acids present. There are indications that arginine and lysine (Kidder and Dewey, 1947b) at high levels are concerned, but such investigations are incomplete.

Hydroxy and keto analogs of valine have also been tested and found to be completely inactive.

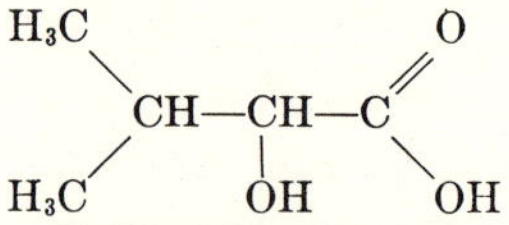

α-Hydroxyisovaleric acid

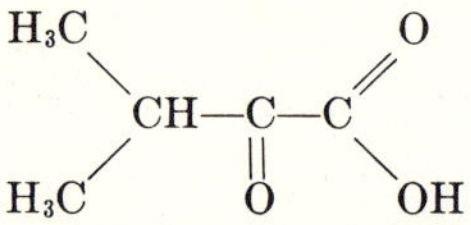

α-Ketoisovaleric acid

Here again a "complete" amino acid mixture was used, and these results were not entirely unexpected. In a mixture of only essential amino acids, these analogs should be expected to permit growth.

L. Serine

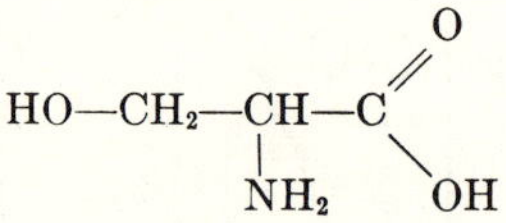

The role played by serine in the metabolism of *Tetrahymena* has been discussed in part under the separate amino acids. It may be summed up by saying that it performs a dual function: (1) detoxification and (2) growth stimulation. Its effect can be partially replaced by other amino acids, but none is quite so effective. Maximum effects were obtained with 10 γ/ml. of either L- or DL-serine.

In the case of serine it is again necessary to state that recent results (Kidder and Dewey, unpublished) are somewhat contradictory to those obtained earlier (Kidder and Dewey, 1947b). Although it is true that the serine "requirement" is dependent to a large extent upon the kinds and concentrations of amino acids used, it is also true that DL-serine is much less than half as active as L-serine in all cases (Fig. 19). In the medium most closely approximating that used earlier (Kidder and Dewey, 1947b), about three times as much DL-serine is required for half-maximal stimulation as L-serine. In other media the amount varied from two to three times as much. The medium which showed the largest requirement was one in which all the amino acids were present. This problem still requires a great deal more investigation to determine how serine is acting. At least part of the effects obtained can be explained by the fact that D-serine is a competitive inhibitor of L-serine. Indices of from 2.5 to 3.0 have been obtained at suboptimal levels of L-serine. At nearly optimal or supraoptimal levels, D-serine has little effect. This appears to be the general case when dealing with inhibitions due to D-isomers of amino acids.

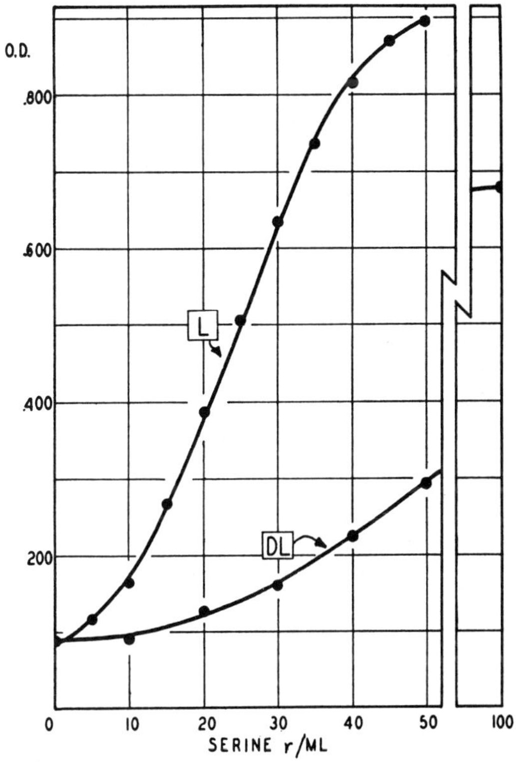

FIG. 19. Dose response of *Tetrahymena geleii* W to serine. Medium S minus serine.

M. Nonessential Amino Acids

It is readily seen that nonessential amino acids may have a stimulatory effect on the growth of *Tetrahymena,* since their addition relieves to some extent the burden of synthesis laid upon the organism, but they also serve as an additional source of nitrogen, sparing that of the essential amino acids. This is a general and nonspecific effect; no single amino acid can be held accountable for it, although proline appears to be somewhat stimulatory alone (Kidder and Dewey, 1949a). This may be an indication of the close relationship between *Tetrahymena* and *Glaucoma scintillans,* which requires proline (Fuller, 1948).

Glycine also deserves mention, since it can replace the serine effect to some extent, although rather inefficiently. Recent work of Sakami (1948, 1949) on serine synthesis may serve to explain this result in part. Sakami has found that in the mammal (rat) serine is formed by the condensation of formate with glycine.

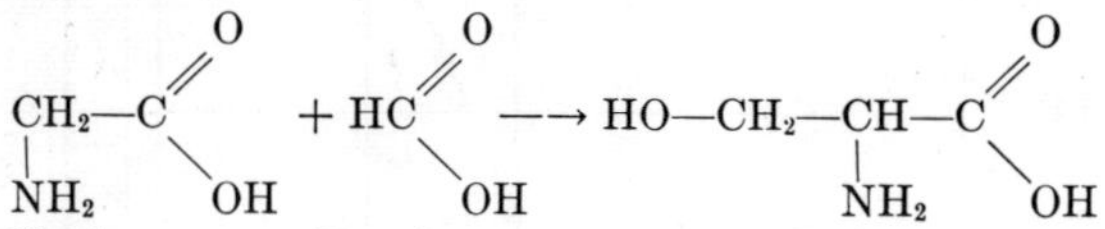

$$CH_2-C\begin{smallmatrix}O\\ \\OH\end{smallmatrix} + HC\begin{smallmatrix}O\\ \\OH\end{smallmatrix} \longrightarrow HO-CH_2-CH-C\begin{smallmatrix}O\\ \\OH\end{smallmatrix}$$
$$\text{Glycine} \qquad \text{Formic acid} \qquad \text{Serine}$$

Therefore, glycine should be more effective in the presence of formate than alone. Some evidence confirming this fact in serine synthesis by *Tetrahymena* has recently (Kidder and Dewey, unpublished) been obtained. It occurs only under certain conditions and in certain media. In any case the effect is small and more work must be done to determine under what conditions this synthesis occurs most readily.

Inasmuch as glycine is a dispensable amino acid, it was impossible to determine whether or not the various analogs of glycine were utilized. As would be expected under these circumstances, N-acetylglycine, N-propionylglycine and N-butyrylglycine had no effect on growth even in high concentrations. Also according to expectations N-laurylglycine was irreversibly inhibitory at low concentrations (50 γ/ml. gave complete inhibition).

It has been possible to show quite definitely that *Tetrahymena* can utilize nitrogen in the form of ammonia (Kidder, 1947). This occurs only under certain very definite conditions. First, the amino acid content of the medium must be very low; secondly, there must also be present an extra source of carbon such as dextrose. The addition of either ammonium salts or dextrose alone has very little effect on growth. In the presence of both together, larger populations are obtained, which persist for longer periods. Rose et al. (1949) have demonstrated that a similar effect may be obtained with rats.

N. Digestion

Tetrahymena is known to possess proteolytic enzymes. According to Lawrie (1937), an extract prepared from the ciliates by precipitation with $MgSO_4$ is capable of digesting gelatin and casein readily and egg albumen less easily. The enzyme has an optimum at pH 6.0 and is partially activated by cyanide at pH 7.0, and is therefore characterized as a papainase. Pepsin-like and trypsin-like enzymes are said to be absent and peptidases were not investigated. The presence of peptidases is indicated by the fact that N-substituted amino acids are utilized (see above) as are glycyl-L-leucine and L-leucylglycine (Kidder and Dewey, unpublished).

It is this ability to digest proteins which has made investigations of the possibility of using *Tetrahymena* as an assay organism for amino acids worth while. This is particularly true in the case of tryptophan, since this amino acid is destroyed during acid hydrolysis of proteins. Such a method, using *T. geleii* H was first reported by Rockland and Dunn (1946). These same authors later (1949a) reported assay procedures for isoleucine, leucine, lysine, and histidine as well as tryptophan. For these assays Rockland and Dunn utilize acid production and titration as a means of estimating growth following incubation times up to 24 days. Since acid production depends to a large extent on the degree of anaerobiosis, the use of this method is open to question, as is the application of the method to determinations of the biological values of proteins (Rockland and Dunn, 1949b).

Lwoff and Roukhelman (1926) first showed that *Tetrahymena* is capable of digesting peptones and peptides, since these materials decreased in concentration in the medium as cultures grew. At the same time there is at first an increase in amino acid nitrogen and later a decrease in free amino nitrogen, indicating utilization of free amino acids. These authors also noted an increase in ammonia nitrogen as cultures grew and aged. Since urea was neither formed nor utilized and uric acid could not be detected in the cultures, it follows that ammonia is the main, if not the only, nitrogenous excretory product. These results were confirmed by Doyle and Harding (1937) using washed, bacteria-fed ciliates. The amount of ammonia-nitrogen excreted was very nearly the same as the amount of protein N calculated to have been ingested. These authors also failed to find urea in their cultures.

In view of the fact that urea does not appear to be a normal metabolite of *Tetrahymena,* it was interesting to find that both urea and certain analogs of urea inhibit the growth of the ciliate (Dewey, Kidder and Parks, unpublished data). Relatively high concentrations of urea are required for one-half maximum inhibition (about 700 γ/ml.), and so far the inhibition has been irreversible. The following urea analogs have thus far been tested: acetylmethylurea, acetylurea, thiourea, ethyl carbamate, nitrourea, semicarbazide and 3-thiosemicarbazide. The acetyl derivatives of urea proved not to be inhibitory, although they were tested only up to a concentration of 2 mg./ml. Urethane (ethyl carbamate) was irreversibly inhibitory with half maximum inhibition at a concentration of about 5 mg./ml. Thiourea was somewhat less inhibitory than urea on a molar basis, but nitrourea was ten times more inhibitory. Semicarbazide resembled nitrourea, but the 3-thiosemicarbazide was approximately 3 times as effective.

In attempts to reverse these inhibitory effects, it was found that a mixture of the vitamins released the inhibition caused by the semicarbazide. Upon testing these separately, the vitamin B_6 group was indicated as the factor responsible. Of the three available members of the

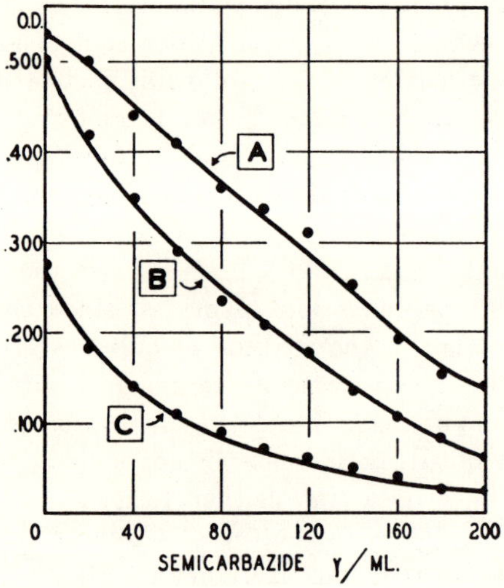

FIG. 20. Inhibition of the growth of *Tetrahymena geleii* W by semicarbazide. The following amounts in γ/ml. of pyridoxamine · HCl were included: A = 0.01; B = 0.005; C = 0.001. Medium A minus vitamin B_6 group.

group, pyridoxamine was more effective than pyridoxal and both were more so than pyridoxine (Fig. 20). An interesting observation in this connection is that autoclaving the pyridoxamine with the regular components of the basal medium (salts, amino acids, vitamins, etc.) makes it effective whereas, when the pyridoxamine is autoclaved only with the semicarbazide, it is inert. Whether this effect is due to an interaction of the pyridoxamine with some component of the basal medium which makes it more effective in counteracting the inhibitor or whether the inhibitor forms an inactive complex with pyridoxamine or in some way destroys it is still under investigation. The other analogs in the series with the exception of 3-thiosemicarbazide, are not released by pyridoxamine under any condition.

The release produced by pyridoxamine is not complete. A phenomenon similar to that observed with folic acid antagonists (see below) is observed here. The amount of semicarbazide required for half maximal inhibition increases more slowly than does the relative amount of pyridoxamine, so that the inhibition index decreases with increasing amounts of pyridoxamine (Fig. 20). It is impossible to say, at the moment, whether or not this effect is due to insufficient production of the active substance from pyridoxamine upon autoclaving, or whether it is due, as suggested in the case of folic acid (see below) to the involvement of more than one enzyme system in the inhibition. The fact that amino acid concentration appears to affect the inhibition may indicate that interactions between several amino acids may be involved. Increasing the concentration of all the amino acids makes the semicarbazide more inhibitory. It has, however, been impossible to relate this effect to any one single amino acid. The situation is a complex one and requires more investigation before it can be clarified.

V. Growth Factors

A detailed description of the various arguments which have been put forth for and against growth factor requirements of ciliates is unnecessary. It is of course now obvious that Hall's doubts of 1943 regarding growth factor requirements of *Tetrahymena* were based on purely theoretical grounds. Our present knowledge of the synthetic abilities of *T. geleii* W makes it possible to review both the qualitative and quantitative aspects of this problem.

So little is known of the requirements for recognized vitamins in ciliates other than *Tetrahymena* that only a short statement is necessary. Thus, Kidder and Dewey (1942) reported that *Glaucoma scintillans* required thiamine and was unable to synthesize the compound even in the presence of Factor S (see below). Tatum and associates (1942; Garnjobst *et al.*, 1943), using a chemically undefined medium, were able to show that *Colpoda duodenaria* requires high levels of thiamine, riboflavin, pantothen, nicotinamide, and pyridoxine, in addition to the amounts of these vitamins contained in their medium. They found that *Colpoda* does not appear to require *p*-aminobenzoic acid, biotin, or inositol, while the status of choline and folic acid is unknown.

Unless otherwise stated, therefore, the following discussion will deal exclusively with the genus *Tetrahymena.*

A. Thiamine

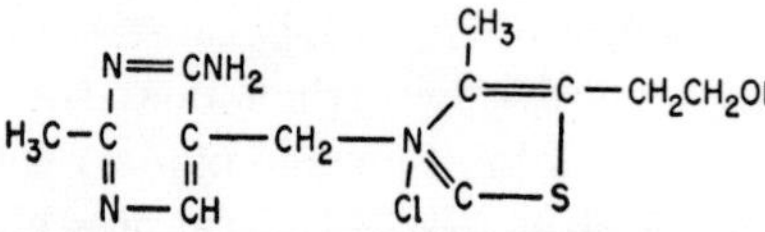

Lwoff and Lwoff (1937) were the first to find a thiamine requirement for a ciliate. They found that *T. geleii* (*G. piriformis*) would not grow in unsupplemented silk peptone, but would grow if thiamine was added. The addition of neither the pyrimidine nor the thiazole portion of the molecule (separately or combined) would suffice. They later showed (1938) that neither the 6-methyl analog of thiamine nor thiochrome could substitute for thiamine.

Several reports have appeared in which crude basal media (casein, peptone, etc.) were used, which were assumed to be free of thiamine or heat and alkali treated to destroy thiamine. A number of strains of *T. geleii* failed to grow in such media in the absence of thiamine (Elliott, 1939; Hall, 1940; Dewey and Kidder, 1941; Hall and Cosgrove, 1944). During this period the nutritional requirements of these ciliates were so imperfectly known that it was difficult to evaluate results, when growth failed, as due to specific limiting factors or to the presence of inhibitory substances. It was shown by Kidder and Dewey (1942), however, that *T. geleii* W and *T. vorax* could be grown in a "vitamin-free" casein medium supplemented with alkali-treated alfalfa meal extract. They recognized the fact that the alfalfa extract contained a number of growth factors required by the ciliates, but that it could not, following their treatment, contain thiamine. They postulated a new factor, "Factor S," which could catalyze the synthesis of thiamine from pyrimidine and thiazole. Later (Kidder and Dewey, 1944) they showed that *T. geleii* W could be grown in serial transfers in casein or gelatin hydrolyzates (plus tryptophan) supplemented with dethiaminized alfalfa extract. They noted that the addition of thiamine to the medium caused a large increase in cell size. After the discovery that the nitrogen requirements of *Tetrahymena* could be met by crystalline amino acids (Kidder and Dewey, 1945b), it became possible to attack the thiamine problem in a slightly different manner. Factor S had been found most abundantly in the leafy portions of higher plants, but not in appreciable amounts in materials of animal origin. In the amino acid medium, growth failed unless supplemented with certain crude fractions, in addition to eleven known "vitamins." However, it was found (Kidder and Dewey, 1945c) that the "unknown factors" could be supplied by a butanol extracted water solution of liver powder which could then be dethiaminized without producing inhibitory materials. Gelatin hydrolyzate medium plus the dethiaminized liver fraction would not support appreciable growth unless thiamine was added. Factor S, in dethiaminized alfalfa extract, added in lieu of thiamine, would again make growth possible. It was found, however, that no alfalfa Factor S was required if crystalline amino acids were used as the nitrogen source. A small amount of the amino acid solution, added to the gelatin hydrolyzate medium, would substitute for Factor S or thiamine. Tests soon showed that it was the unnatural isomers of a number of the amino acids (added in the racemic form) which were responsible for the growth of the ciliates without alfalfa extract or thiamine. It appeared at this time, therefore, that *Tetrahymena* required some D-configuration if an exogenous source of thiamine was not present. The authors suggested that the D-configurations could have come from racemization of amino acids in the alfalfa extracts during heat and alkali treatment, in the earlier experiments, and that there appeared to be substances which specifically block the biosynthesis of thiamine in gelatin hydrolyzate and in the liver extract. This block is counteracted by Factor S or high levels of D-amino acids.

It is interesting to note that the announcement by Snell and Guirard (1943) that D-alanine could replace pyridoxine in the growth of *Lactobacillus casei* and the last paper on *Tetrahymena* just discussed were in press simultaneously. Recent work on the D-alanine and pyridoxine relationships (Holden, Furman and Snell, 1949; Holden and Snell, 1949) have indicated that the D-amino acid appears to be an end product of a reaction in which pyridoxal is a cofactor to a racemase. There is still no clue as to the use to which *L. casei* puts the D-alanine. A similar enzyme may be concerned in the conversion of D- to L-methionine by *L. arabinosus* (Camien and Dunn, 1950).

Recently the question of the role of thiamine in the metabolism of *Tetrahymena* has been reopened by the Amherst group (Parks, Dewey and Kidder, unpublished data). With the biochemical requirements of *Tetrahymena* known to the extent that all components (with the exception of protogen, to be discussed later) of the medium are chemically defined, a more nearly adequate evaluation of the thiamine problem can be made. It has been found that in the synthetic medium (Dewey *et al.*, 1950) now employed, thiamine must be present for growth. This medium contains many of the amino acids (in the racemic form) which were earlier shown to make growth possible in the absence of thiamine. It has become evident that an unknown factor or factors must be present in addition to the D-amino acids (in the absence of thiamine), and this factor was present in the liver and alfalfa extracts. The addition of dethiaminized extracts from various substances to the synthetic medium allows growth to take place.

One striking result of growing the ciliates in the absence of thiamine is the accumulation of acid at about the fourth day of incubation. In the presence of very active fractions (minus thiamine), the accumulation of acid results in death and cytolysis of the cells. The acid has not been identified, but it is probably lactic acid or pyruvic acid (Thomas, 1943), which is to be suspected in view of the known function of thiamine pyrophosphate as the co-carboxylase of pyruvate. It has been proposed (Parks, Dewey and Kidder, unpublished) that, in *Tetrahymena,* pyruvic acid is decarboxylated according to the following scheme:

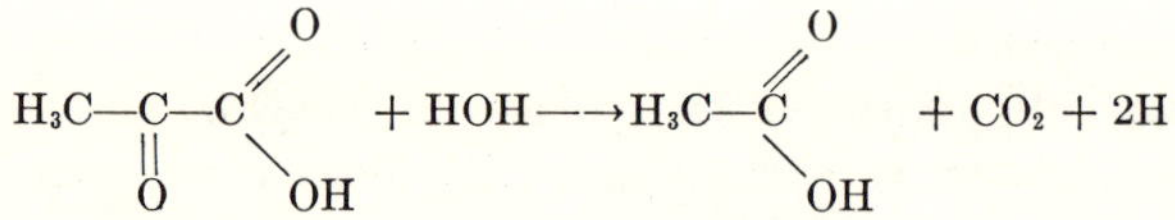

Pyruvic acid Acetic acid

In this reaction CO_2 is split out from the carboxyl of the pyruvic acid leaving one hydrogen. Simultaneously a hydroxyl ion from water is coupled to the keto carbon to produce the carboxyl group of acetic acid. This releases a second hydrogen which, with the first, is transferred to a hydrogen acceptor. This hydrogen acceptor may be protogen, as it has been shown that acetate spares protogen (Kidder *et al.,* 1950). Protogen has more than this one function for *Tetrahymena,* as acetate will not abolish its requirement. It has been shown by Snell and Broquist (1949) that protogen can replace acetate for *L. casei.* It is therefore the acetate factor of Guirard, Snell, and Williams (1946a,b) and is also the "pyruvate oxidation factor" of O'Kane and Gunsalus (1948).

It appears, therefore, that the decarboxylation of pyruvate in *Tetrahymena* can be dispensed with, provided a second reaction, normally mediated through thiamine, is carried out by a nonthiamine unknown compound and certain D-amino acids. It may be that we will find that this unknown compound is the product of a thiamine-controlled reaction.

That thiamine normally plays a role in the metabolism of *Tetrahymena* is shown by the fact that competitive growth inhibition results upon the addition of neopyrithiamine. Neopyrithiamine is an analog wherein the thiazole portion of the molecule has been replaced by a pyridine ring.

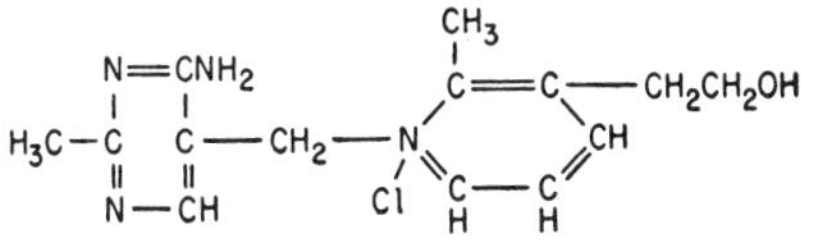

It is not usual to find competitive inhibition resulting where the normal metabolite is synthesized by the organism. Figure 21 shows the results of experiments where the inhibition produced by neopyrithiamine is released by thiamine.

To sum up our present knowledge of the role of thiamine in the metabolism of *Tetrahymena,* we can say that the vitamin is a requirement

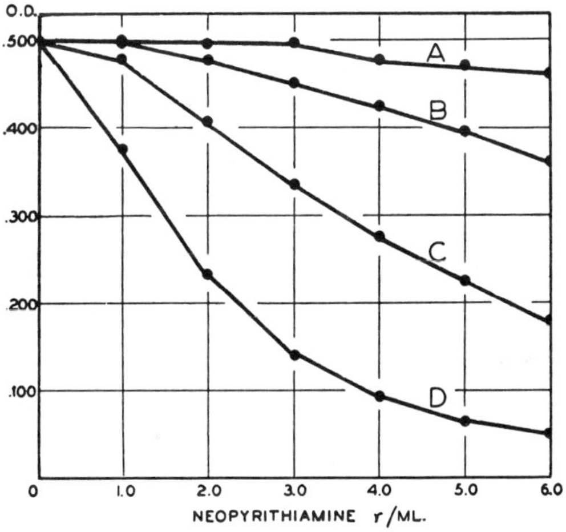

FIG. 21. Inhibition of the growth of *Tetrahymena geleii* W with neopyrithiamine and its release by thiamine. The following amounts, in γ/ml. of thiamine HCl are indicated: A = 0.04; B = 0.02; C = 0.01; D = 0.005. Medium A minus thiamine.

for normal metabolism, functioning in at least two reactions. One reaction appears to be the decarboxylation of pyruvate to acetate and CO_2, a reaction which is not necessary for growth. The second is an unknown reaction which may be dispensed with provided compounds (D-configurations and an unknown substance), which may be end products of the reaction, are provided. In any case, it is perfectly possible to obtain excellent growth of *Tetrahymena* [six strains of *T. geleii* and two strains of *T. vorax* (Kidder and Dewey, 1945a)] in a medium devoid of thiamine, but it is also true that exogenous thiamine, when present, functions in the normal metabolism of these ciliates.

B. Riboflavin

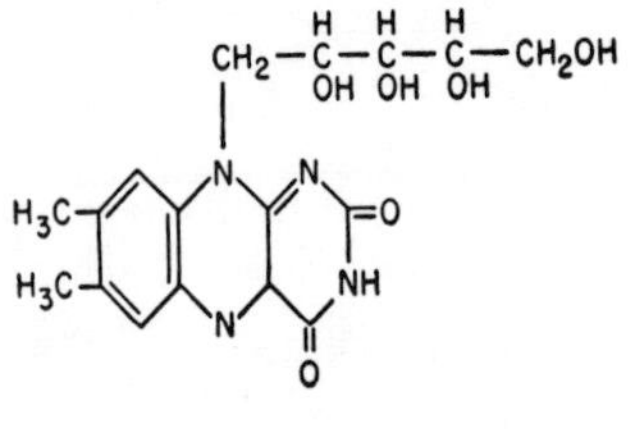

Hall and Cosgrove (1944) and Hall (1944), working with casein and gelatin media, report that *Tetrahymena geleii* GP (*Glaucoma piriformis*) requires riboflavin for optimum growth. In their media, however, some growth was obtained without its addition. Essentially similar results were obtained by Kidder and Dewey (1945a) on *T. geleii* W in an amino acid medium, to which were added the then "unknown factors," necessary for the growth of the ciliate, in the form of a treated lead acetate filtrate of liver fraction L. Assays of this supplement for riboflavin with *Lactobacillus casei* indicated that it was free of riboflavin. The ciliates were found to grow in the "riboflavin-free" supplemented amino acid medium through serial transplants with a generation time of 5.21 hours and a maximum yield at 96 hours of 67,000 cells/ml. Upon the addition of riboflavin, the generation time was reduced to 4.37 hours, and the maximum yield was raised to 164,000 cells/ml. Although it was clear that exogenous riboflavin was exerting a favorable influence, it also seemed clear that the ciliates were capable of synthesizing a portion of their riboflavin. Later (Kidder and Dewey, 1949c), when the necessity for adding liver fraction was removed, it was immediately evident that *T. geleii* W was incapable of riboflavin synthesis, for no growth resulted in the synthetic medium unless riboflavin was added. Linear response to riboflavin is shown in Fig. 22. This ciliate, then, is dependent upon an exogenous source of riboflavin and, under the conditions of the tests, half maximal growth results with the addition of 0.001 γ/ml. The requirement is increased when a more concentrated medium is used.

Recently Elliott (1949a) published, in abstract form, the statement that, for *T. geleii* E, riboflavin was neither essential nor stimulatory. This statement is incorrect, as it has been shown in the Amherst Laboratories that the growth of *T. geleii* E fails in the second transplant unless

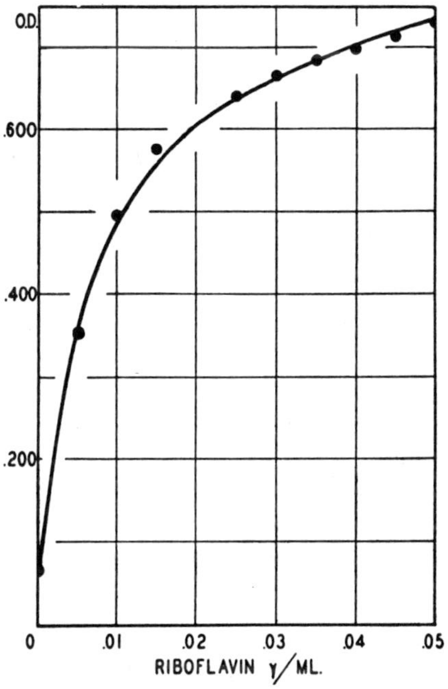

FIG. 22. Dose response of *Tetrahymena geleii* W to riboflavin. Medium D minus riboflavin.

riboflavin is present (Kidder and Dewey, unpublished). One can only conclude that Elliott was not using a riboflavin-free medium.

Isoriboflavin, a riboflavin analog, shows competitive inhibition, reversed by riboflavin, when tested with *T. geleii* W. The inhibition index is high (400–500), meaning that this analog has a rather low affinity for the enzyme concerned.

C. VITAMIN B_6 GROUP

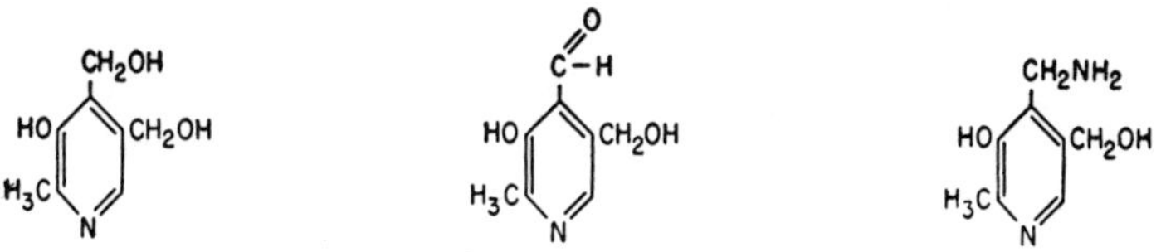

Tetrahymena geleii W is dependent upon an exogenous source of vitamin B_6 (Kidder and Dewey, 1949b). In the earlier experiment (Kidder and Dewey, 1945a), it appeared that the ciliate could grow in a pyridoxine-free medium, but again the results of the *Lactobacillus casei* assays on the liver fraction used in the medium were misleading. Using a synthetic medium it was shown that high levels of pyridoxine were required and the activity depended on the length of time the medium, containing pyridoxine, was autoclaved (Kidder and Dewey, 1949b). It is known that amination of pyridoxine to pyridoxamine takes place upon standing with amino acids, however, and this would account for the growth observed even when filtered pyridoxine was added. Pyridoxal and pyridoxamine were found to be approximately equal in activity and as much

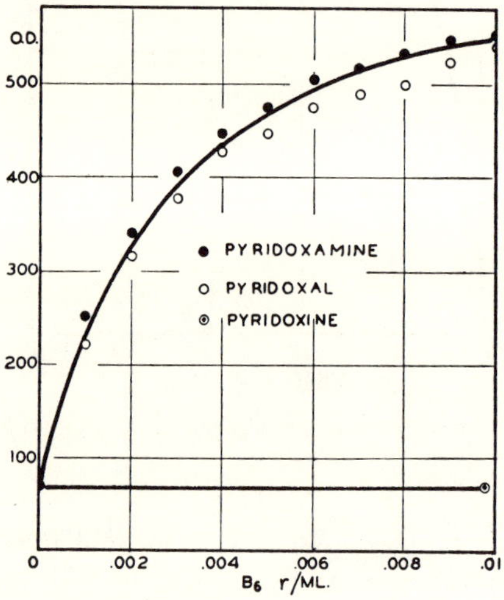

Fig. 23. Dose response of *Tetrahymena geleii* W to three members of the Vitamin B_6 group. Medium A minus vitamin B_6 group.

as 500 times as active as pyridoxine. Half maximum growth resulted with the addition of 0.0007 γ/ml. in the less concentrated base medium. Figure 23 represents the growth response to three of the members of the vitamin B_6 group in medium A.

It has been shown by Lichstein *et al.* (1945) that pyridoxal phosphate functions as a coenzyme for certain decarboxylases and transaminases. It was of interest, therefore, to test pyridoxal phosphate for growth-promoting activity. The sample of pyridoxal phosphate used was of 45% purity. This was determined by heating the compound at 100° C. in 0.1*N* HCl for 3 hours and comparing it on a molar basis with pyridoxal·HCl similarly treated (Fig. 24). When filtered pyridoxal phosphate was compared to filtered pyridoxal · HCl, the activity of the phosphate was found to be only 65% that of pyridoxal (Fig. 25). It would appear that *Tetrahymena* must expend energy in dephosphorylation before a useful compound is formed. There appears to be no reason to believe that the phosphorylated compound cannot be absorbed.

This result need not be taken to mean that pyridoxal phosphate does not function in the enzyme systems of *Tetrahymena,* but only that other

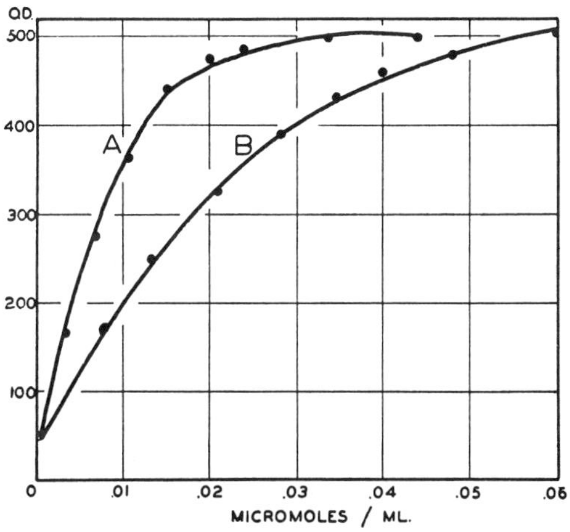

FIG. 24. Test of activity of a sample of pyridoxal phosphate. Curve A represents the response of *Tetrahymena geleii* W to pyridoxal·HCl treated at 100° C. for 3 hours. Curve 3 represents the response to a similarly treated sample of pyridoxal phosphate, plotted as if it were a pure sample. The treatment yields pyridoxal and the response indicates that the sample actually contained 45% pyridoxal phosphate. See Fig. 25. Medium A minus vitamin B_6 group.

reactions mediated through pyridoxal are important. It may still be that some of the pyridoxal phosphate is used as such, but growth is limited by a requirement for pyridoxal which is only forthcoming, under these conditions, following dephosphorylation.

On the other hand, desoxypyridoxine is not inhibitory to *Tetrahymena* (Kidder and Dewey, unpublished) even at high levels and when tested with suboptimal pyridoxal (Mushett *et al.,* 1947; Porter *et al.,* 1947). If the hypothesis for the action of desoxypyridoxine proposed by Umbreit and Waddell (1949) is correct (that desoxypyridoxine is phosphorylated

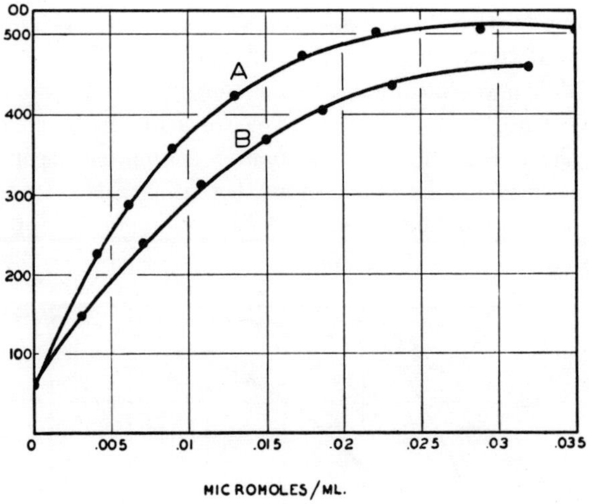

FIG. 25. Comparison of the activities of pyridoxal and pyridoxal phosphate (adjusted for activity). These results indicate that pyridoxal phosphate is only 65% as active as pyridoxal. Medium A minus vitamin B_6 group.

and then saturates the tyrosine decarboxylating enzyme in place of pyridoxal phosphate), then it would appear that *Tetrahymena* may not possess decarboxylating enzymes using pyridoxal phosphate as a cofactor.

D. PANTOTHEN

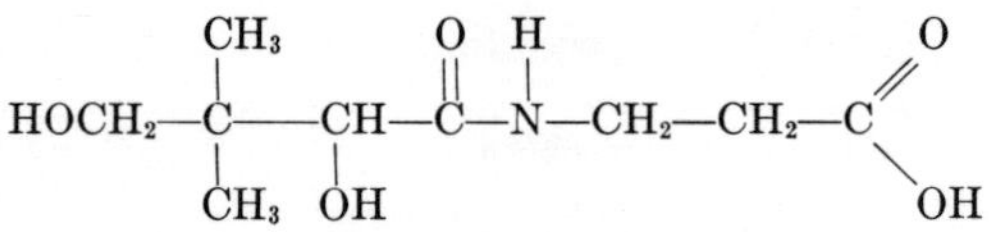

Tetrahymena geleii W was found to be dependent upon an exogenous source of pantothenic acid (Kidder and Dewey, 1949c), which was in contrast to the earlier work (Kidder and Dewey, 1945a) where unreliable assay procedures had indicated that pantothen-free supplements were being used. Half maximum growth resulted when 0.01 γ of Ca pantothenate per milliliter of medium was added and maximum growth resulted with the addition of 0.1 γ/ml. (Fig. 26).

The recent work of Lipmann and his associates (1947; Hegsted and Lipmann, 1948) has implicated pantothenic acid as an important constituent of a cofactor for certain acetylating enzymes. One of the characteristics of the pantothen-containing cofactor (coenzyme A) is its stability. Methods for assaying coenzyme A involve either the use of higher animals (birds) or treatment of the coenzyme with liver and kidney enzymes to release the pantothen which is then assayed microbiologically. Inasmuch as *Tetrahymena* possesses an animal-like metabolism and requires pantothen, it was thought likely that it would be able to utilize the pantothen from coenzyme A. This was found, however, not to be the case. Coenzyme A* is not attacked by *Tetrahymena,* nor is it used as such. As a consequence, no pantothenic acid activity was exhibited, although approximately 75% of the calculated pantothen could be made

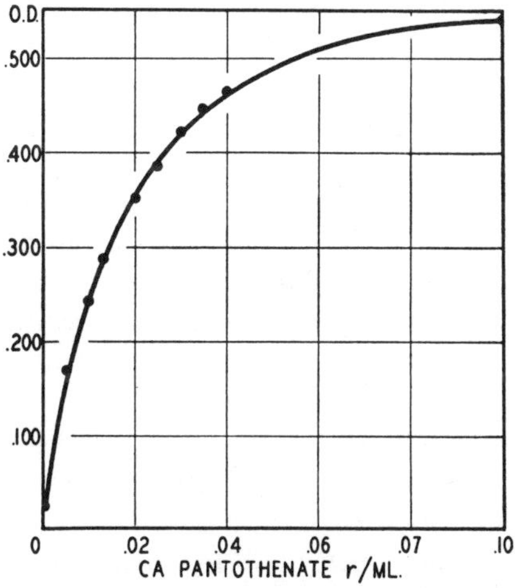

FIG. 26. Dose response of *Tetrahymena geleii* W to Ca pantothenate. Medium D minus pantothen.

available to *Tetrahymena* by the appropriate coenzyme A-splitting enzyme treatment. It must be concluded from these results that probably not all pantothen in nature is found in the form of coenzyme A (Novelli *et al.,* 1949), that certainly *Tetrahymena* does not possess the necessary enzymes for the liberation of pantothen from coenzyme A, and that coenzyme A is not utilized as such by the ciliates as a source of pantothen. It is possible that coenzyme A cannot be absorbed by the ciliate, as it may not be by the lactic acid bacteria. This does not seem likely, however, as *Tetrahymena* is perfectly capable of ingesting particles through its well-

* The sample of coenzyme A and the liver and kidney enzymes used were generously supplied by Dr. Fritz Lipmann.

formed mouth. The report of King *et al.* (1948) indicates that coenzyme A is active for *Acetobacter suboxydans*.

E. Pteroylglutamic Acid

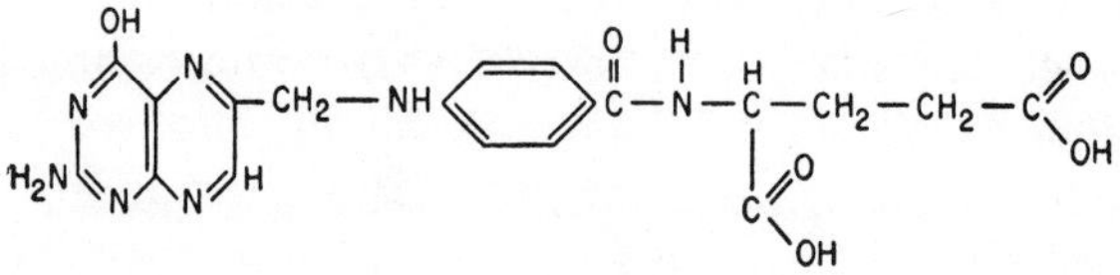

Due to the fact that pteroylglutamic acid (PGA) is readily adsorbed on activated charcoal at low pH, it was early found possible (Kidder, 1946) to prepare "folic acid-free" liver fractions for supplements to the amino acid media used at that time. In media so prepared, it was found that *Tetrahymena geleii* W could not grow in the absence of added folic acid. However, there was enough storage of folic acid in the cells so that inoculations into "folic acid-free" medium always resulted in almost optimum growth in the first transplant. Subsequent transplants failed unless folic acid was added. It was later shown (Kidder and Fuller, 1946) that *T. geleii* responded nearly identically to the Lederle folic acid crystals and to Vitamin B_c, later shown to be the same compound (Binkley *et al.*, 1944; Jukes and Stokstad, 1948). When the chemical identity of folic acid had been established (Mowatt *et al.*, 1946) and it was recognized that there were various conjugates, further studies were conducted to determine the enzyme pattern of *Tetrahymena* in relation to PGA. It was found (Kidder and Dewey, 1947a) that *Tetrahymena* possesses a "conjugase" since it responds to pteroyldiglutamylglutamic acid (the fermentation factor, PTGA) and pterolyhexaglutamylglutamic acid (folic acid conjugate, PHGA). With the medium then used (containing folic acid-free liver extract), the activities on a molar basis were found to increase in proportion to the glutamic acid residues. Thus, half maximum growth resulted from the addition of 0.00032 γ/ml. of PGA in the conjugated form of PHGA: 0.0005 γ/ml. of PGA in the conjugated form of PTGA: 0.00064 γ/ml. in the free PGA state. Later work (Kidder and Dewey, 1949c) indicated, however, that the liver supplement contained inhibitors, for when tested in the synthetic medium, free PGA was equal to PTGA and PHGA on a molar basis.

By using various parts of the PGA molecule, certain enzyme deficiencies were noted. *Tetrahymena* does not possess the enzymes necessary for joining the pteridine portion of the molecule to the *p*-aminobenzoic acid. This is shown by the failure of the ciliates to grow when either xanthopterine or 7-methylxanthopterine plus *p*-aminobenzoylglutamic acid were added to the medium. Likewise, the addition of pteroic acid plus glutamic acid failed to promote growth. This tells us that the dehydrolytic enzymes necessary for completing the peptide linkage between the PAB carboxyl radical and the glutamic amino group are absent (Kidder and Dewey, 1947a). These blocks are illustrated in the following scheme:

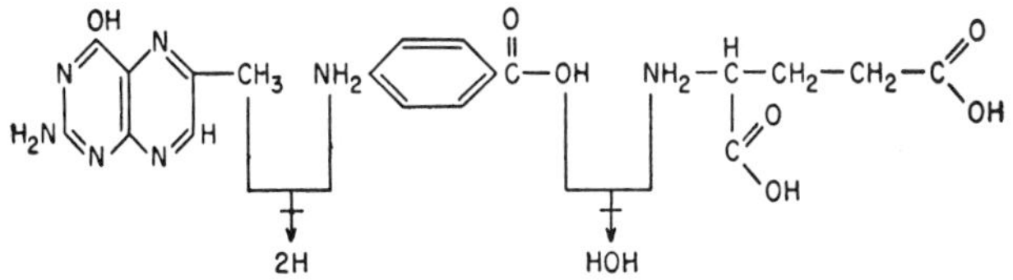

None of these combinations was inhibitory as evidenced by the fact that the addition PGA always resulted in a normal response. These results place *Tetrahymena* in the same category as the higher animals, as in all cases fragments of the PGA molecule have proved inactive as substitutes for the intact molecule.

On the other hand, unlike the lactic acid bacteria and higher animals, *Tetrahymena* can use the 4-aminopteroylglutamic acid (Aminopterin) in place of PGA. Oxidative deamination of position 4 may be carried out. Methopterin (N^{10} methyl PGA) can likewise be used, but demethylation appears to be more difficult as this compound is less active. Deamination of position 4 and demethylation of position 10 simultaneously appear to put too great a strain on the metabolic machinery, however, and 4-amino N^{10} methyl PGA (Amethopterin) cannot be used instead of PGA and is inhibitory to growth.

Tests on a number of the antifolics demonstrated that, in addition to Amethopterin, Adenopterin (4-amino-9,10-dimethyl PGA) and Aninopterin (4-amino-9-methyl PGA) were also inhibitory. Inhibition with Aninopterin (released by PGA) appears to be strictly competitive, as the inhibition index remains approximately constant over wide ranges of inhibitor concentrations. The inhibition index shifts, however, with the concentrations of PGA used in the case of both Amethopterin and Adenopterin. As the concentration of Amethopterin doubles, the amount of PGA needed for release increases eight- to tenfold. As the concentration of Adenopterin doubles, the amount of PGA required increases four- to fivefold. Shifting inhibition indices may indicate that the inhibitor is affecting more than one system.*

Both pteroylaspartic acid and 4-aminopteroyl aspartic acid (Amino-an-fol) spare PGA but will not replace it. All the pteridines and substituted pteroic acids tested proved to be entirely inert.

F. Nicotinamide

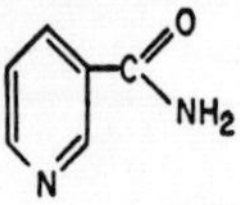

Although an earlier report (Kidder and Dewey, 1945e) stated that nicotinic acid (or nicotinamide) was only stimulatory to *Tetrahymena,* it was found that, in the synthetic medium (Kidder and Dewey, 1949c), growth was impossible when niacin was omitted. Half maximum growth resulted with the addition of 0.0055 γ of nicotinamide per milliliter of medium and maximum growth occurred with the addition of 0.1 γ/ml. (Fig. 27).

It has been shown that many organisms can synthesize niacin from tryptophan. The synthetic pathways have been worked out using *Neurospora* mutants (Bonner and Beadle, 1946; Beadle *et al.,* 1947; Bonner, 1948; Mitchell and Nyc, 1948) and involve anthranilic acid, indole, and serine for tryptophan synthesis and kynurenine and 3-hydroxyanthranilic acid as intermediates in the formation of niacin.

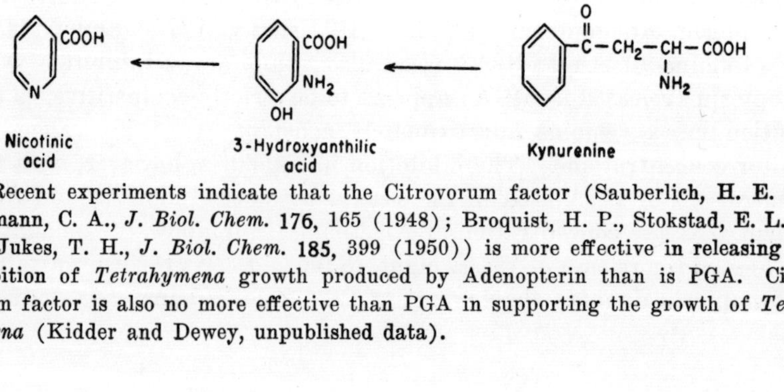

* Recent experiments indicate that the Citrovorum factor (Sauberlich, H. E. and Baumann, C. A., *J. Biol. Chem.* **176**, 165 (1948); Broquist, H. P., Stokstad, E. L. R., and Jukes, T. H., *J. Biol. Chem.* **185**, 399 (1950)) is more effective in releasing the inhibition of *Tetrahymena* growth produced by Adenopterin than is PGA. Citrovorum factor is also no more effective than PGA in supporting the growth of *Tetrahymena* (Kidder and Dewey, unpublished data).

Part of this series of reactions is possible for mammals (tryptophan to niacin), and it has recently been shown that the feeding of hydroxyanthranilic acid to the rat increased significantly the excretion of N′-methylnicotinamide (Mitchell *et al.*, 1948). Experiments were performed with *Tetrahymena,* using media calculated to simulate the rat diets which had been used, to determine whether or not tryptophan levels would affect the niacin requirement (Kidder *et al.,* 1949). It was found that both tryptophan and niacin are absolute requirements for the ciliate and,

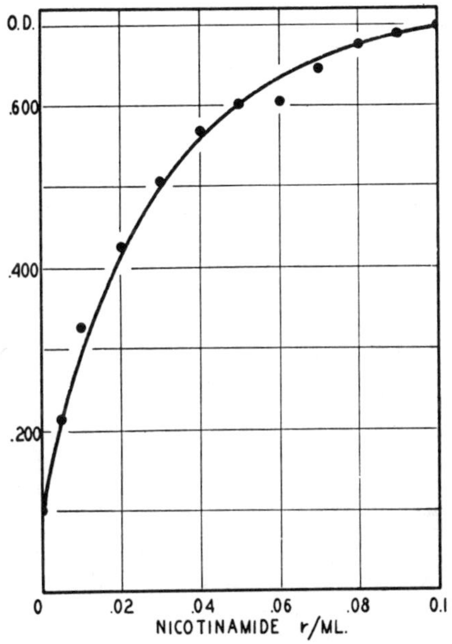

FIG. 27. Dose response of *Tetrahymena geleii* W to nicotinamide. Medium D minus nicotinamide.

moreover, higher levels of tryptophan have no sparing action on niacin. It must be concluded, therefore, that *Tetrahymena* has lost enzyme systems, still retained by the mammal, for one or more steps in the conversion of tryptophan to niacin. Intermediates between these two compounds (kynurenine and hydroxyanthranilic acid) have not, as yet, been tested. In respect to niacin synthesis, then, *Tetrahymena* has progressed further along the road of physiological evolution than the mammal.

Pyridine-3-sulfonic acid, an analog of niacin, is not inhibitory to *Tetrahymena* up to 1 mg./ml. (Kidder and Dewey, unpublished).

G. Biotin

It was reported in 1945 (Kidder and Dewey, 1945e) that exogenous biotin was not required by *Tetrahymena.* Some stimulation was noted when the vitamin was added to a biotin-free medium. It was also found that the addition of sterile raw egg white or avidin produced no inhibitory effects. When biotin was re-examined in the synthetic medium (Kidder and Dewey, 1949c), again no requirement was demonstrated. Indefinitely transplantable growth was obtained in a medium lacking biotin and the various compounds usually associated with biotin metabolism (fatty acids, aspartic acid, acetate). Slight stimulation resulted when biotin was added to such a medium, but this was considered of doubtful significance. As might have been expected under these conditions, desthiobiotin (a biotin analog) exerted no inhibitory effects upon growth.

H. Choline, Inositol, *p*-Aminobenzoic Acid, Vitamin B_{12}

It has been determined that choline is not necessary as a growth factor for *Tetrahymena* (Kidder and Dewey, 1948a). When choline-less *Neurospora* was employed as the assay organism, it was shown that choline increased in the medium as the ciliates multiplied. It has likewise been shown that neither inositol nor *p*-aminobenzoic acid are dietary requirements nor are they stimulatory to growth (Kidder and Dewey, unpublished). The indications are that *Tetrahymena* is able to synthesize adequate amounts of vitamin B_{12}. The addition of B_{12} to the medium does not affect the growth of the ciliate and tests with *Euglena gracilis* (Stokstad, communication) indicate an increase in the B_{12} in the medium following the multiplication of the cells.

I. Purines and Pyrimidines

It was found early in the work on the biochemistry of *Tetrahymena* (Dewey, 1941, 1944) that growth was possible only if certain extracts of natural materials were present. Two factors were at first recognized, Factors I and II, on the basis of heavy metal precipitation. When plant extracts, for instance, were treated with lead acetate, Factor I was precipitated while Factor II remained in the filtrate. Growth in a casein medium (containing salts and vitamins) was impossible unless both factors were present. Later (Kidder and Dewey, 1945d) it was found that, when the crystalline amino acids were substituted for the casein, growth failed. A third unknown factor (Factor III) was therefore postulated as being present in proteins and protein hydrolyzates as well as in crude materials. On the basis of the chemical characteristics of Factors I and III, components of nucleic acid were suspected. Factors I and III could be entirely replaced by yeast nucleic acid or thymus nucleic acid (Kidder and Dewey, 1945f). Hydrolysis of the nucleic acid by heat and alkali

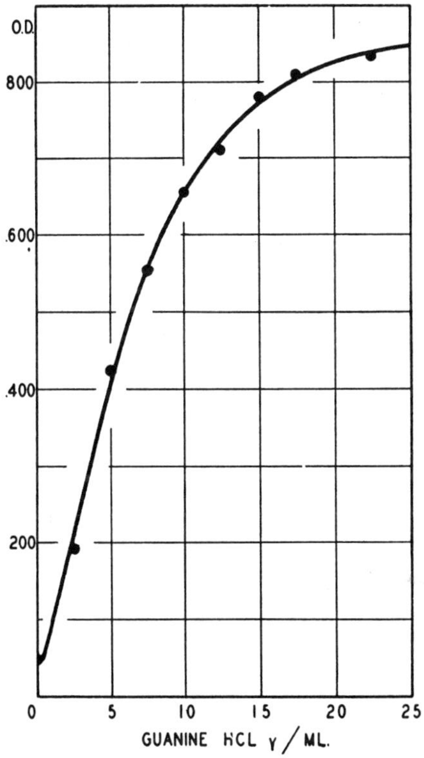

FIG. 28. Dose response of *Tetrahymena geleii* W to guanine·HCl. Medium A minus purines, containing 10 mg./ml. Tween 80 instead of Tween 85.

resulted in more readily available materials. In a series of tests it was demonstrated that *Tetrahymena* requires exogenous sources of guanine and uracil (Kidder and Dewey, 1948b). No natural purine could replace guanine (Fig. 28), although guanosine and guanylic acid were approximately twice as active on a molar basis. Adenine and hypoxanthine were active in sparing guanine while xanthine and uric acid were entirely inert. No natural pyrimidine base could replace uracil for

Tetrahymena (Fig. 29), but the block in this case was found to be the lack of enzymes responsible for performing the riboside linkage. While cytosine was inert, cytidine and cytidylic acid were as active on a molar basis as uracil, uridine, or uridylic acid for the ciliate. Neither thymine nor thymidine replaced or spared uracil (Kidder *et al.*, 1950b).

It is well established that higher animals (rat, pigeon) are able to synthesize both purines and pyrimidines and build them into nucleosides,

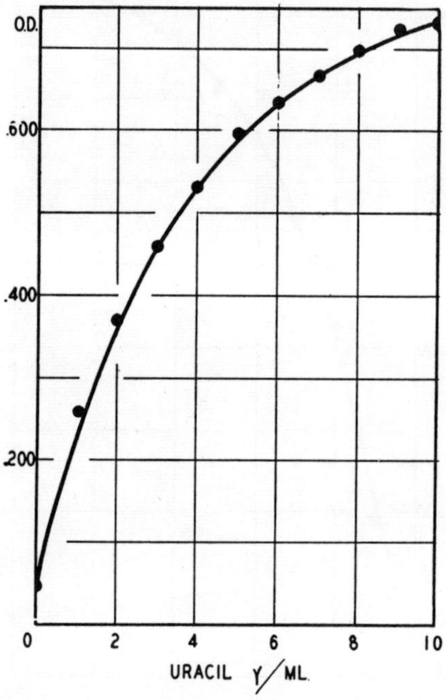

FIG. 29. Dose response of *Tetrahymena geleii* W to uracil. Medium A minus pyrimidine, containing 10 mg./ml. Tween 80 instead of Tween 85.

nucleotides, and nucleic acids. *Tetrahymena* does not possess certain enzymes for these reactions, so again it appears that this ciliate is further along in its physiological evolution than the mammal. In other words, the ciliate, during the course of its evolution, has lost enzymes for carrying out very important syntheses, and these enzymes have been retained in the vertebrate line.

By using many substituted purines it was demonstrated that the substituent groups characteristic of guanine were of prime importance

(Kidder and Dewey, 1949e). In only two cases were compounds found which could replace guanine and which did not have the substituent groups characteristic of guanine. In both cases the activity was low (2.3% for 2,6-diaminopurine and 15% for 1-methylxanthine), and it was apparent that aminating and deaminating mechanisms were at work. It will be recalled that xanthine was inert, and it is of interest that the ciliate can synthesize guanine from it only if position 1 is filled. This was taken to mean that the methyl substitution at position 1 prevented enolization and "froze" the keto group at position 6. The keto group

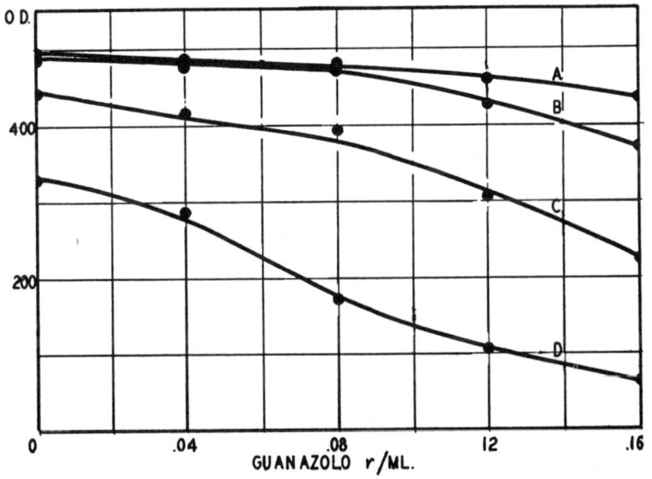

FIG. 30. Inhibition of the growth of *Tetrahymena geleii* W by guanazolo (5-amino-7-hydroxy-1-v-triazolo(d)pyrimidine) and its release by guanylic acid. The following amounts of guanylic acid, in γ/ml. are represented as follows: A = 40; B = 30; C = 20; D = 10. Medium A minus purine.

appears to be less reactive than the hydroxyl group, so amination at position 2 could proceed at a rapid enough rate to produce some 1-methylguanine which is about 75% as active as guanine.

A number of substituted purines proved to be inhibitory, and it was demonstrated that those most closely resembling guanine in chemical configuration were the most active inhibitors. The triazolo analog of guanine (5-amino-7-hydroxy-1-v-triazolo (d) pyrimidine) was the most active inhibitor encountered. One mole of the triazolo compound would completely block twelve to thirteen moles of guanine. The inhibition is antagonized by guanine, guanosine, and by guanylic acid (Fig. 30), and adenine and hypoxanthine were also effective. Evidence was produced which indicated that the enzyme system responsible for the metabolism of guanine also metabolized this analog. The triazolo analog of adenine (7-amino-1-v-triazolo (d) pyrimidine) was found to be much less inhibitory to *Tetrahymena* than the guanine analog (Kidder *et al.*, 1950b). In this case the release of the inhibition was confined to adenine and its ribosides (Fig. 31). Table I summarizes the data obtained from use of substituted purines and purine analogs.

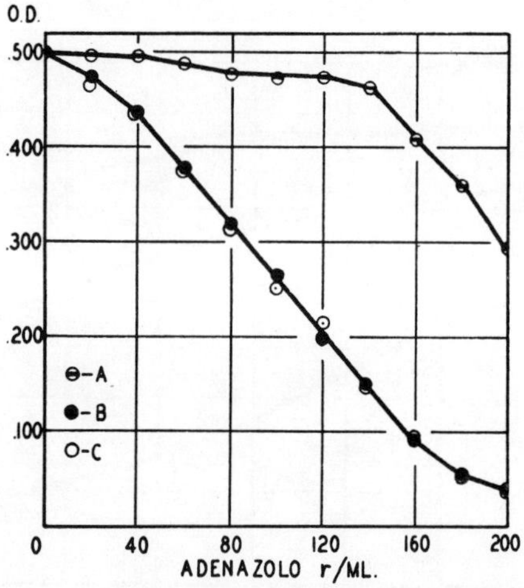

FIG. 31. Inhibition of the growth of *Tetrahymena geleii* W by adenazolo (7-amino-1-v-triazolo(d)pyrimidine). This inhibitor is not released by guanine, guanosine, guanylic acid, or hypoxanthine, but is released by adenine or its ribosides. In the curves shown here adenine and guanylic acid are included in the following amounts: A contains 30 γ/ml. guanylic acid and 50 γ/ml. adenine; B contains 30 γ/ml. guanylic acid and 20 γ/ml. adenine; C contains 100 γ/ml. guanylic acid and 20 γ/ml. adenine. Medium A minus purine.

A similar set of studies on substituted pyrimidines (Kidder and Dewey, 1949d) showed that any substitution on the uracil molecule reduced the activity, while certain substituted pyrimidines were inhibitory. Table II presents this data.

Continuation of the work on pyrimidine metabolism of *Tetrahymena* (Kidder *et al.*, 1950b) has brought to light several more points of interest. Orotic acid (4-carboxyuracil), which has been shown to be a pyrimidine precursor in the rat (Arvidson *et al.*, 1949; Bergström *et al.*, 1949) and to function in the metabolism of certain pyrimidine-less *Neu-*

TABLE I. Activity of Substituted Purines or Purine Analogs on *Tetrahymena geleii* W

Replacement	Per cent activity*	Sparing†	Inhibitory	Inhib. Index‡	Inert
1-Methylguanine	75	Adenine	5-Amino-7-hydroxy-1-v-triazolo(d)pyrimidine	0.02	7-Hydroxy-1-v-triazolo(d) pyrimidine
					5-hydroxy-7-amino-1-v-triazolo(d)pyrimidine
1-Methylxanthine	15	Hypoxanthine	5,7-Diamino-1-v-triazolo (d)pyrimidine	27	Xanthine
2,6-Diaminopurine	2.3	7-Methylguanine (poor)	7-Amino-1-v-triazolo(d) pyrimidine	5	7-Methylxanthine
8-Methylguanine	4	1,7-Dimethylguanine (poor)	3,7-Dimethylxanthine (theobromine)	280§	3-Methylxanthine
			1,3-Dimethylxanthine (theophylline)	260§	2-Thio-6-ketopurine
			1,7-Dimethylxanthine (paraxanthine)	280§	Uric acid
			1,3,7-Trimethylxanthine (caffeine)	200§	8-Ethylguanine
			Benzimidazole	350§	8-Methyladenine
			o-Phenylenediamine	20§	4,6-Diaminobenzimadazole

* Calculated from guanine, which was taken as 100%.
† Growth increases in the presence of suboptimal guanine.
‡ Inhibition index is the smallest ratio between the amount of inhibitor and antagonist at which half maximum growth occurs.
§ Not released, or only at low levels of inhibitor, by any purine. The numbers indicate the amount in γ/ml. required for half maximum inhibition.

TABLE II. Activity of Substituted Pyrimidines on *Tetrahymena geleii* W

Replacement	Per cent activity*	Sparing†	Inhibition	Inhib. index‡	Inert
1-Methyluracil	1.00	4-Methyluracil	2-Thio-4-Aminopyrimidine	⟨10	2-Amino-4-hydroxypyrimidine (isocytosine)
3-Methyluracil	0.84		5-Hydroxyuracil (isobarbituric acid)	10	
5-Bromouracil (?)	0.5		2-Thio-6-ketopyrimidine (2-thiouracil)	20	1,3-Dimethyluracil
5-Nitouracil (?)	0.12		2,4-Diaminopyrimidine	100	5-Methylcytosine
			5-Aminouracil	100	4-Hydroxyuracil (barbituric acid)
			2,4-Diamino-6-methyl pyrimidine	170	
			2,4-Diamino-5-formyl-amino-6-hydroxypyrimidine	170	4-Hydroxypyrimidine
			2-Aminopyrimidine	⟩1000	4,5-Dihydroxypyrimidine
			5,5-Diethylbarbituric acid (barbital)	1000§	5-Methyl-6-hydroxy-pyrimidine
			5-Ethyl-5-phenyl-barbituric acid (phenobarbital)	275§	2,4,6-Triamino-5-formyl-aminopyrimidine
			5-Ethyl-5-isopropyl barbituric acid	730§	2,4-Diamino-5-acetylamino-6-hydroxypyrimidine
			5-Ethyl-5-n-butyl barbituric acid (neonal)	200§	2,4-Diamino-5-propionyl-amino-6-hydroxypyrimidine
			5-Ethyl-5-(1-methyl butyl) barbituric acid (nembutal)	140§	5-Aminobarbituric acid (uramil)
			5-Ethyl-5-cyclohexenyl barbituric acid (phanodorm)	330§	6-Methylthiouracil
			3,5-Dimethyl-5-cyclohexenyl barbituric acid (evipal)	200§	1-desaza uracil
			6-n-Propylthiouracil	500§	4-carboxy uracil (orotic acid)
			5-Methylthiouracil	450§	
			bis-Desazouracil (resorcinol)	375§	

* Calculated from uracil, which was taken as 100%.
† Growth increase in the presence of suboptimal uracil.
‡ Inhibition index is the smallest ratio between the amount of inhibitor and antagonist at which half maximum growth occurs.
§ Not released by any pyrimidine. The numbers indicate the amount in γ/ml. required for half maximum inhibition.

rospora mutants (Mitchell and Houlahan, 1947; Mitchell *et al.,* 1948), is inactive for *Tetrahymena.*

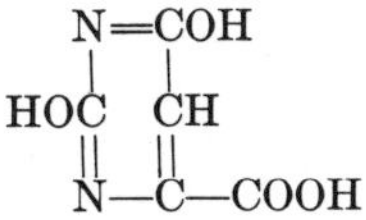

Orotic acid

This means that the ciliate does not possess the necessary decarboxylating enzymes for this compound. It was also found that while thymine and thymidine did not spare uracil, they did spare pteroylglutamic acid. This was interpreted to mean that thymine must be synthesized from non-pyrimidine precursors, not by methylation of uracil, nor by transforma-

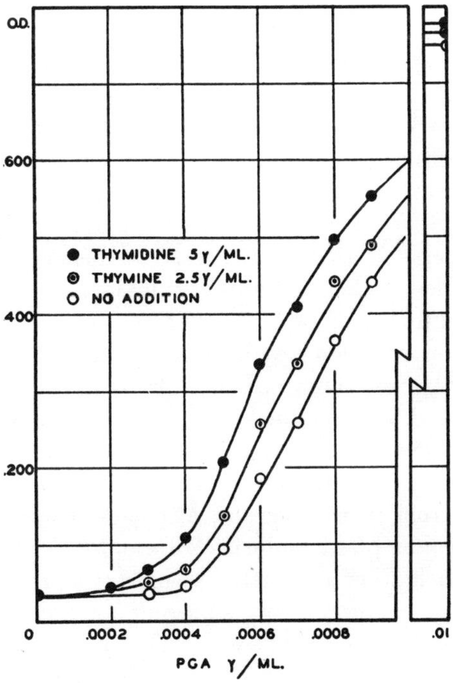

FIG. 32. Sparing action of thymidine and thymine on PGA as tested with *Tetrahymena geleii* W. The amounts of thymidine and thymine shown here represent the optimum for sparing. The inoculum for these experiments were PGA-depleted organisms. Medium A, minus PGA, containing 10 mg./ml. Tween 80 instead of Tween 85.

tion of orotic acid, and that PGA must function in one or more of the steps leading to thymine synthesis. Inasmuch as thymidine is more active than thymine in sparing PGA (Fig. 32), it appears that the ribose linkage enzyme may also involve PGA as a cofactor. On the basis of the known metabolic reactions involving pyrimidines and purines, the pathways are outlined and compared to what is known in the higher animals (Kidder *et al.*, 1950b).

J. Protogen

In two early studies on the nutrition of *Tetrahymena* it was found that two fractions of crude materials must be added to a casein containing medium to support growth (Dewey, 1941, 1944). These fractions were designated Factors I and II. Later, using an amino acid-containing medium (Kidder and Dewey, 1945d), a third factor (Factor III) was discovered. Factors I and III were subsequently identified as nucleic acid derivatives (Kidder and Dewey, 1945f), but Factor II had to be added as the lead acetate filtrate of natural materials. In 1946 Dr. E. L. R. Stokstad began work on the concentration of Factor II, which was soon found to be a complex of substances. By appropriate methods the old Factor II was fractionated into what was called Factors IIA and IIB (Stokstad *et al.*, 1949). Factor IIB proved to be a combination of known substances (Kidder and Dewey, 1949a). It was possible to replace all Factor IIB activity with increased pyridoxine, iron, and amino acids together with high levels of copper.

Factor IIA was concentrated by chromatographic adsorptions until one unit of activity was contained in a fraction of a microgram of dry weight (Kidder and Dewey, 1949a; Stokstad *et al.*, 1949). The name "protogen" was provisionally given to this factor until its chemical identity could be used to suggest a more appropriate term. Concentration was carried out from liver fractions, although it occurs in higher concentrations in certain plant materials (Cerophyl). Protogen has been concentrated to such an extent that it seems quite probable that it represents a single chemical entity. In its present state of purity it is entirely free of all other known growth factors, including vitamin B_{12}.

Protogen represents the only chemically unknown component of a medium which will support excellent growth of eight strains of *Tetrahymena geleii* and two strains of *T. vorax*.

Protogen has been identified with the "acetate factor" and with the "pyruvate oxidation factor" of lactic acid bacteria (Snell and Broquist,

1949). It has been found in most materials of plant origin and in the various tissues of the body of mammals. It is liberated from tissues following autolysis, but very little occurs in a free form in the heart or skeletal muscle. Enzymatic treatment releases large quantities of it from these muscles, however (Keevil, 1950). In view of its ubiquitous nature, it is likely that protogen will eventually be found to function in the metabolism of all organisms.

K. General

The results reported by Kline (1943) have been reviewed under the discussion of amino acids given above. This paper deals with the growth requirements of what was obviously another strain of *Tetrahymena* (referred to at that time as *Colpidium striatum*). Kline claims to have obtained transplantable growth in a medium devoid of eight components, the omission of any *one* of which causes growth failure in *Tetrahymena.* Four of these missing components are amino acids (histidine, leucine, lysine, and threonine). The other four are pteroylglutamic acid, pyrimidine (e.g., uracil), purine (guanine, guanosine, or guanylic acid), and protogen. In addition, the pyridoxine was included at a level (0.2 γ/ml.) far below requirement (Kidder and Dewey, 1949a). These omissions of biologically important compounds, which the ciliate is incapable of synthesizing, make the claims of Kline regarding the requirements of his organism dubious indeed. If these claims were true, then this one ciliate would seem to violate the general pattern of the animal kingdom so far as PGA, leucine, lysine, and threonine are concerned, and as far as histidine is concerned except for the human (Rose, 1949). In addition to these omissions, other conditions of his medium (mentioned above) are so far out of line with what we know about *Tetrahymena* as to cast serious doubt on the reliability of the report.

The fact that Kline's report has been given space in at least two general discussions (Anon., 1948; Scheer, 1948) makes it seem important to state our views on this work. We have pointed out on other occasions (Kidder and Dewey, 1945b,e, 1949a) that his results were not in accord with known facts, and it must be stated emphatically here that the most charitable thing we can say of Kline's results is that his cultures were bacterially contaminated, for even with the relatively impure commercial chemicals of the period in which the work was done, one could hardly expect eight or nine chemically different compounds to have been inadvertantly included as chemical contaminants.

VI. General Conditions of Growth

As an aid to those who may wish to use *Tetrahymena* for assays or other studies, a section on general techniques and methods of handling may be useful. Stock cultures of the organism may be kept in 2% proteose peptone, in which the organism will live for at least one month. In practice it is safer to make weekly transplants of the stock cultures and for use to keep several daily transplants going. This insures a supply of organisms of a suitable age for inoculation of experimental media each day. The cultures are kept in an incubator at 25° C. in the dark. While the stock cultures are kept in an upright position, it is better to grow experimental cultures in an inclined position to insure adequate aeration. Better aeration conditions are obtained by the use of flasks, but these are cumbersome when a large number of cultures is necessary.

In making up the basal medium it is convenient to prepare concentrated solutions of the various components or groups of components of the medium and to mix them in the proper proportions as needed. These are kept in the refrigerator under toluene. In this way a medium lacking any desired component may be prepared at will. Table III gives a list of the solutions used in this laboratory. Under certain conditions it may be more convenient to prepare large volumes of the complete medium as desired at double strength, if only one or a few different bases are to be used. In such a case, the medium may be made up without the vitamins, protogen, Tween, and dextrose and kept indefinitely under toluene. Fresh solutions of the components omitted may be added as needed.

Considerable time and effort have been spent in trying to obtain a medium to be used in assay procedures which would show no nonspecific stimulation upon the addition of raw or crude materials. It was at first believed that raw materials contain a stimulatory substance. Later it was found that by increasing all the components of the medium simultaneously it was possible to obtain growth even better than that obtained in the presence of crude materials. However, if one continues to increase all components, a point is reached at which certain of them become inhibitory. If, then, these are held constant at a lower level and the other components increased further, further improvements in growth can be obtained. A great many variations of the original medium have been tested. Table IV lists a number of these. It has been found that medium 2C gives the best growth (Dewey *et al.*, 1950). Medium D gives about the same growth although it contains somewhat less SNA.

TABLE III. Stock Solutions of Ingredients of Basal Media*

1. = IVA 3:100	g./l.
L-Arginine	4.0
L-Histidine	1.5
DL-Isoleucine	5.0
L-Leucine	5.0
L-Lysine	4.0
DL-Methionine	6.0
L-Phenylalanine	2.0
DL-Serine	8.0
DL-Threonine	5.0
L-Tryptophan	1.6
DL-Valine	1.0
2. = SCH 2:100	
DL-Alanine	5.50
L-Arginine	4.30
L-Aspartic acid	6.10
Glycine	0.50
L-Glutamic acid	11.65
L-Histidine	2.10
DL-Isoleucine	6.30
L-Leucine	9.70
L-Lysine	7.60
DL-Methionine	3.40
L-Phenylalanine	5.00
L-Proline	8.75
DL-Serine	7.70
DL-Threonine	8.80
L-Tryptophan	1.20
DL-Valine	6.60
3. = 8 K.F. 1:100	mg./l.
Ca Pantothenate	10.00
Nicotinamide	10.00
Pyridoxine·HCl	100.00
Pyridoxal·HCl	10.00
Pyridoxamine·HCl	10.00
Riboflavin	10.00
Pteroylglutamic acid	1.00
Thiamine·HCl	100.00
Biotin (free acid)	0.05
Choline chloride	100.00
4. = sulfates 1:100	g./l.
$MgSO_4 \cdot 7H_2O$	10.00
$Fe(NH_4)_2(SO_4)_2 \cdot 6H_2O$	2.50
$MnCl_2 \cdot 4H_2O$	0.05
$ZnCl_2$	0.005
5. = chlorides 1:100	
$CaCl_2 \cdot 2H_2O$	5.00
$CuCl_2 \cdot 2H_2O$	0.50
$FeCl_3 \cdot 6H_2O$	0.125
6. = phosphate 1:100	
K_2HPO_4	100.00
KH_2PO_4	100.00
7. 1:100	
Dextrose	250.00
8. 1:50	
Tween 85	35.00
Tween 80	undiluted
9. 1:100	
Sodium acetate	100.00
10. = SNA 1:100	
Guanylic acid	3.00
Adenylic acid	2.00
Cytidylic acid	2.50
Uracil	1.00
11. 1:1000	U./ml.
Protogen	1000

* Proportions given are for medium A (see Table IV). Other media are prepared simply by varying the proportions used.

In order to make comparisons of these various media, it was found to be necessary to dilute the cultures after growth with an equal volume of water. Growth was so dense that turbidimetric determinations were at the very high end of the scale where readings are too inaccurate to be utilized. In certain of such media growth is so profuse as to require the use of rather large amounts of buffer. If conditions are allowed to be-

TABLE IV. Basal Media (Amounts in γ/ml.)

Designation of Medium	III	IV	V	A	S	2A	4A	C	2C	D
DL-alanine	None	55.00	55.00	110.00	275.00	220.00	440.00	550.00	1100.00	1100.00
L-arginine	125.00	83.00	163.00	206.00	215.00	412.00	824.00	430.00	860.00	860.00
L-aspartic acid	None	61.00	61.00	122.00	305.00	244.00	488.00	610.00	1220.00	1220.00
Glycine	None	5.00	5.00	10.00	25.00	20.00	40.00	50.00	100.00	100.00
L-glutamic acid	None	233.00	233.00	233.00	582.50	466.00	932.00	1165.00	2330.00	2330.00
L-histidine	125.00	36.00	66.00	87.00	105.00	174.00	348.00	210.00	420.00	420.00
DL-isoleucine	125.00	113.00	213.00	276.00	315.00	552.00	1104.00	630.00	1260.00	1260.00
L-leucine	250.00	147.00	247.00	344.00	485.00	688.00	1376.00	970.00	1940.00	1940.00
L-lysine	250.00	116.00	196.00	272.00	380.00	544.00	1088.00	760.00	1520.00	1520.00
DL-methionine	500.00	94.00	214.00	248.00	170.00	496.00	992.00	340.00	680.00	680.00
L-phenylalanine	350.00	70.00	110.00	160.00	250.00	320.00	640.00	500.00	1000.00	1000.00
L-proline	None	175.00	175.00	250.00	437.50	500.00	1000.00	877.00	1754.00	1754.00
DL-serine	250.00	157.00	317.00	394.00	385.00	788.00	1576.00	770.00	1540.00	1540.00
DL-threonine	125.00	138.00	238.00	326.00	440.00	652.00	1304.00	880.00	1760.00	1760.00
L-tryptophan	50.00	28.00	60.00	72.00	60.00	144.00	288.00	120.00	240.00	240.00
DL-valine	125.00	76.00	96.00	162.00	330.00	324.00	648.00	660.00	1320.00	1320.00
Calcium pantothenate	0.10	0.10	0.10	0.10	0.10	0.20	0.40	0.50	0.50	0.60
Nicotinamide	0.10	0.10	0.10	0.10	0.10	0.20	0.40	0.50	0.50	0.60
Pyridoxine hydrochloride	2.00	2.00	2.00	1.00	None	2.00	4.00	5.00	5.00	6.00
Pyridoxal hydrochloride	None	None	None	0.10	0.10	0.20	0.40	0.50	0.50	0.60
Pyridoxamine hydrochloride	None	None	None	0.10	0.10	0.20	0.40	0.50	0.50	0.60
Riboflavin	0.10	0.10	0.10	0.10	0.10	0.20	0.40	0.50	0.50	0.60
Pteroylglutamic acid	0.01	0.01	0.01	0.01	0.01	0.02	0.04	0.05	0.05	0.06
Thiamine hydrochloride	1.00	1.00	1.00	1.00	1.00	2.00	4.00	5.00	5.00	6.00
Biotin (free acid)	0.0005	0.0005	0.0005	0.0005	0.0005	0.001	0.002	0.0025	0.0025	0.003
Choline chloride	1.00	1.00	1.00	1.00	1.00	2.00	4.00	5.00	5.00	6.00
$MgSO_4 \cdot 7H_2O$	100.00	100.00	100.00	100.00	100.00	200.00	400.00	100.00	140.00	140.00
$Fe(NH_4)_2(SO_4)_2 \cdot 6H_2O$	25.00	25.00	25.00	25.00	25.00	50.00	100.00	25.00	62.50	62.50
$MnCl_2 \cdot 4H_2O$	0.05	0.05	0.05	0.50	0.50	1.00	2.00	0.50	1.25	1.25
$ZnCl_2$	0.05	0.05	0.05	0.05	0.05	0.10	0.20	0.05	0.125	0.125
$CaCl_2 \cdot 2H_2O$	50.00	50.00	50.00	50.00	50.00	100.00	200.00	50.00	30.00	30.00
$CuCl_2 \cdot 2H_2O$	5.00	5.00	5.00	5.00	5.00	10.00	20.00	5.00	3.00	3.00
$FeCl_3 \cdot 6H_2O$	1.25	1.25	1.25	1.25	1.25	2.50	5.00	1.25	0.75	0.75
K_2HPO_4	100.00	100.00	100.00	1000.00	1000.00	1000.00	1000.00	500.00	500.00	500.00
KH_2PO_4	None	None	None	1000.00	1000.00	1000.00	1000.00	500.00	500.00	500.00
Guanylic acid	None	None	None	30.00	30.00	60.00	120.00	150.00	300.00	150.00
Adenylic acid	None	None	None	20.00	20.00	40.00	80.00	100.00	200.00	100.00
Cytidylic acid	None	None	None	25.00	25.00	50.00	50.00	125.00	250.00	125.00
Uracil	None	None	None	10.00	10.00	20.00	40.00	50.00	100.00	50.00
Hydrolyzed YNA	100.00	100.00	100.00	None	None	None	None	None	None	None
Dextrose	1000.00	1000.00	1000.00	2500.00	2500.00	2500.00	2500.00	2500.00	2500.00	2500.00
Sodium acetate	1000.00	1000.00	1000.00	1000.00	1000.00	1000.00	1000.00	None	None	1000.00
Tween 85	500.00	500.00	500.00	700.00	None	1400.00	2800.00	None	None	None
Tween 80	None	None	None	None	10,000.00	None	None	10,000.00	24,000.00	24,000.00
Protogen	None	1.00*	1.00*	1.00*	1.00*	2.00*	4.00*	2.00*	2.00*	2.00*
Factor II (Cerophyl)	1:5	None	None	None	None	None	None	None	None	None

* Units per milliliter.

References: Medium III (Kidder and Dewey, 1947b, 1948a); Medium IV (Kidder and Dewey, 1948b, 1949b, c, d, e); Medium V (Kidder and Dewey, 1949a); remainder (Dewey *et al.*, 1950).

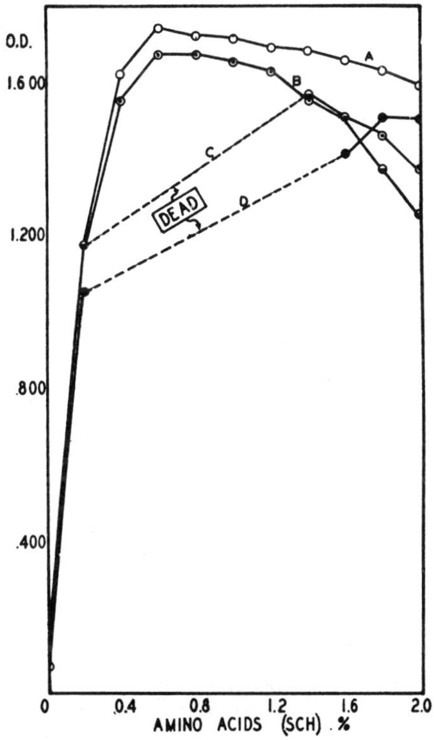

FIG. 33. Response of *Tetrahymena geleii* to increasing amounts of amino acids. All tubes were diluted before reading. Tubes represented by curve A contained 2,000 γ/ml. Na acetate, were diluted with water and kept in a slanted position until read. Under these conditions the pH did not drop, and the cells remained in good condition. Tubes represented by curve B also contained 2,000 γ/ml. Na acetate. They were diluted with phosphate buffer and held in an upright position until read. Slight cytolysis occurred due to rapid fermentation of dextrose while standing. Tubes represented by curve C contained 2,000 γ/ml. Na acetate. They were diluted with water and held in an upright position until read. Rapid death and cytolysis occurred in the middle ranges while the high concentrations of the amino acids toward the upper end prevented to some extent the drastic drop in pH due to fermentation. The tubes represented by curve C contained no acetate. They were diluted with water and kept slanted until read. The lack of acetate buffering in the middle ranges of amino acid concentration had allowed enough acid to accumulate to cause the death and cytolysis of the cells even before dilution. It should be noted that the dextrose of the basal medium is high (5,000 γ/ml.). Medium C minus amino acids (dextrose doubled).

come even slightly anaerobic, the organisms produce enough acid to cause cytolysis (Fig. 33). In medium 2C or D part of the function of the high amino acid concentration is to buffer the medium. This becomes apparent when high dextrose concentrations are used. In order to substitute a somewhat less expensive buffer, sodium acetate was used and the amino acid concentration was correspondingly reduced as the dextrose

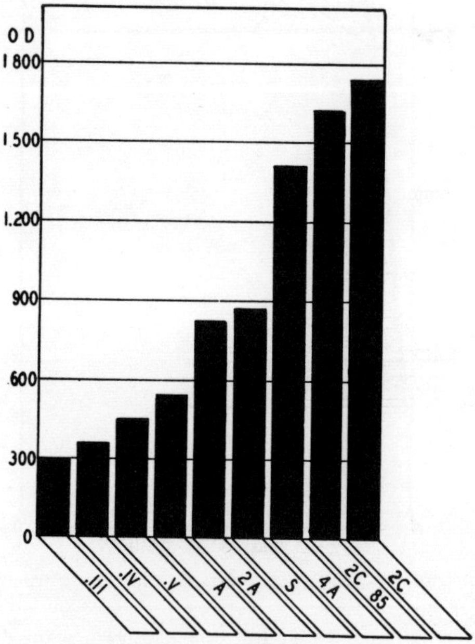

FIG. 34. Schematic representation of growth yields in various media. Media designations correspond to those given in Table IV. ''2C 85'' means that 7 mg./ml. Tween 85 is substituted for 24 mg./ml. Tween 80 in medium 2C.

concentration was raised. Figure 34 represents schematically the growth yields which are obtained in the various media given in Table IV.

In use the media are made up in double concentration. The variable is placed in the tubes, also in double the desired concentration in a volume of 2 ml. After autoclaving the tubes and the base separately, sterile dextrose solution in the desired concentration, plus the inoculum, is added to the base medium. One milliliter of a 3-day-old culture per 100 ml. of double strength base is a convenient amount for the inoculum. The tubes are then inoculated and loaded at the same time by the addition of 2 ml. of the basal medium, using the Brewer Automatic Pipetting Machine

(Kidder *et al.*, 1951). The latter procedure is carried out aseptically. After allowing four days for growth, the tubes are read in a photoelectric colorimeter using a red filter. Growth curves may be constructed by making readings upon the same tubes at various intervals.

VII. Applications

During the past several years the field of microbiology has taken its place as an important branch of investigation contributing to our knowledge of biochemistry. Consequently, although it seems unnecessary to point out possible applications to biochemical thinking that any microbiological study may contribute, it nevertheless seems wise to review briefly what one may expect from the studies of the biochemistry of animal microorganisms.

Biological Assays. It now seems clear that *Tetrahymena* (and probably ciliates in general) possess enzyme systems somewhat comparable to those of higher animals. This being so, it should be possible to adapt this organism for assay purposes and to obtain information regarding quantitative amounts of many biologically important compounds in natural materials. There are good reasons for believing that *Tetrahymena* will respond to even bound forms of a number of vitamins and of many amino acids (in whole proteins), thereby reducing the inaccuracies caused by hydrolysis, either by chemical or enzymic methods.

From a consideration of the requirements of *Tetrahymena,* it can be seen that successful assay procedures should be possible for the following amino acids: histidine, leucine, lysine, isoleucine, methionine, phenylalanine, threonine, and tryptophan. Assays for arginine, serine, and valine would be less exact. The following vitamins could be quantitatively detected: vitamin B_6 (pyridoxal and/or pyridoxamine), nicotinic acid and/or nicotinamide, pteroylglutamic acid (either free or conjugated), and riboflavin. Assays for thiamine would be unsafe due to the ability of *Tetrahymena* to respond to certain non-thiamine materials in natural products in a thiamine-free medium. Pantothenic acid assay would be unreliable in relation to the amount bound in the form of coenzyme A. *Tetrahymena* is currently used for protogen assays (Stokstad *et al.*, 1949; Keevil, 1950). Assay procedures for guanine or its ribosides should be useful, provided the basal medium contained adenine.

From our experiences, it is indicated that growth response as measured by turbidity is the method of choice for the evaluation of results. Acidimetric methods (Rockland and Dunn, 1946, 1949a, 1949b) do not necessarily reflect growth response, as mentioned earlier.

It is always sound practice, before embarking on microbiological assay procedures, to investigate not only the quantitative requirements of the microorganism, but also stimulatory substances in natural materials. Only then may one be reasonably sure of devising a basal medium which will not give nonspecific stimulation. Only recently have we felt that we are approaching that point with *Tetrahymena* (see above).

Cellular Metabolism. Tetrahymena is singularly well suited for investigations of many phases of cellular metabolism. It is now possible to study, under controlled conditions, many enzymatic reactions which are common to animals. It should be possible to extract enzymes from ciliates grown under a variety of conditions. Adequate amounts of *Tetrahymena* protoplasm can be obtained and extraction of enzymes is facilitated by the softness of the organisms.

Natural limitations are imposed on studies of this sort, of course, by the level of the animal organization. Many interesting metabolic phenomena peculiar to higher animals could not be investigated. Among those which are obvious can be mentioned blood studies, hematopoeiesis, hormones, and bone formation.

Animal Nutrition. Much valuable information can be attained in a short time regarding animal nutrition by the use of *Tetrahymena.* Applications to higher animals, of information gained in this way, regarding interactions of basic materials and proportional relations of constituents of foods should prove very useful. Basic metabolic pathways can be worked out on this type of animal micro model.

Chemotherapy. The response of *Tetrahymena* to many compounds of potential value in connection with chemotherapy makes it possible to use this organism as a tool. For instance, if and when the enzyme system or systems affected by the numerous barbiturates (see Table II), now used as medications, can be discovered, then more rational use of such drugs can be made.

It is in the field of chemotherapy that enzyme differences between *Tetrahymena* and mammals offer most promise. It has been shown above that the patterns of purine and pyrimidine metabolism differ greatly. These compounds are being more and more recognized as of prime importance to living organisms. *Tetrahymena* has already been used as a pilot organism for the discovery of a possible difference between certain types of mouse neoplastic cells and the normal mammalian cells (Kidder *et al.,* 1949), by the use of a guanine analog of unusual inhibitory activity in a guanine requiring system. Further investigations along the lines opened up by these observations may prove useful in the field of cancer chemotherapy.

REFERENCES

Anonymous (1948). *Nutrition Revs.* **6**, 38.

Arvidson, H., Eliasson, N. A., Hammarsten, E., Reichard, P., Ubisch, H. v., and Bergström, S. (1949). *J. Biol. Chem.* **179**, 169.

Baker, E. G. S., and Baumberger, J. B. (1941). *J. Cellular Comp. Physiol.* **17**, 285.

Beadle, G. W., Mitchell, H. K., and Nyc, J. F. (1947). *Proc. Natl. Acad. Sci. U. S.* **33**, 155.

Bergström, S., Arvidson, H., Hammarsten, E., Eliasson, N. A., Reichard, P., and Ubisch, H. v. (1949). *J. Biol. Chem.* **177**, 495.

Binkley, F. (1944). *J. Biol. Chem.* **155**, 39.

Binkley, S. B., Bird, O. D., Bloom, E. S., Brown, R. A., Calkins, D. G., Campbell, C. J., Emmett, A. D., and Pfiffner, J. J. (1944). *Science* **100**, 36.

Bonner, D. (1948). *Proc. Natl. Acad. Sci. U. S.* **34**, 5.

Bonner, D., and Beadle, G. W. (1946). *Arch. Biochem.* **11**, 319.

Camien, M. N., and Dunn, M. S. (1950). *J. Biol. Chem.* **182**, 119.

Chaix, P., Chauvet, J., and Fromageot, C. (1947). *Antonie van Leeuwenhoek. J. Microbiol. Serol.* **12**, 145.

Colas-Belcour, J., and Lwoff, A. (1925). *Compt. rend. soc. biol.* **93**, 1421.

Dewey, V. C. (1941). *Proc. Soc. Exptl. Biol. Med.* **46**, 482.

Dewey, V. C. (1944). *Biol. Bull.* **87**, 107.

Dewey, V. C., and Kidder, G. W. (1941). *Biol. Bull.* **81**, 285.

Dewey, V. C., Parks, R. E., Jr., and Kidder, G. W. (1950). *Arch. Biochem.* **29**, 281.

Doyle, W. L., and Harding, J. P. (1937). *J. Exptl. Biol.* **14**, 462.

Elliott, A. M. (1935a). *Arch. Protistenk.* **84**, 472.

Elliott, A. M. (1935b). *Arch. Protistenk.* **84**, 156.

Elliott, A. M. (1939). *Physiol. Zoöl.* **12**, 363.

Elliott, A. M. (1949a). *Anat. Record* **105**, 47.

Elliott, A. M. (1949b). *Physiol. Zoöl.* **22**, 337.

Fuller, R. C., III (1948). Thesis, Amherst College.

Furgason, W. H. (1940). *Arch. Protistenk.* **94**, 224.

Garnjobst, L., Tatum, E. L., and Taylor, C. V. (1943). *J. Cellular Comp. Physiol.* **21**, 199.

Genghof, Dorothy S. (1949). *Arch. Biochem.* **23**, 85.

Glaser, R. W., and Coria, N. A. (1935). *Am. J. Hyg.* **21**, 111.

Guirard, B. N., Snell, E. E., and Williams, R. J. (1946a). *Arch. Biochem.* **9**, 361.

Guirard, B. N., Snell, E. E., and Williams, R. J. (1946b). *Arch. Biochem.* **9**, 381.

Hall, R. H. (1938). *Biol. Bull.* **75**, 395.

Hall, R. H. (1941). *Physiol. Zoöl.* **14**, 193.

Hall, R. P. (1933). *Anat. Record* **57** (suppl.), 95.

Hall, R. P. (1939). *Anat. Record* **75** (suppl.), 150.

Hall, R. P. (1940). *Anat. Record* **78** (suppl.), 164.

Hall, R. P. (1942). *Physiol. Zoöl.* **15**, 95.
Hall, R. P. (1943). *Vitamins and Hormones* **1**, 249.
Hall, R. P. (1944). *Physiol. Zoöl.* **17**, 200.
Hall, R. P., and Cosgrove, W. B. (1944). *Biol. Bull.* **86**, 31.
Hall, R. P., and Cosgrove, W. B. (1945). *Physiol. Zoöl.* **18**, 425.
Hall, R. P., and Elliott, A. M. (1935). *Arch. Protistenk.* **85**, 443.
Hegsted, D. M., and Lipmann, F. (1948). *J. Biol. Chem.* **174**, 89.
Holden, J. T., Furman, C., and Snell, E. E. (1949). *J. Biol. Chem.* **178**, 789.
Holden, J. T., and Snell, E. E. (1949). *J. Biol. Chem.* **178**, 799.
Horowitz, N. H. (1947). *J. Biol. Chem.* **171**, 255.
Jahn, T. L. (1936). *Proc. Soc. Exptl. Biol. Med.* **33**, 494.
Johnson, D. F. (1935). *Anat. Record* **64** (suppl.), 106.
Johnson, W. H., and Baker, E. G. S. (1942). *Science* **95**, 333.
Johnson, W. H., and Tatum, E. L., (1945). *Arch. Biochem.* **8**, 163.
Jukes, T. H., and Stokstad, E. L. R. (1948). *Physiol. Revs.* **28**, 51.
Keevil, C. S., Jr. (1950). Thesis, Amherst College.
Kidder, G. W. (1941). *Biol. Bull.* **80**, 50.
Kidder, G. W. (1941). *In* Protozoa in Biological Research, p. 448. G. N. Calkins and F. M. Summers, eds. Columbia University Press, New York.
Kidder, G. W., (1946). *Arch. Biochem.* **9**, 51.
Kidder, G. W., (1947). *Ann. N. Y. Acad. Sci.* **49**, 99.
Kidder, G. W., and Dewey, V. C. (1942). *Growth* **6**, 405.
Kidder, G. W., and Dewey, V. C. (1944). *Biol. Bull.* **87**, 121.
Kidder, G. W., and Dewey, V. C. (1945a). *Physiol. Zoöl.* **18**, 136.
Kidder, G. W., and Dewey, V. C. (1945b). *Arch. Biochem.* **6**, 425.
Kidder, G. W., and Dewey, V. C. (1945c). *Biol. Bull.* **89**, 131.
Kidder, G. W., and Dewey, V. C. (1945d). *Arch. Biochem.* **6**, 433.
Kidder, G. W., and Dewey, V. C. (1945e). *Biol. Bull.* **89**, 229.
Kidder, G. W., and Dewey, V. C. (1945f). *Arch. Biochem.* **8**, 293.
Kidder, G. W., and Dewey, V. C. (1947a). *Proc. Natl. Acad. Sci. U. S.* **33**, 95.
Kidder, G. W., and Dewey, V. C. (1947b). *Proc. Natl. Acad. Sci. U. S.* **33**, 347.
Kidder, G. W., and Dewey, V. C. (1948a). *Proc. Natl. Acad. Sci. U. S.* **34**, 81.
Kidder, G. W., and Dewey, V. C. (1948b). *Proc. Natl. Acad. Sci. U. S.* **34**, 566.
Kidder, G. W., and Dewey, V. C. (1949a). *Arch. Biochem.* **20**, 433.
Kidder, G. W., and Dewey, V. C. (1949b). *Arch. Biochem.* **21**, 58.
Kidder, G. W., and Dewey, V. C. (1949c). *Arch. Biochem.* **21**, 66.
Kidder, G. W., and Dewey, V. C. (1949d). *J. Biol. Chem.* **178**, 383.
Kidder, G. W., and Dewey, V. C. (1949e). *J. Biol. Chem.* **179**, 181.
Kidder, G. W., Dewey, V. C., Andrews, M. B., and Kidder, R. R. (1949). *J. Nutrition* **37**, 521.
Kidder, G. W., Dewey, V. C., and Parks, R. E., Jr. (1951). *Physiol. Zoöl.* **24**, 69.
Kidder, G. W., Dewey, V. C., and Parks, R. E., Jr. (1950a). *Arch. Biochem.* **27**, 463.
Kidder, G. W., Dewey, V. C., Parks, R. E., Jr., and Heinrich, M. R. (1950b). *Proc. Natl. Acad. Sci. U. S.*, **36**, 431.
Kidder, G. W., Dewey, V. C., Parks, R. E., Jr., and Woodside, G. L. (1949). *Science* **109**, 511.
Kidder, G. W., and Fuller, R. C., III (1946). *Science* **16**, 160.

Kidder, G. W., Stuart, C. A., McGann, V. G., and Dewey, V. C. (1945). *Physiol. Zoöl.* **18**, 415.

King, T. E., Locher, L. M., and Cheldelin, V. H. (1948). *Arch. Biochem.* **17**, 483.

Kline, A. P. (1943). *Physiol. Zoöl.* **16**, 405.

Lawrie, N. R. (1937). *Biochem. J.* **31**, 789.

Lichstein, H. C., Gunsalus, I. C., and Umbreit, W. W. (1945). *J. Biol. Chem.* **161**, 311.

Lipmann, F., Kaplan, N. O., Novelli, G. D., Tuttle, L. C., and Guirard, B. (1947). *J. Biol. Chem.* **167**, 869.

Loefer, J. B. (1938). *J. Exptl. Zoöl.* **79**, 167.

Lwoff, A. (1923). *Compt. rend.* **176**, 928.

Lwoff, A. (1924). *Compt. rend. soc. biol.* **91**, 344.

Lwoff, A. (1932). Monographie de l'institut Pasteur, Paris.

Lwoff, A., and Lwoff, M. (1937). *Compt. rend. soc. biol.* **126**, 644.

Lwoff, A., and Lwoff, M. (1938). *Compt. rend. soc. biol.* **127**, 1170.

Lwoff, A., and Roukhelman, N. (1926). *Compt. rend.* **183**, 156.

Lwoff, M. (1934). *Compt. rend. soc. biol.* **115**, 237.

MacLeod, R. A., and Snell, E. E. (1947). *J. Biol. Chem.* **170**, 351.

Mitchell, H. K., and Houlahan, M. B. (1947). *Federation Proc.* **6**, 506.

Mitchell, H. K., Houlahan, M. B., and Nyc, J. F. (1948). *J. Biol. Chem.* **172**, 525.

Mitchell, H. K., and Nyc, J. F. (1948). *Proc. Natl. Acad. Sci. U. S.* **34**, 1.

Mitchell, H. K., Nyc, J. F., and Owen, R. D. (1948). *J. Biol. Chem.* **175**, 433.

Mowat, J. H., Boothe, J. H., Hutchings, B. L., Stokstad, E. L. R., Waller, C. W., Angier, R. B., Semb, J., Cosulich, D. B., and SubbaRow, Y. (1946). *Ann. N. Y. Acad. Sci.* **48**, 279.

Mushett, C. W., Stebbins, R. B., and Barton, M. N. (1947). *Trans. N. Y. Acad. Sci.* **9**, 291.

Novelli, G. D., Kaplan, N. O., and Lipmann, F. (1949). *J. Biol. Chem.* **177**, 97.

O'Kane, D. J., and Gunsalus, I. C. (1948). *J. Bact.* **56**, 499.

Ormsbee, R. A. (1942). *Biol. Bull.* **82**, 423.

Ormsbee, R. A., and Fisher, K. C. (1944). *J. Gen. Physiol.* **27**, 461.

Pace, D. M., and Ireland, R. L. (1945). *J. Gen. Physiol.* **28**, 547.

Pace, D. M., and Lyman, E. D. (1947). *Biol. Bull.* **92**, 210.

Peterson, R. E. (1942). *J. Biol. Chem.* **146**, 537.

Phelps, A. (1936). *J. Exptl. Zoöl.* **72**, 479.

Porter, C. C., Clark, I., and Silber, R. H. (1947). *J. Biol. Chem.* **167**, 573.

Rockland, L. B., and Dunn, M. S. (1946). *Arch. Biochem.* **11**, 541.

Rockland, L. B., and Dunn, M. S. (1949a). *J. Biol. Chem.* **179**, 511.

Rockland, L. B., and Dunn, M. S. (1949b). *Food. Technol.* **3**, 289.

Rose, W. C. (1938). *Physiol. Revs.* **18**, 109.

Rose, W. C. (1949). *Federation Proc.* **8**, 546.

Rose, W. C., Oesterling, M. H., and Womack, M. (1948). *J. Biol. Chem.* **176**, 753.

Rose, W. C., Smith, L. C., Womack, M., and Shane, M. (1949). *J. Biol. Chem.* **181**, 307.

Ryan, F. J., Tatum, E. L., and Giese, A. C. (1944). *J. Cellular Comp. Physiol.* **23**, 83.

Sakami, W. (1948). *J. Biol. Chem.* **176**, 995.

Sakami, W. (1949). *J. Biol. Chem.* **178**, 519.

Scheer, B. T. (1948). Comparative Physiology. Wiley, New York.

Seaman, G. R. (1949). *Anat. Record* **105**, 42.

Snell, E. E., and Broquist, H. P. (1949). *Arch. Biochem.* **23**, 326.

Snell, E. E., and Guirard, B. M. (1943). *Proc. Natl. Acad. Sci. U. S.* **29**, 66.

Stokstad, E. L. R., Hoffmann, C. E., Regan, M., Fordham, D., and Jukes, T. H. (1949). *Arch. Biochem.* **20**, 75.

Tatum, E. L., Garnjobst, L., and Taylor, C. V. (1942). *J. Cellular Comp. Physiol.* **20**, 211.

Taylor, C. V., and van Wagtendonk, W. J. (1941). *J. Cellular Comp. Physiol.* **17**, 349.

Thomas, J. O. (1943). *Stanford Univ. Bull.* **18**, 20.

Umbreit, W. W., and Waddell, J. G. (1949). *Proc. Soc. Exptl. Biol. Med.* **70**, 293.

Van Niel, C. B., Thomas, J. O., Ruben, S., and Kamen, M. D. (1942). *Proc. Natl. Acad. Sci. U. S.* **28**, 157.

van Wagtendonk, W. J., and Hackett, P. L. (1949). *Proc. Natl. Acad. Sci. U. S.* **35**, 155.

van Wagtendonk, W. J., and Taylor, C. V. (1942). *J. Cellular Comp. Physiol.* **19**, 95.

Author Index

Numbers in italics refer to the page on which the reference is listed in the bibliography at the end of each article.

D

E

N

S

T

U

V

W

Y

Z

Subject Index

B

C

Q

R